Industrial Inorganic Chemicals: Production and Uses

Industrial Inorganic Chemicals: Production and Uses

Edited by
R. Thompson, CBE, FEng
Consultant

THE ROYAL
SOCIETY OF
CHEMISTRY

ISBN 0–85404–514–7

A catalogue record of this book is available from the British Library.

Published by The Royal Society of Chemistry,
Thomas Graham House, The Science Park, Cambridge CB4 4WF, UK

Typeset by Vision Typesetting, Manchester, UK
Printed by Redwood Books Ltd., Trowbridge, Wiltshire, UK

12.7.96

Preface

In 1977 'The Modern Inorganic Chemicals Industry' appeared as The Royal Society of Chemistry Special Publication No. 31, being the Proceedings of a two-day conference which formed part of that year's Annual Congress at University College, London. At the time there was no comprehensive collected textbook on the contemporary inorganic chemicals industry, and the conference content was specifically designed such that the delivered papers be publishable in book form to fill a recognized gap. As convener and editor I was also concerned that the treatment should be *authoritative*, so consequently recognized authors were invited to contribute.

SP31 was an almost instant success, becoming a standard text in university and polytechnic chemistry and chemical engineering courses. The first print quickly sold out and the Proceedings were later to be printed twice more.

Nearly two decades have passed since SP31 was begun. The inorganic chemicals industry, like others, has moved on; albeit perhaps more slowly in production methods as is the way with capital-intensive commodity industries. It was decided that the time had come to prepare what is in effect a second edition, again with the main criterion that each chapter be an authoritative contribution from an industry expert. Several of the authors are in fact the same as in the earlier work, while others had contributed to the almost equally popular companion volume SP40, 'Speciality Inorganic Chemicals'. That recorded papers from a Salford University symposium run by the Inorganic Chemicals Group of the Society's Industrial Division in 1980. Yesterday's specialities can become today's commodities, and the importance of alumina and silica (debatably then in the commodities league) in their various forms is, for example, sufficient to merit inclusion here alongside other high-tonnage chemicals. Three chapters not included this time relate to water, the chlor-alkali industry, and soluble silicates, for accounts of which the reader is referred to SP31.

Chapters which have been extended over the earlier volumes are on aluminium, fluorine, and phosphorus chemicals, where the production and uses of derived products (many of which are themselves sold in large tonnages) are described in some detail.

Inorganic chemicals manufacture is highly integrated: a product from one part of the industry is often a feed-stock of another. This is no better illustrated than by sulfuric acid (Chapter 4) which finds major outlets *inter alia* in the production of boron, phosphorus, and titanium products described later in the book. The whole interplay across the industry is reviewed in Chapter 1, a further purpose of which is to place the various production and consumption rates in relative as well as absolute perspective.

The chemicals industry is often perceived (with some past justification) as being dirty and environmentally unacceptable. Air was polluted, rivers contaminated, and tracts of land laid waste, as well as the health and well-being both of employees and people living near to the plants being affected. The last two decades have seen considerable changes both in practical action being taken and the tightening of legislation requiring it. An overview given in Chapter 2 deals with these aspects, relating them also to metals smelting and the physical treatment and use of production by-products.

One of the features which heightened the demand for SP31 was the inclusion of an educational chapter that addressed some of the shortcomings of contemporary chemistry courses, at least insofar as degree courses in 'pure' chemistry were concerned. Graduate chemical engineers are often more fitted to enter manufacturing industry than their chemistry counterparts. It was therefore decided that we should again include a primer plus bibliography on at least the elements of chemical engineering. While this book is not designed primarily for student use, Chapter 3 could usefully be read first by those unfamiliar with the design and manufacturing procedures of the chemicals industry.

Finally, I must pay tribute to the individual authors, all busy people, who have contributed to this book; not least to those who finished their chapters on time and had to await completion by others. All were written within a span of about two years and thus none, bearing in mind any proof-stage alterations, will become outdated for some years to come.

Ray Thompson CBE, FEng

Contents

Contributors

S.P.S. Andrew, FRS, FEng, *Formerly ICI Plc., Agriculture Division*

A.K. Barbour, OBE, *Consultant, Environmental Science and Regulation*

B. Bertsch-Frank, *Degussa AG, Postfach 1345, D-6450, Hanau 1, Germany*

A. Dorfer, *Degussa AG, Postfach 1345, D-6450, Hanau 1, Germany*

T.A. Egerton, *Tioxide Group Services Ltd., West Site, Haverton Hill Road, Billingham, Cleveland TS23 1PS, UK*

K.A. Evans, *Alcan Chemicals Ltd., Chalfont Park, Gerrards Cross, Buckinghamshire SL9 0QB, UK*

D.C. Freshwater, FEng, *Louisiana State University, Baton Rouge*

K. Gilbert, *British Sulphur Corporation,*

G. Goor, *Degussa AG, Postfach 1345, D-6450, Hanau 1, Germany*

K.W.A. Guy, FEng, *Air Products Plc., Hersham Place, Molesey Road, Walton-on-Thames, Surrey KT12 4RZ, UK*

P. Kleinschmit, *Degussa AG, Postfach 1345, D-6450, Hanau 1, Germany*

J.C. McCoubrey, *Aston University, Aston Triangle, Birmingham B4 7ET, UK*

R.L. Powell, *Research and Technology Department, ICI Chemicals and Polymer Ltd., PO Box 8, The Heath, Runcorn, Cheshire WA7 4QD, UK*

T.A. Ryan, *Research and Technology Department, ICI Chemicals and Polymer Ltd., PO Box 8, The Heath, Runcorn, Cheshire WA7 4QD, UK*

H.U. Süss, *Degussa AG, Postfach 1345, D-6450, Hanau 1, Germany*

A. Tetlow, *Tioxide Group Services Ltd., West Site, Haverton Hill Road, Billingham, Cleveland TS23 1PS, UK*

R. Thompson, CBE, FEng, *Consultant, Formerly RTZ Borax*

A.A. Trickett, *Acid Technology, Chemetics International Company Ltd., 1818 Cornwall Avenue, Vancouver, British Columbia, Canada V6J 1C7*

R.D.A. Woode, *Brunner Mond and Company Limited, PO Box 4, Mond House, Northwich, Cheshire CW8 4DT, UK*

CHAPTER 1

Economic Importance of the Inorganic Chemical Sector

K. GILBERT

1 INTRODUCTION AND DEFINITIONS

As noted in the column 'On This Day' in *Chemistry and Industry*, 5 October 1992, Dumas and Liebig presented a joint memoir to the French Academy on 23 October 1837 asserting that 'in inorganic chemistry the radicals are simple; in organic chemistry they are compounds – that is the sole difference. The laws of combination [and] reaction are the same in the two branches of chemistry'. As a result of their work on radicals, and Wohler's urea synthesis from ammonium cyanate, the sharp distinction between organic and inorganic compounds, which had existed for centuries, was gradually lost.

Not entirely lost, however. The definition of 'inorganic' in the recently published 'Collins English Dictionary', 3rd Edition, 1991, defines 'inorganic' as relating to or denoting chemical compounds that do not contain carbon. When I first encountered chemistry at school the entry to the subject was by way of inorganic chemistry, the supposedly more difficult subject of organic chemistry being deferred for a year. There were separate textbooks for the two subjects and I still have 'General and Inorganic Chemistry' by Partington, and 'Organic Chemistry' by Fieser and Fieser. Many people of my generation will remember the Siamese cats that decorated the preface to the Fieser and Fieser.

Whilst the definition of inorganic is clear enough for the purposes of this book the inorganic sector of the chemical industry is in many respects inextricably tied up with the organic sector. For example, about 30% of the chlorine produced by the chloralkali industry ends up in the commodity plastic, poly(vinyl chloride). Chlorine is also used to produce propylene oxide, isocyanates, and other chemicals in which it does not appear in the molecule. There are many other examples of these connections between industrial inorganic and organic chemistry. Some of the more important examples are dealt with under the individual chemical classes described below.

This introductory review concentrates on commercial aspects of the inorganic chemicals business as technical matters are dealt with in detail in subsequent chapters.

Table 1 *USA: Production of Main Inorganic Chemicals in Terms of Volume*

>10 million tonnes a^{-1}

chlorine	ammonia
caustic soda	phosphoric acid
sulfuric acid	diammonium phosphate

>5 million tonnes a^{-1} < 10 million tonnes a^{-1}

soda ash	ammonium nitrate
sulfur	nitric acid

>1 million tonnes a^{-1} < 5 million tonnes a^{-1}

hydrochloric acid	ammonium sulfate
alumina	monoammonium phosphate
aluminium sulfate	

>500 000 tonnes a^{-1} < 1 million tonnes a^{-1}

titanium dioxide	sodium sulfate
aluminium hydroxide	calcium phosphate monobasic
sodium silicate	calcium phosphate dibasic

>100 000 tonnes a^{-1} < 500 000 tonnes a^{-1}

hydrocyanic acid	calcium phosphate tribasic
hydrofluoric acid	hydrogen peroxide
potassium sulfate	ferric chloride
sodium chlorate	magnesium chloride
sodium tripolyphosphate	phosphorus
bromine	phosphorus trichloride
calcium chloride	sulfur dioxide

---

2 THE SIZE OF THE INDUSTRY

The industrial inorganic chemicals industry manufactures a very large number of products ranging in volume from many millions of tonnes to a few kilograms per year and ranging in value from as little as $25 t^{-1} to $25 kg^{-1} or more.

The annual report produced by the US Department of Commerce, Bureau of the Census, entitled 'Inorganic Chemicals' lists 130 chemicals or groups of chemicals of which, in 1991, 35 individual inorganic chemicals were produced in volumes greater than 100 000 tonnes (Table 1).

According to the European Chemical Industry Council (CEFIC), inorganics accounted for 7.2% of the total value of sales of the West European chemical industry in 1990 or 23.4 billion electrochemical unit (ECU) (about $29.8 billions) (see Figure 1). However, this excludes fertilizers (almost all of which will be mineral fertilizer and therefore falling within the definition of inorganics adopted for this review) which account for a further 3.7% or 12.0 billion ECU making 35.4 billion ECU (about $45.1 billions) in all. Industrial gases, which are included in the review below, are not included in the CEFIC analysis.

CEFIC has given an estimate of world chemicals output by product sector group for 1989 (Table 2). Inorganic chemicals represent 8.1% and fertilizers and agrochemicals a further 7.3%.

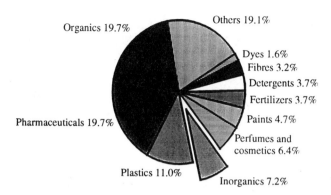

Figure 1   *West Europe: Chemical Industry by Sector Sales Value – 1990*
            (from Facts and Figures: West European Chemical Industry, September 1991)

Table 2   *World: Chemicals Output*

Product Sector Group	Estimated Product Sector Group Breakdown 1989 (US $billions)					
	West Europe	*North America*	*Japan*	*Other Far East*	*All Other*	*Total*
Inorganics	31	19	12	2	27	91
Fertilizers and agrochemicals	24	18	10	4	27	83
Petrochemicals	119	105	62	10	118	414
Man-made fibres	14	15	10	8	28	75
Pharmaceuticals	60	44	38	8	42	192
Other chemicals	92	74	58	7	44	275
Total	340	275	190	39	286	1130
Percentages of Regional Totals						
Inorganics	9.1	6.9	6.3	5.1	9.4	8.1
Fertilizers and agrochemicals	7.1	6.5	5.3	10.3	9.4	7.3
Petrochemicals	35.0	38.2	32.6	25.6	41.3	36.6
Man-made fibres	4.1	5.5	5.3	20.5	9.8	6.6
Pharmaceuticals	17.6	16.0	20.0	20.5	14.7	17.0
Other chemicals	27.1	26.9	30.5	17.9	15.4	24.3
Total	100.0	100.0	100.0	100.0	100.0	100.0
Percentages of Product Sector Group Totals						
Inorganics	34.1	20.9	13.2	2.2	29.7	100.0
Fertilizers and agrochemicals	28.9	21.7	12.0	4.8	32.5	100.0
Petrochemicals	28.7	25.4	15.0	2.4	28.5	100.0
Man-made fibres	18.7	20.0	13.3	10.7	37.3	100.0
Pharmaceuticals	31.3	22.9	19.8	4.2	21.9	100.0
Other chemicals	33.5	26.9	21.1	2.5	16.0	100.0
Total	30.1	24.3	16.8	3.5	25.3	100.0

Source: Chemical Industry Main Markets, 'An Assessment by Area and Product Group 1989–2000'
(CIA, 1990).

Table 3 *World: Chemical Output*

| | Indicative Growth Rates 1979–89 (% pa) | | | | | |
	West Europe	North America	Japan	OECD Area	All Other Areas	Total
Inorganics	−1.0	0.0	2.0	0.0	1.5	0.5
Fertilizers and agrochemicals	0.5	1.0	−1.0	0.5	4.5	2.0
All sectors	2.5	3.6	4.2	3.2	4.5	3.5

Source: Chemical Industry Main Markets, 'An Assessment by Area and Product Group 1989–2000' (CIA, 1990).

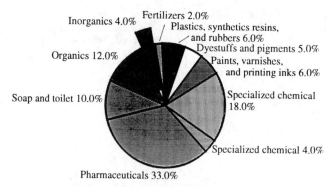

Figure 2 *UK Chemical Industry by Sector Gross Value Added – 1990*
[from UK Chemical Industry Facts, July 1992 (Chemical Industries Association)]

CEFIC has also estimated indicative growth rates for the period 1979–89 (Table 3).

According to the Chemical Industries Association (CIA) sales of principal products of the UK chemical industry were $27.27 billion in 1990. In terms of gross value added, inorganics represented 4% and fertilizers 2% of the whole (Figure 2).

Japan's Ministry of International Trade and Industry (MITI) publishes production figures for the Japanese chemical industry on a monthly basis. The analysis covers 70 inorganic chemicals (not including fertilizers or industrial gases) of which about 26 are produced in amounts exceeding $100\,000\,t\,a^{-1}$.

3 THE MAIN SECTORS OF THE INORGANIC CHEMICALS INDUSTRY

3.1 Alkalis and Chlorine

The chemicals reviewed under this heading are sodium hydroxide (caustic soda), soda ash, and chlorine. These three chemicals are produced on a very large scale –

about 41 million $t\,a^{-1}$, 31 million $t\,a^{-1}$, and 38 million $t\,a^{-1}$ respectively on a world-wide basis.

3.2 Chlorine and Caustic Soda

These chemicals are considered together as they are produced together, in a fixed weight relationship, from the electrolysis of sodium chloride brine. Small quantities of co-product chlorine arise in the production of sodium metal, potassium hydroxide, potassium nitrate, and magnesium metal. Small quantities of caustic soda are produced from soda ash using the soda lime process (see below).

Chlorine provides one the main links between commercial inorganic and organic chemistry. An analysis of the world market carried out in 1991 showed that 71.3% of total chlorine consumption was in the production of organic chemicals including poly(vinyl chloride) and solvents. Of the remaining 28.7% only 14.6% was consumed in the production of other inorganic chemicals (including titanium dioxide, see below) with the rest going into water treatment and pulp and paper.

The most important feature of the chlorine and caustic soda market is that the production level is determined by chlorine consumption. Electrolysis of brine produces about 1.1 t caustic soda to 1 t chlorine. In practice the ratio of products for sale or use is about 1.08:1 according to the production statistics published by the Chlorine Institute.

The chloralkali industry traditionally operates at a rate that satisfies the demand for chlorine. The amount of co-produced caustic soda may or may not satisfy the demand. If caustic soda production falls short of demand prices move upwards; if production exceeds demand prices move downwards. The range of movement of caustic soda prices can be very large. Over the 1980s the low point was about $40 t^{-1} for caustic liquor on a 100% basis and the high point was over $500 t^{-1}. High prices for caustic soda are not an indication of high capacity utilization and the need for new capacity as they would be for other commodity chemicals. Indeed, high caustic soda prices may be an indication of low or falling capacity utilization.

Producers of caustic soda and chlorine by electrolysis calculate their production costs in terms of the electrochemical unit (ECU) which is 1 short ton of chlorine and 1.1 short tons of caustic soda (or 0.91 tonnes chlorine and 1 tonne of caustic soda). The object of the producer is to obtain a netback for the ECU which exceeds the cost of the ECU by a margin sufficient to provide an adequate return on investment. If a netback to the manufacturer of $350 ECU^{-1} was considered satisfactory it could be generated by a combination of chlorine and caustic soda prices as shown in Figure 3.

A main element of chloralkali production costs is the cost of electricity. Depending on the type of cell used, electricity consumption ECU^{-1} ranges from about 2200 kWh for membrane cells to 3100 for mercury cells. A 2 cent $(kWh)^{-1}$ difference in electricity cost can make a difference of $44–62 ECU^{-1} in production cost. Most chlorine is used at or very near the point of production. To be uncompetitive in chlorine production can mean, in all probability, being

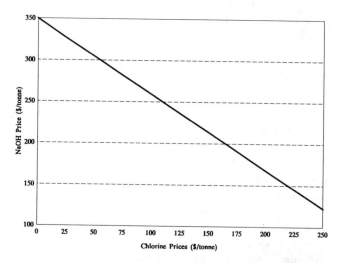

Figure 3 *Chlorine Price Necessary to Recover $350 ECU^{-1} at Specified Caustic Soda Prices*

uncompetitive in downstream chlorine derivatives including poly(vinyl chloride). The UK chlorine industry has made strong representations about the high level of domestic electricity prices to the power generators and the government, as it feels that some of its overseas competitors receive favoured electricity prices.

The main trend in chloralkali technology is that towards the use of membrane cells on the grounds of overall energy efficiency. Whilst mercury cells require more electricity they produce caustic soda liquor at 50% and so there is no steam requirement for evaporation. Diaphragm cells use about 2500 kWh ECU^{-1} but produce low strength caustic soda solution containing salt. Membrane cells use the least electricity and produce 35% caustic soda and so steam demand is low.

For most of the period since 1987, caustic soda demand has exceeded caustic soda production ex-brine electrolysis. The market for caustic soda has been brought back into balance through the price mechanism. High prices for caustic soda discourage usage, lead to substitution by soda ash and other alkalies, and encourage alternative production methods for caustic soda without co-product chlorine production. Regarding the production of soda-lime caustic soda, this has been introduced by the natural soda ash industry based in Wyoming (see Chapter 5). Three plants, with a total capacity of 218 000 t a^{-1}, are in operation. The viability of these plants is entirely dependant on the market price of caustic soda on the west coast of the US which in turn depends on the strength of the chlorine market.

The reason for the imbalance has been a weak chlorine market. Chlorine demand has been adversely affected by environmental factors in the pulp and paper industry, in water treatment, and in the use of chlorinated solvents and chlorofluorocarbons.

On a world-wide basis, the decline has been most marked in the pulp and paper industry which consumes about 3 million t^{-1} in bleaching processes. The concern arises from the environmental impact of chlorinated organics which are formed in the bleaching process. The chemical mechanisms which lead to the formation of chlorinated organics are not fully understood but it has been established that the highest proportion occur in the chlorination and extraction effluents of conventional kraft pulp bleaching. These compounds are not completely decomposed by the biological wastewater treatments commonly employed. The concern over chlorine has led to a rapid growth in alternative pulping processes and bleaching processes. The reduction of chlorine usage through chemical substitution and process modification is being accomplished by the following method:

- changed chlorination conditions;
- replacing chlorine with chlorine dioxide;
- use of hydrogen peroxide;
- oxidative extraction;
- use of ozone; and
- biobleaching.

What has been bad for chlorine has been good for sodium chlorate, the precursor for chlorine dioxide. Consumption of sodium chlorate has grown rapidly in the last ten years. World sodium chlorate capacity exceeds 1.6 million t a^{-1} with most of the capacity being concentrated in the pulp and paper producing areas of Scandinavia and North America.

Chlorine is used as a bactericide in the primary treatment of supply water and wastewater. Chlorine also serves as an oxidant for H_2S and Fe^{II} and various other impurities in raw water. As in the pulp and paper industry there has been considerable concern about chlorinated organics arising from the chlorination of water. This has led to a reduction in chlorine usage in some places. The trend in West Europe, North America, and Japan is for a continuing decline in chlorine consumption for water treatment but that this loss of market will just about be made up, in the period to 2000, by growth in demand in the rest of the world. The chlorine industry has stressed the importance to human health of clean drinking water which is not available to many millions of the world's inhabitants. The balance of benefit is very much in favour of chlorination.

Ozone, chlorine dioxide, and chloramine can be used in place of chlorine. As with chlorine, all have advantages and disadvantages.

The use of chlorine in the production of chlorinated solvents and chloro-fluorocarbons (CFCs) is connected as the former are raw materials for the latter. The main chlorinated solvents are:

- C1   methyl chloride, methylene chloride, chloroform, and carbon tetrach-loride (with 1, 2, 3, and 4 atoms of chlorine respectively);
- C2   ethyl chloride, perchloroethylene, 1,1,1-trichloroethane (methyl chloro-form), and trichloroethylene (with 1, 4, 3, and 3 atoms of chlorine respectively); and
- C3   propylene dichloride (with 2 atoms of chlorine).

A large proportion of the world capacity for chlorinated solvents is on sites on which chlorine is produced and from which by-product hydrochloric acid can be recycled.

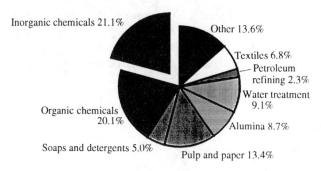

Figure 4   *World: Estimated Caustic Soda Consumption by End-use – 1991*

The main existing and potential starting materials for chlorofluorocarbons (CFCs), hydrofluorochlorocarbons (HCFCs), and hydrofluorocarbons (HFCs) are:

- carbon tetrachloride   CFC-11, CFC-12
- chloroform   HCFC-22
- perchloroethylene   CFC-113, CFC-114, HCFC-123, HCFC-124, HFC-134a
- 1,1,1-trichloroethane   HCFC-134b, HCFC-141b
- trichloroethylene   HCFC-123, HCFC-124, HFC-134a
- vinylidene chloride   HCFC-141b, HCFC-142b.

Under the Montreal Protocol 1987 and its subsequent revisions all the CFCs will be phased out as will carbon tetrachloride and 1,1,1-trichloroethane on the grounds of their effects on the ozone layer. The HCFCs, which have a much smaller ozone depleting potential (ODP) than CFCs (if CFC-11 has an ODP of 1 then HCFCs are in the range 0.03–0.05), will replace CFCs in some applications but are likely to be replaced in turn by the HFCs which have ODP's of zero. It is generally considered that the future market for these classes of products will be smaller in tonnage terms than it has been in the past. This will have an impact on chlorine demand, hydrofluoric acid demand, and by-product hydrochloric acid production.

The caustic soda market is affected by the general level of industrial activity and, as we have seen, by its selling price. Caustic soda has a wide range of end-uses of which the most important are within the chemical industry (Figure 4).

The high prices which have prevailed in the caustic soda market in the last few years have led some users in the chemical industry to switch to soda ash, *e.g.* in the manufacture of sodium phosphates.

The pulp and paper industry is also a very important user of caustic soda and as the industry also uses large quantities of chlorine this justified the installation of chloralkali capacity by companies in the industry. The declining use of chlorine has destroyed the rationale of these plants in some cases, as the pulp and paper industry has no other use for chlorine which fetches very low prices on the open market. The pulp and paper industry is reported to be interested in processes for the electrolysis of sodium sulfate which gives caustic soda and sulfuric acid, both of which are used in the industry.

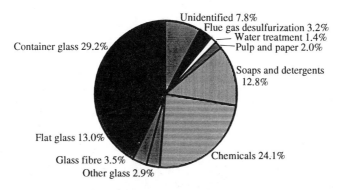

Figure 5   *USA: Soda Ash End-use Breakdown 1990 (1000 t)*
(from US Bureau of Mines)

The production of alumina for the manufacture of aluminium metal is an important consumer of caustic soda and, because of the location of alumina plants near deposits of bauxite, is the most important market for deep-sea traded caustic soda. The alumina business is very sensitive to caustic soda prices which at current levels represent the biggest cost element of production. Alumina producers have shown great interest in the soda-lime production of caustic soda and one plant using this technology exists although it is not operating at present (see Chapter 11).

3.2.1   Soda Ash (Sodium Carbonate). The soda ash business is a battle between natural and synthetic ash. Natural soda ash is made from deep-mined trona in Wyoming and from dry lake deposits in California, Kenya, and Botswana. Synthetic soda ash is made, using the Solvay process and its variants, from salt and limestone. The Solvay process produces large volumes of effluents, mainly of calcium chloride, and is energy intensive. The growing costs of effluent disposal and energy have put synthetic soda ash producers under increased pressure from the lower cost natural soda ash. North America only has one small synthetic soda ash plant in operation.

Europe is steadily taking greater volumes of imported natural soda ash. It is interesting to note that a significant proportion of the Wyoming industry is owned by the European synthetic soda ash producers, Solvay and Rhône–Poulenc.

Soda ash has a wide range of end-uses but is very dependent on the glass industry (Figure 5). Glass making is the largest market for soda ash. The manufacture of most types of glass require it, the most important sectors for consumption being containers, flat glass, and glass fibre. More and more container glass is being recycled, which depresses soda ash demand although without recycling glass containers would probably lose more market share to plastics. Caustic soda can be used as a replacement for soda ash in glass but at the present price-relationship between the alternatives, this is an academic consideration.

An important chemical derivative of soda ash is sodium bicarbonate which is used in household products such as carpet deodorizers, animal feeds, baked products, and fire extinguishers. Flue gas desulfurization is an end-use for sodium bicarbonate with potential for the future.

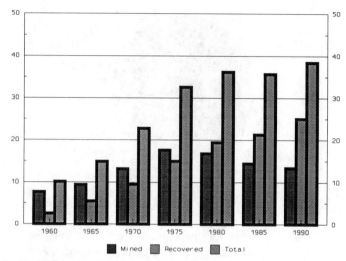

Figure 6 *World Brimstone Production by Source 1960–1990 (million t S)*

3.3 Mineral Acids

3.3.1 Sulfur and Sulfuric Acid. Elemental sulfur, generally still referred to as brimstone in the industry (Figure 6), and sulfuric acid have to be considered together as about 90% of sulfur is used in the form of sulfuric acid, *i.e.* elemental sulfur is converted to sulfuric acid or the sulfur arises as sulfuric acid, for example from non-ferrous smelters (see Chapter 2) and is used as such.

The two main features of the market thus are:

(i) Sulfur arises predominantly as a by-product.
(ii) A significant proportion of sulfuric acid production is of by-product origin.

The growth of by-product sulfur production has paralleled the growth of treatment of sour natural gas and the removal of sulfur during oil refining. The production of sulfur by straightforward mining (the exploitation of native sulfur) and by Frasch mining has declined as a proportion of the total.

Sulfur is a commodity and its market price reflects the balance between supply and demand. By-product sulfur, by definition, is produced whether it is required or not. The cost of production of by-product sulfur, at the point of production, can be considered to be zero. The production of sulfur by Frasch mining (underground melting) has running costs which are largely made up of the energy cost for water heating. The energy cost can be lessened by co-generation, but even so Frasch mining becomes marginal at sulfur prices below about $80 t^{-1}$, a level to which prices can be driven by oversupply.

Modern sulfur-burning sulfuric acid plants also co-generate electricity from high-pressure steam. As a rule of thumb, one tonne of sulfur is equivalent to two barrels of oil as far as energy content is concerned. A large proportion of sulfur-burning sulfuric acid capacity is installed on sites where phosphoric acid and fertilizer phosphates are also produced. The sulfuric acid plant can usually produce all the electric power and low pressure steam required on the site.

This energy aspect of sulfuric acid production is important in relation to the sales value of by-product sulfuric acid. In order to be competitive, by-product sulfuric acid has to be delivered to a site on which there is a sulfur-burning sulfuric acid plant at no more than cost of the on-site plant less energy credit. This can result in the supplier receiving a negative netback. This is acceptable if the cost of disposal is greater than the negative netback, or if the costs of plant closure due to an inability to move the sulfuric acid are high as they are with non-ferrous smelters.

The smelting of sulfides of copper, zinc, and lead is generating larger quantities of acid as the markets for these metals increase in volume but more importantly, as more sulfur values are captured for environmental reasons. Fortunately, more sulfuric acid is being put back into this industry as heap-leaching and *in-situ*-leaching grow in importance. For example, the quantity of sulfuric acid produced by the non-ferrous metals industry of Chile is rapidly increasing (from an estimated 0.57 million tonnes in 1990 to an estimated 2.19 million tonnes in 1995) but overall consumption in that country is also increasing (from an estimated 0.94 million tonnes in 1990 to 1.74 million tonnes in 1995).

Apart from the very important use for sulfuric acid in the production of phosphate fertilizers, the acid is used for non-ferrous metal leaching (primarily for copper) as noted above, in steel pickling, and in the production of titanium dioxide, sodium sulfate, potassium sulfate, hydrofluoric acid, boric acid, and many other inorganic chemicals. It is also used in petroleum refining (mainly as an alkylation catalyst), in pulp and paper, the manufacture of anionic surface active agents, the manufacture of rayon, and in the manufacture of many organic chemicals.

3.3.2 Hydrochloric Acid. Hydrochloric acid (HCl) provides a good illustration of the close relationship between inorganic and organic chemistry. HCl can be made, and is made, by burning chlorine in hydrogen, a convenient synthesis as both these starting materials are available from brine electrolysis (see above). However, only relatively small quantities are made in this way. Far greater quantities arise in chlorination and phosgenation reactions.

For the USA, statistics are available which show total production of hydrochloric acid including anhydrous, production from chlorine and hydrogen and production as by-product. In 1991 the production figures were 3.30 million short tons, 0.40 million short tons, and 2.90 million short tons respectively with deliberate production representing 12.3% of total production.

The by-product derivation of much HCl means that the acid has a very low value in those circumstances where the only alternative to a low-priced sale is to neutralize in order to dispose to waste.

Producers of large quantities of by-product HCl can recycle the acid by combining chlorination processes with oxychlorination processes or by using the acid to esterify an alcohol. An example of the first is the practice by which producers of vinyl chloride monomer combine the direct chlorination of ethylene and the oxychlorination of ethylene, both of which give ethylene dichloride, with the HCl being recycled from the ethylene dichloride pyrolysis stage. An example of the second is the combination of the production of methyl chloride from methanol and HCl, with the HCl being returned from reactions involving further chlorination of methyl chloride to give methylene chloride and chloroform; or from chlorination of methane to give methyl chloride, methylene chloride, and chloroform.

The vinyl chloride monomer synthesis can be balanced so that all the HCl produced in the ethylene diamine pyrolysis stage is recycled. It is rather more difficult to balance the chloromethanes processes as the market demand for the three or more products is not necessarily the same as the proportions produced in the reactions.

This chemistry is often carried out in complexes in which chlorine is produced and where there are process options for recycling the HCl arising from all the units on site, including the hydrochloric acid produced in the incineration of chlorinated wastes.

In the future it is expected that hydrochloric acid will be produced from the incineration of household wastes. In Germany, for example, the household waste incineration plants that are planned have an estimated potential hydrochloric acid production of $300\,000$–$400\,000\,t\,a^{-1}$ in total.

The other major source of by-product HCl is the manufacture of isocyanates. Isocyanates are generally made by phosgenating amines according to the equation:

$$RNH_2 + COCl_2 \rightarrow RNCO + 2HCl$$

In this case the HCl cannot be recycled other than through the regeneration of chlorine. Some makers of isocyanates have integrated schemes with the by-product HCl going to other reactions on site. Others are able to sell all or part of the by-product HCl. In some cases chlorine is regenerated from the HCl by using electrolysis, but this chlorine costs more to produce than does chlorine ex-brine electrolysis under the conditions, presently prevailing in the industry. The economics of HCl electrolysis have to be seen more in terms of an environmentally-acceptable waste treatment process for these companies. Alternatives to electrolysis are processes such as Kel-Chlor, Shell, MT Chlor (Mitsui), and, the latest, the catalytic carrier process. All are more expensive, under present conditions, than virgin chlorine.

The production of chlorofluorocarbons, hydrofluorochlorocarbons, and similar products, also leads to by-product HCl production. Where these processes are on integrated sites the HCl can usually be recycled. On non-integrated sites the by-product HCl has either to be sold (which may or may not be possible), converted to chlorine, or put to waste.

There are also completely inorganic sources of by-product HCl. Three of commercial importance are the Mannheim process production of sodium and potassium sulfates, the production of potassium nitrate from potash (potassium chloride) and nitric acid, and the production of high quality magnesia from magnesium chloride.

The Mannheim process involves the high temperature reaction of salt or potash with sulfuric acid. The process is usually integrated with an HCl-consuming process such as the manufacture of dicalcium phosphate as in Belgium.

The production of potassium nitrate by the reaction of potash and nitric acid with by-product HCl production is only practiced in Israel. Synthetic potassium nitrate is made in the USA from the same raw materials but chlorine is the by-product. In Israel the by-product HCl is used to acidulate phosphate rock to give phosphoric acid.

The production of high quality magnesia from magnesium chloride is also unique to Israel. By-product magnesium chloride solution from the potash industry is thermlly decomposed at high temperature. The by-product HCl solution is used to acidulate phosphate rock to give phosphoric acid. It is noteworthy that Israel has indigenous potash and phosphate rock and was able to develop the technology to separate phosphoric acid from water-soluble calcium chloride.

The markets for commercial hydrochloric acid (solutions in water of concentrations of 31.4%, 35.2%, and 37.1%) include:

steel pickling;
manufacture of calcium chloride;
oil and gas well acidizing;
production of high-fructose corn syrup (HFCS);
regeneration of ion-exchange resins;
manufacture of gelatin;
production of soy sauce; and
production of magnesium metal.

The common wisdom in the chemical industry is that HCl does not travel, the limit of deliveries being about 300 km. However, the cost of disposal to waste is rising and so by-product material can be delivered over greater distances and still provide a positive netback over the cost of disposal.

3.3.3 Nitric Acid. Nitric acid is primarily a fertilizer intermediate and this aspect is covered in the next section and in Chapter 6. It is also used in the manufacture of explosives [the major product, explosive grade ammonium nitrate (EGAN) is a special version of the fertilizer, ammonium nitrate (AN)], adipic acid, nitro-compounds (including nitrobenzene, most of which is converted to aniline), potassium nitrate (see under fertilizers), and synthetic sodium nitrate. A recent investigation into the industrial uses of nitric acid gave a world consumption of 12.5 million tonnes in 1991 and the market sector breakdown as shown in Figure 7.

3.4 Mineral Fertilizers

The three main plant nutrients are nitrogen, phosphorus, and potassium. The secondary nutrients are calcium, magnesium, and sulfur and the main micronut-rients are boron, chlorine, copper, iron, manganese, molybdenum, and zinc. Plants require soluble forms of these nutrients.

Fertilizers are categorized by their N, P, and K content (actually N, P_2O_5, and

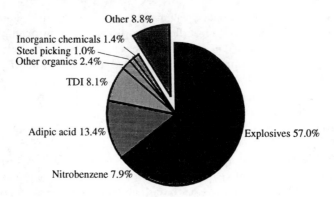

Figure 7   *World: Non Fertilizer Nitric Acid Consumption 1991 (1000 t)*
             (from British Sulphur Estimates)

Table 4   *Typical Content of Plant Material (kg tonne⁻¹ dry material basis)*

nitrogen	20	iron	0.100
phosphorus	2	manganese	0.080
potassium	15	zinc	0.040
calcium	8	boron	0.030
magnesium	2	copper	0.005
sulfur	1	molybdenum	0.001

K_2O content), *e.g.* a fertilizer designated 15.15.15 contains 15 units of each nutrient. Plants require all the nutrients (main, secondary, and micro-) in the correct proportions. An indication of the relative requirements is given by the typical content in plants (Table 4).

An examination of the usage of mineral fertilizers by particular countries shows that the ratio of N, P, and K can vary widely (Table 5).

The fertilizer industry is passing through a difficult period. A high proportion of world production capacity is state owned (probably over 70%) and commercial disciplines are not always observed. There has been over-building of new capacity leading to poor capacity utilization. Some fertilizer markets have been weak due to overproduction of agricultural commodities. Farm incomes have been declining. Nitrogen fertilizers have been blamed for contaminating sources of drinking water. The manufacture of phosphate fertilizers can produce discharges of fluorine and heavy metals. Finally, the economic chaos in Eastern Europe has led to a collapse of demand in that region.

Despite these recent setbacks the importance of mineral fertilizers to the provision of food for a growing world population will lead to a recovery of demand and continued growth in the future.

3.4.1   Nitrogen. The fertilizer nitrogen industry is based on ammonia production. Ammonia is considered to be a petrochemical in some circles as it is derived from hydrocarbons such as methane and naphtha. We take the view that a discussion of

Table 5   *World: Fertilizer NPK Ratios of Specified Countries*

Country	N	Ratio P₂O₅	K₂O
		Ratio	
Country	N	P_2O_5	K_2O
Australia	1:	1.32:	0.33
Brazil	1:	1.52:	1.52
Canada	1:	0.51:	0.31
China	1:	0.30:	0.09
Egypt	1:	0.26:	0.07
France	1:	0.54:	0.74
Germany	1:	0.37:	0.54
India	1:	0.40:	0.16
Iran	1:	1.05:	0.03
Italy	1:	0.75:	0.38
Japan	1:	1.13:	0.88
Mexico	1:	0.28:	0.06
Nigeria	1:	0.41:	0.45
Pakistan	1:	0.26:	0.02
Poland	1:	0.50:	0.87
Spain	1:	0.52:	0.37
Turkey	1:	0.52:	0.05
UK	1:	0.27:	0.33
USA	1:	0.37:	0.45
World average	1:	0.47:	0.32

Source: British Sulphur fertilizer consumption data-base.

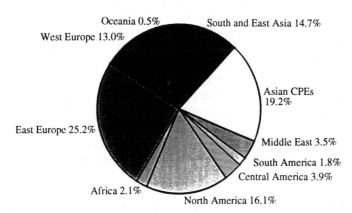

Figure 8   *World: Ammonia Production 1990 (million t N)*

inorganic chemicals and mineral fertilizers would not be complete without including ammonia (see also Chapter 6).

Ammonia is one of the most important inorganic chemicals in volume terms. In 1990 96.75 million tonnes (as N) was produced. The regional breakdown was as shown in Figure 8.

Production of ammonia in 1991 in the UK was 0.866 million tonnes N. This was 7.7% of West European production and 0.9% of world production.

Table 6 *World: Changes in Ammonia Production*

Region/Country	Specified Regions and Countries (million t N)		
	1980	1985	1990
West Europe	12.13	12.43	12.62
Middle East	1.41	1.92	3.37
Latin America/Caribbean	2.94	4.33	5.48
Africa	1.10	1.92	2.07
Saudi Arabia	0.17	0.44	0.94
Venezuela	0.36	0.40	0.56
Trinidad and Tobago	0.46	1.09	1.52
Nigeria	0.00	0.00	0.33
Indonesia	1.04	2.05	2.74

The ammonia process has been improved in efficiency over the last ten years. A plant installed today will use about 27–30 million BTU t^{-1} of natural gas compared with *ca.* 40 million BTU t^{-1} ten years ago. As the cost of natural gas is the main element of production cost, this is an important consideration. The dependence of ammonia production on low cost natural gas has resulted in a growth of production in areas such as the Arabian Gulf, Trinidad, Venezuela, Indonesia, and Nigeria. At the same time production in Western Europe has stagnated (Table 6).

As far as the fertilizer industry is concerned, ammonia is usually converted to a derivative before use. Urea, ammonium nitrate, and ammonium phosphates are the common forms, with the ammonium nitrate often being in the form of calcium ammonium nitrate or combined with urea in the form of urea–ammonium nitrate solutions. The manufacture of ammonium nitrate requires nitric acid which is made from ammonia. Ammonium nitrate is generally preferred to urea in high latitudes and urea is preferred to ammonium nitrate in low latitudes, but these preferences can be overturned if the costs per unit N clearly favour one or other product.

Ammonium sulfate is also available to the fertilizer industry. Its manufacture from ammonia and sulfuric acid is rarely economic but by-product material is available from several processes of which the most important is the manufacture of caprolactam.

The main derivative of ammonia is urea, an important nitrogen fertilizer in its own right. Urea is most conveniently made in plants adjoining ammonia plants in order to take advantage of the by-product carbon dioxide from the ammonia plant. If ammonia is in the grey area between inorganic and organic, urea must be considered organic and outside the scope of this discussion.

Diammonium phosphate (DAP) is a very important commodity fertilizer chemical but it is more important as a source of phosphorus than nitrogen. It contains 64 units of nutrient. It is made by reacting anhydrous ammonia with a solution of phosphoric acid using the heat of reaction to evaporate most of the water (see Chapter 14).

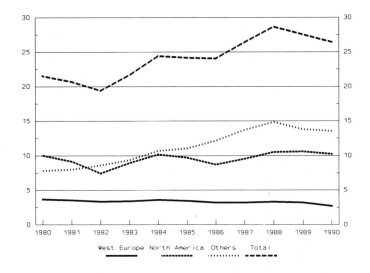

Figure 9   *World: Phosphoric Acid Production 1980–1990 (million t P_2O_5)*

Monoammonium phosphate (MAP), with 53–68 units of nutrients, is less important than diammonium phosphate, not for any reason of technical or agronomic deficiency, but merely due to the fact that diammonium phosphate was commercialized first as a world-scale product. MAP and DAP can be made on the same plant from the same raw materials.

3.4.2   Phosphorus. Mineral phosphate fertilizers are derived from phosphate rock which occurs widely throughout the world. The principal sources of phosphate rock are the southeast part of the USA, North Africa, Northwest Africa, South Africa, Middle East, Russia, and Kazhakstan.

Plant-life requires soluble phosphorus and so phosphate rock as such cannot be used except for certain grades of 'soluble' rock on acid soils. In general the rock has to be converted to a soluble form and this is commonly accomplished by acidulation. Acidulation with sulfuric acid without separation of calcium sulfate gives single superphosphate, a low analysis fertilizer (16–22 units of nutrient). Acidulation of phosphate rock with sulfuric acid with calcium sulfate removal (wet-process phosphoric acid) gives a solution of phosphoric acid of concentration of 27–30% P_2O_5 which can be concentrated to 52–54% P_2O_5 (so-called 'merchant acid') or 70–72% (so-called 'superacid'). Both are traded internationally in large quantities.

Phosphoric acid is not used as such as a fertilizer but is converted to a solid form such as triple superphosphate (containing 44–48 units of nutrient), MAP or DAP (see under nitrogen above).

The production of phosphoric acid has increased markedly over recent years in regions other than West Europe and North America (Figure 9).

The notable development of the phosphoric acid industry has been its growth in

Table 7 *Phosphoric Acid Production – Specified Countries* $(1000 \, t \, P_2O_5)$

Country	1980	1985	1990
Morocco	574	1290	2500
Tunisia	625	738	1047
Jordan	0	273	268
Israel	76	219	240

phosphate rock rich countries, particularly Morocco, and its decline in regions dependant on phosphate rock imports (Table 7).

The UK lost its last phosphoric acid production in 1992 with the closure of a plant at Whitehaven, although this closure had more to do with environmental factors than with straight economic considerations.

3.4.3 Potassium. 'Potash' (potassium chloride) production is classified as industrial mineral rather than inorganic chemicals production. For the bulk of production of potash no chemical transformation takes place. Deposits of potash are found in many parts of the world but relatively few are exploited commercially. The most important producers are Canada (Saskatchewan and New Brunswick), Russia, Israel, Jordan, Germany, and France. The UK has one producer and is largely self-sufficient.

Natural potassium chloride is the starting material for potassium chemicals including the important fertilizers potassium sulfate (see under sulfuric acid) and potassium nitrate (see under nitric and hydrochloric acids).

3.5 Industrial and Food Phosphates

This sector is closely related, in chemical production terms, with the manufacture of phosphate fertilizers. This is because the key intermediate is phosphoric acid. The industrial and food phosphate industry is only about one-twentieth the size of the fertilizer phosphate industry when measured in terms of volume of P_2O_5.

Three technologies are used by the industry (more fully described in Chapter 14). All start from phosphate rock.

Firstly, it can be converted to elemental phosphorus by smelting in an electric furnace with coke and silica. The elemental (yellow) phosphorus can be burned to give phosphorus pentoxide which dissolves in water to give 'thermal' phosphoric acid. Thermal acid is of good quality and is suitable for most applications.

Secondly, phosphate rock can be acidulated with sulfuric acid to give an impure phosphoric acid (as in the fertilizer industry). This impure acid can be converted into phosphate salts of acceptable purity for industrial applications using chemical purification techniques by which impurities are precipitated and are removed by filtration or settling. This method does not produce phosphoric acid for industrial sale.

The third method also starts with impure phosphoric acid. The acid is pre-treated by a combination of physical and chemical methods and is then purified by solvent-extraction. This process is widely used in Europe and has recently come on-line in the USA where it is competing with the well-established phosphorus-based route.

The main product which the industrial phosphates industry makes for sale is sodium tripolyphosphate (STPP). STPP is a detergent builder and whilst it is generally accepted as the most effective product available it is in declining demand in its main markets, North America and West Europe, for environmental reasons. Some markets are closed to phosphates in laundry detergents; restrictions of the level of phosphorus apply in others. The scientific arguments against phosphates in detergents are weak, but they prevail as public opinion favours 'non-phosphate'.

Apart from the important detergents/cleaners end-use, industrial phosphates have a very wide range of applications including metal treatment, water treatment, food and beverage, catalysts, cement manufacture, ceramics, drilling muds, excipients, fire-fighting compositions, fire retardants, paper processing, pigments, refractories, toothpaste, and yeast production. Most uses are mature and overall growth, leaving aside detergents, is similar to that of gross national product (GNP) growth. The use of technical grade phosphoric acid in oil recovery and in fuel cells may become important in the long-term.

Phosphorus is also converted to its direct derivatives, the most important of which are phosphorus trichloride, phosphorus oxychloride, and phosphorus pentasulfide. Phosphorus trichloride is made by the direct reaction of phosphorus and chlorine. Phosphorus oxychloride is made from phosphorus trichloride. Phosphorus pentasulfide is made from the direct reaction of phosphorus and sulfur.

Apart from phosphorus oxychloride manufacture, phosphorus trichloride is used as a pesticide intermediate (the herbicide glyphosate being particularly important), phosphite stabilizers, alkyl aryl phosphates, phosphonates, and acid chlorides. Phosphorus oxychloride is an intermediate for phosphate esters (used as plasticizers, flame retardants, and functional fluids), pesticides, and lubricating oil additives. Phosphorus pentasulfide is used to make the family of zinc dialkyl-dithiophosphate lubricant additives and is a key intermediate for many organophosphorus insecticides including malathion, chlorpyrifos, and terbuphos.

3.6 Aluminium Chemicals

The most important aluminium chemical is alumina, the key intermediate in the production of aluminium metal. The consumption of metallurgical alumina in 1991 in the Western World was about 28 million tonnes. Primary aluminium production in 1991 in the Western World was 14.3 million tonnes. Alumina is also used in refractories, in chemicals manufacture, and in the manufacture of catalysts (see also Chapter 11).

Most of the world's alumina is made by the two-stage, Bayer, process. The dissolution of bauxite in caustic soda is followed by hydrolysis of the sodium aluminate to give alumina and regenerate caustic soda. It has been calculated that the industry has 40 million tonnes of caustic soda circulating within its plants.

Other important aluminium chemicals are listed here.

(i)   Aluminium sulfate, which is usually derived from bauxite by sulfuric acid attack although it can also be made from aluminium hydroxide by reaction with sulfuric acid. Aluminium sulfate is important in the pulp and paper industry where it is used in sizing and water treatment. It is also used in municipal and industrial water and waste treatment.

(ii) Aluminium chloride anhydrous, which is made from aluminium metal by chlorine attack. The majority of commercial uses are catalytic and include the manufacture of ethyl benzene, detergent alkylates, ethyl chloride, and dyestuffs. The main non-catalytic use is the nucleation of titanium dioxide in the chloride process.

(iii) Aluminium chloride hydrous, which is made from aluminium hydroxide and hydrochloric acid. Polyaluminium chloride, a rapidly growing water treatment chemical, is a version of this product.

(iv) Aluminium hydroxide trihydrate, which is the Bayer process product before calcination to alumina. It is an important flame retardant and may also be used as a toothpaste abrasive, an additive for paper, and for the manufacture of synthetic marble.

(v)   Aluminium sodium silicate (synthetic zeolite), which is produced commercially by the crystallization of sodium aluminosilicate gels prepared from pure sodium aluminate, sodium silicate, and sodium hydroxide solutions. The sodium aluminate being made by the dissolution of alumina trihydrate in aqueous sodium hydroxide.

(vi) Aluminium fluoride, which is made from aluminium hydroxide and hydrofluoric acid. It is an important part of the electrolyte for aluminium metal production.

Synthetic zeolite has important uses in catalysts and as a builder in the detergent industry (see Chapter 12). In heavy duty laundry detergent powders zeolite has replaced sodium tripolyphosphate (STPP) in those geographical areas which have phosphate bans or restrictions (see next section).

3.7   Silicates and Zeolites

Sodium silicates, made from silica sand and soda ash (and/or caustic soda) are important ingredients in detergents, catalysts, pigments, and paper adhesives. They are also used in water, paper, and ore treatment.

The technology for the manufacture of sodium silicates is very similar to glass-making technology. The process involves fusing soda ash and high quality silica sand at high temperature. The proportions of the two ingredients determine the type of silicate produced and its alkalinity.

The market for sodium silicates has been somewhat weak in recent years but the growing use of peroxide bleaching in pulp and paper is boosting demand in that sector of the market. Waste treatment applications are also growing.

The decline of sodium tripolyphosphate (mentioned above under industrial phosphates) as a builder in detergents has coincided with a period of strong growth

Table 8   *World: Detergent Zeolite Capacity* (1000 t)

Region	In-place	Projects
West Europe	549	390
East Europe	310	0
North America	448	23
Latin America	0	0
Africa	0	35
Asia	319	71
Oceania	0	0
Total	1626	519

for synthetic zeolite builders. From their commercial introduction in 1978 world capacity has increased to about 1.6 million a^{-1} (Table 8).

These sodium aluminium silicates deal effectively with calcium but not with magnesium, and so they are used in conjunction with co-builders such as polyacrylates. There have been objections to these co-builders on the grounds that they are not biodegradable but the zeolites themselves are environmentally benign.

If the zeolites have a disadvantage, it is that they are insoluble. The laminar sodium silicate builders, which are water-soluble, are under intense study. If the promise of these novel sodium silicates is confirmed, and if they can be made at an economic cost, then they may well replace zeolites in the powder detergents of the future.

3.8   Hydrogen Peroxide and Persalts

Hydrogen peroxide has been one of the stars of the inorganic chemicals industry over recent years. The particular merit of hydrogen peroxide as an oxidizing agent is that when it has fulfilled its function only water is left. It is the ultimate in environmental acceptability.

Growth in consumption has been rapid. In the USA the average growth rate for the period 1980–90 was 5.8%. The most important use is in pulp and paper bleaching where hydrogen peroxide has benefited from the pressure on the industry to reduce chlorine consumption. Also important are environmental applications, chemical synthesis, and textile bleaching. Hydrogen peroxide is also used in the leaching of uranium ores and for detoxifying the sodium cyanide used in gold leaching operations.

All modern hydrogen peroxide plants use auto-oxidation of anthraquinone derivatives such as ethyl or t-butyl anthraquinone (see Chapter 7). Previously, hydrogen peroxide was produced by the electrolysis of ammonium or potassium bisulfate solution, a process which is still used for the production of persulfates. Another route, which was used by Shell in the US, involved the oxidation of isopropanol to hydrogen peroxide and acetone.

Du Pont is developing a liquid-phase process that directly combines hydrogen and oxygen over a proprietary catalyst. It has been suggested that this process could have a much lower capital cost that the auto-oxidation process.

In Western Europe, a major use for hydrogen peroxide is the manufacture of sodium perborate, a solid bleaching agent which can be incorporated in household powder laundry detergents. Sodium perborate is made from sodium borate and hydrogen peroxide, as described in Chapter 7.

Sodium perborate is a very effective bleach and stain remover at elevated wash temperatures and at lower temperatures in the presence of an activator. Prior to 1975, white cotton textiles were laundered at or near the boil. Under these conditions sodium perborate was particularly effective. At the lower wash temperatures recommended for synthetic and synthetic/cotton blend fabrics, sodium perborate was less effective. Its position as a major ingredient in European formulations was maintained, as average wash temperatures fell, by the development of activators such as TAED (tetraacetoxy-ethylenediamine).

The trend to concentrated detergents has led to the use of sodium perborate monohydrate rather than tetrahydrate which was the standard product for many years.

Sodium perborate has not made the same impact in US detergents. US washing machines often do not have heaters, and washing temperatures are low compared to Europe. The other significant difference is that the concentration of detergent in the wash water is much lower in the US than it is in Europe. Under these conditions sodium perborate is not effective. Bleaching was accomplished by soaking the washing in chlorine bleach (sodium hypochlorite solution). The situation today is that activators have been developed that render sodium perborate effective at the temperatures used in the US and products containing the system are on the market. Their market share is limited, however.

Sodium percarbonate, made from soda ash and hydrogen peroxide, is an alternative to sodium perborate.

Hydrogen peroxide itself is also used as a bleaching agent in pulp and paper, and textiles. Its environmental uses (municipal and industrial waste treatment, detoxifying the cyanide used in gold leaching) are showing strong growth. It is also used in the manufacture of organic peroxides, in hydrometallurgical operations, de-inking waste paper, and as a cleaning solvent. Organic peroxides are used as catalysts and curing agents for unsaturated polyesters, polyethylene, PVC, polystyrene, and other plastics and rubbers.

3.9 Fluorine, Hydrofluoric Acid, and Fluorides

Inorganic fluorine chemistry is largely based on fluorspar, a mineral form of calcium fluoride which is found in commercial quantities in many countries of which the most important are China, Mongolia, Mexico, the CIS, France, Italy, UK, and Morocco. Fluorine values, usually in the form of fluorosilicic acid, are also isolated during the manufacture of phosphoric acid, superphosphates, and defluorinated phosphate rock (tricalcium phosphate) from phosphate rock which, typically, contains 4% F (see also Chapter 8).

Hydrofluoric acid is made from acid-grade fluorspar, containing 97–98%

calcium fluoride, which is attacked with sulfuric acid to give hydrofluoric acid gas and gypsum. The HF is absorbed in water to give a solution from which various concentrations can be made. Anhydrous HF is made by concentrating the aqueous solution by distillation.

Hydrofluoric acid is used to make fluorocarbons, fluorine-containing plastics, fluoride salts, and speciality chemicals including herbicides and pharmaceuticals. It is used to make synthetic cryolite for the aluminium industry. It is an important alkylating agent in the petroleum industry where it competes with sulfuric acid. It finds outlets as a pickle for stainless steel and in uranium chemical production, and a glass etchant.

The most important end-use is the manufacture of fluorocarbons. The rapid changes in this sector of the organic chemical industry, which also affect the chlorine and hydrochloric acid markets, with the phasing out of CFCs -11, -12, and -113 and the development of new HCFCs and HFCs, will reduce HF consumption, at least in the short-term.

About 75% of sodium fluoride is consumed in water fluoridation.

3.10 Bromine and Iodine

Bromine is made from brines which are pumped from underground deposits such as those in Arkansas, or which arise as by-products of potash refining operations such as those based on the Dead Sea. It is released from the heated brines by reaction with chlorine.

Bromine is very toxic and it is corrosive, and is expensive to transport as it requires special purpose-built tanks which are costly to make and maintain, and involve the supplier in a large capital outlay. For this reason it makes sense for the producer of bromine to also make the major bromine derivatives. Brominations produce hydrobromic acid as a by-product and this has to be recycled for good production economics. The recycling can be done in the bromine plant.

These constraints mean that there are very few bromine manufacturers. In the US there are three, all in Arkansas. There is a major producer in Israel, Dead Sea Bromine, and one each in Germany, France, Italy, Spain, and UK. Bromine is also produced in the CIS, Japan, and India. World bromine consumption is about $400\,000\,t\,a^{-1}$.

Well over 50% of the production of bromine is converted to organic compounds, *e.g.* flame retardants, methyl bromide (an agricultural fungicide), and ethylene dibromide (a component of anti-knock compositions and a fumigant). Inorganic compounds are used in oil and gas completion fluids, photography, pharmaceuticals, hair-waving treatments, and swimming pool disinfection.

The most important group of derivatives of bromine are the flame retardants. These include tetrabromobisphenol A, tribromophenol, ethylenebistetrabromophthalimide, and polybrominated diphenyl ethers.

Iodine production is in even less hands than bromine, not because it is a hazardous chemical but because there are fewer commercially viable sources and the market is much smaller in tonnage terms. World iodine production is about $16\,000\,t\,a^{-1}$ with Japan accounting for about 47%.

Most iodine producers process iodine-rich brines which occur naturally in some locations, sometimes in association with bromine. In Chile, iodine is found in association with sodium nitrate and other salts in deposits which are mined by open-pit and underground methods.

Iodine is isolated from the brine sources by acidification followed by chlorination. The liberated iodine is absorbed in a solution of sulfuric acid and hydroiodic acid from which it can be recovered by a further chlorination. Iodine can also be recovered by ion-exchange.

The process used in Chile involves leaching the ore with an alkaline solution to dissolve iodine and sodium iodate. The iodate is converted to iodide by reaction with sodium bisulfite. Iodine is precipitated from the sodium iodide solution by reaction with fresh leachate. The iodine is filtered off and is purified by sublimation.

Iodine is used in disinfectants (such as iodophors), catalysts, animal feed supplements, in the manufacture of dyes and pigments, in photography, in *X*-ray contrast media, and for iodized table salt. A deficiency of iodine leads to thyroid enlargement (goitre) in humans which is common in some parts of the world including India.

3.11 Titanium Dioxide

Titanium dioxide is the primary white pigment for paints, plastics, and paper. It is manufactured throughout the world, but there are relatively few major manufacturers each of which has plants in a number of locations. The leading manufacturers have not been prepared to licence their technology preferring to enter in local manufacture or joint-venture manufacture. Plants range in size up to about $300\,000\,t\,a^{-1}$.

The titanium minerals rutile and ilmenite are found in beach sands in association with zircon and other heavy minerals. The deposits are usually on or near the surface and mining is straightforward being usually accomplished by dredging. Separation of the minerals is based on density differences.

There are two processes for titanium dioxide manufacture. The older process, which is still widely used, uses sulfuric acid to attack the titanium-bearing raw material. The newer process uses chlorine to attack the raw material. Most new plants use the chlorine process which is cleaner from an environmental standpoint. The sulfate process produces large volumes of waste products. The chlorine process requires a higher quality raw material than does the sulfate process. Chlorine is recycled. Titanium dioxide production is more fully described in Chapter 13.

Environmental pressures have resulted in sulfate plants being closed, with replacement by chlorine process units. Sulfate plants have also been modified in order to recover and recycle sulfuric acid, a development which has had a significant impact on sulfuric acid demand.

Despite the small number of producers, the market fluctuates between tightness and strong prices and slackness and weak prices. The industry has found it difficult to schedule additions of new capacity in order to prevent over-capacity at times of weak demand.

3.12 Borates

The production of refined borates and boric acid depends on a very limited number of sources of borate minerals (see Chapter 10). The main minerals are the sodium borates: borax (also known as tincal) and kernite; colemanite (calcium borate); and ulexite (sodium calcium borate). Commercially viable deposits of borate minerals are extremely rare, the principal ones being in California and Turkey. Deposits are also exploited in Argentina and Chile, while some in Mexico may be used in the future.

This scarcity of mineral resources has resulted in supply being concentrated in very few hands. The two US producers do not offer minerals but concentrate on the production and marketing of refined products. The Turkish producer began as a supplier of mineral (specifically colemanite, which was used in several plants in Europe to make boric acid) and has become a supplier of refined products based on both colemanite and tincal.

The principal refined products are borax (sodium borate) and boric acid. In the US both are made from sodium borate minerals. In Turkey, France, and Italy, boric acid is made from (Turkish) colemanite by acidulation with sulfuric acid, separation of the boric acid solution from calcium sulfate by filtration, and crystallization. In Chile boric acid is produced from ulexite, and in Argentina borax and boric acid are made from tincal.

Glass-making is the main outlet for borax and boric acid. Heat resistant borosilicate glass is widely used for lighting, ovenware, tableware, laboratory equipment, chemical plant, and sealed-beam headlamps. Glass wool and continuous glass fibres for insulation and reinforcement contain boron, as do lime–soda glass for specialist containers for pharmaceuticals and cosmetics. Boron is also an important ingredient in many ceramic glazes. It is an essential trace element for plant nutrition, and where deficiencies occur is added to compound fertilizers or is applied in solution to the growing crop. Borax is used in corrugating adhesives and as a corrosion inhibitor. High purity boric acid is used in the safety systems of nuclear power stations. Mixtures of borax and boric acid are used as timber preservatives and flame retardants. These and other uses are also described in Chapter 10.

In Europe sodium perborate, made from borax and hydrogen peroxide, is an important ingredient in powder laundry detergents for household use. Sodium perborate is a major derivative of both borax and hydrogen peroxide but it is produced by hydrogen peroxide makers rather than by borax makers.

3.13 Industrial Gases

The main industrial gases oxygen and nitrogen are not included in the US Bureau of the Census returns for inorganic chemicals production. The European Chemical Industry Council and the UK Chemical Industries Association do not include industrial gases in their sectoral analyses of the chemical industry. Nevertheless, the products of the industrial gases business are used as intermediates in chemical production and so a brief description of the industry is given here (see also Chapter 9).

The main industrial gases are oxygen, nitrogen, acetylene, argon, hydrogen, and carbon dioxide. Oxygen, nitrogen, and argon are produced in air-separation plants. The size of cryogenic air-separation plants has increased substantially since the mid-sixties. In terms of oxygen production, the change has been 10-fold from about $200\,t\,d^{-1}$ to $2500\,t\,d^{-1}$. Over the same period non-cryogenic technologies have been developed using the differential rates of absorption or permeation for the components of air. These newer technologies are particularly useful for users in remote areas with limited gas consumptions where gas purity is not critical.

Oxygen represents about 29% of the industrial gases market by value. Nitrogen is sold in larger volume than oxygen but it is less expensive and accounts for about 21% of the market by value.

Oxygen is used in welding and primary metals manufacturing. In petrochemicals, it competes with air in several oxidation processes. There is a trade-off between capital investment and operating cost which can favour the use of oxygen. Oxygen is favoured on cost grounds in the direct oxidation of ethylene to ethylene oxide, the TBA-route propylene oxide process, the production of vinyl acetate monomer, and the BASF acetylene process. New plants for acrylic acid and acrylonitrile could be oxygen-based. In general, the substitution of oxygen for air in chemical processes allows the process to be performed in much smaller equipment with significant fuel savings and cleaner flue gases.

The pulp and paper industry is using more oxygen instead of chlorine to remove lignin from chemical pulp. The oxygen delignification step reduces by a factor of two the total production of chlorinated organics in the bleaching step.

It is claimed that the sulfuric acid regeneration process can be improved by using pure oxygen. Oxygen-enhancement (the use of a mixture of air and oxygen) is already in commercial use in regeneration plants. It allows for a greater throughput in existing plants, or smaller plants to treat the same amount of waste acid.

Apart from its very large scale use in ammonia production (see above) nitrogen is widely used as an inerting agent providing a protective atmosphere from oxidation and moisture control in petroleum refining, chemicals, and glass manufacturing. Nitrogen atmospheres are also used for food storage.

Argon is inert and is used as a protective atmosphere in primary smelting and reduction of non-ferrous metals. It enhances oxygen decarburization of stainless steel. It is a shielding gas in arc welding.

Hydrogen for use as a chemical in, *e.g.* methanol and ammonia synthesis, is usually captively produced on site. That required for hydrochloric acid production is usually taken from chloralkali cells. As a merchant product, hydrogen represents about 5% of the industrial gases market by value, and is used for the production of medium-tonnage chemicals such as hydrogen peroxide and aniline. It is also used as an oxygen scavenger in many processes, and as a rocket propellant.

Carbon dioxide arises in large quantities as a by-product in the manufacture of ammonia. Much of this carbon dioxide goes into urea production. Merchant carbon dioxide represents about 9% of the value of the merchant gas market. It is used in the food industry in such applications as crusting, chilling, freezing, firming, and blending, and to carbonate soft drinks. It is also used in heat treatment, welding, aerosols, lasers, and tertiary enhanced oil recovery.

3.14 Miscellaneous

The inorganic chemicals which are commercially important but less so than those covered above include the following.

- **Calcium chemicals**, principally calcium chloride, calcium nitrate, and calcium hypochlorite. Calcium chloride occurs naturally but is generally produced as a by-product, *e.g.* of synthetic soda ash production, or as a derivative of hydrochloric acid which would otherwise go to waste. Calcium chloride is used in road de-icing, for de-icing coal in transit, in dust suppression on roads and in mines, and in oil- and gas-well drilling fluids. Over 700 000 tonnes of calcium chloride were made in the US in 1990. Calcium nitrate is a by-product of the production of nitrophosphate fertilizers, the calcium being derived from that in phosphate rock. It is used as a source of calcium (and nitrate) in agriculture. Calcium hypochlorite is important in the USA where it is widely used in residential swimming pool sanitation.
- **Chromium chemicals**, principally chromic acid and sodium and potassium dichromate. Chromium chemicals are derived from chromite ores which are mainly used for the production of the ferroalloy, ferrochrome, and in refractories. The main uses for chromic acid are wood preservation and metal finishing.
- **Copper chemicals**, principally copper sulfate. Copper sulfate is used in the production of agricultural fungicides and as an essential trace nutrient for plants and animals. Industrial uses include activator in froth-flotation of sulfide ores, wood preservation, electroplating, and leather tanning; and in petroleum refining as a sweetener for sulfur removal. These are all in addition to its major use as the electrolyte in copper production.
- **Iron chemicals**, principally ferric chloride and ferrous sulfate. Most commercial ferric chloride is obtained from steel mill pickling liquors (where hydrochloric acid is used) or as a by-product of the chloride route to titanium dioxide. Almost all applications are in the water treatment area. Technical grades of ferrous sulfate are also produced from steel mill pickling liquors (where sulfuric acid is used) and from by-product copperas from sulfate-process titanium dioxide manufacture. Pure grades are made from iron and sulfuric acid. Ferrous sulfate is used in water treatment, animal feeds, lawn fertilizers, and in the production of iron oxide pigments.
- **Lithium chemicals**, principally lithium carbonate and lithium hydroxide. Lithium values are extracted from brines and from concentrates of pegmatite ores. Brines are concentrated by solar evaporation which is conducted in such a way that sodium and potassium salts crystallize to leave a brine concentrated in lithium. Lime is used to remove magnesium and the lithium is precipitated, as its carbonate, with soda ash. The pegmatite ores are concentrated using flotation and the concentrate is leached with sulfuric acid to give lithium sulfate. The sulfate is converted to carbonate using soda ash. In recent years Chile has become an important source of lithium carbonate, which is the starting material for other lithium chemicals and for lithium metal (used in batteries and aluminium alloys). Lithium carbonate is also

used in glass making, in thermal shock-resistant ceramics, and in aluminium production as an electrolytic bath additive. Lithium hydroxide is used in the manufacture of lithium soaps for incorporation in multi-purpose greases.

- **Magnesium chemicals**, principally magnesia, magnesium chloride, and magnesium sulfate. Magnesium chemicals are derived from sea water, brines and minerals such as magnesite, dolomite, and kieserite. Magnesia is used in refractories, animal feeds, as a secondary plant nutrient, and in water treatment. Magnesium chloride is a raw material for magnesium metal production and is also used in sugar refining, drilling muds, flooring compositions, fluxes, and de-icers. Magnesium sulfate is used in consumer and pharmaceutical products, in the production of high fructose corn syrup, as a soluble source of magnesium as a secondary plant nutrient as well as in pulp bleaching and animal feeds.

- **Manganese chemicals**, principally manganese sulfate used as a source of the micronutrient manganese in agricultural and animal nutrition; and potassium permanganate, which is mainly used in water treatment and as an oxidizing agent. Manganese chemicals are derived from manganese ores (dioxide and carbonate) large quantities of which are used in the production of manganese metal, electrolytic manganese dioxide, and ferroalloys.

- **Potassium chemicals**, the fertilizer aspects of which are dealt with above. The main chemicals are potassium hydroxide and potassium carbonate. Potassium hydroxide is made by the electrolysis of pure potassium chloride solution, the process giving chlorine as a by-product. Potassium hydroxide is the starting point for the manufacture of potassium carbonate, potassium phosphates, and other potassium salts. Potassium carbonate is used in the manufacture of glass for colour TV tubes and in optical glasses.

- **The rare earths** group comprises seventeen elements of which fifteen occur together in nature. The separation and purification of these elements requires sophisticated chemical technology. However, mixtures of rare earths are used by several sectors of industry and these mixtures are produced by less demanding methods. Consequently, the rare earth supply business lies partly in the mining and metallurgy sector and partly in the speciality chemicals sector.

The most important source minerals in 1990 were bastnasite with 57.5% on a rare earth oxide (REO) basis and monazite with 27.1% on a REO basis. The remaining 15.4% was made up of ion adsorption concentrates, xenotime, and other minerals. The importance of monazite will continue to decline owing to the difficulty with the disposal of radioactive thorium.

The most important rare earths in volume terms are cerium, lanthanum, neodymium, yttrium, and praseodymium. World consumption by product is shown in Figure 10. On an REO basis the total in 1990 was about 33 000 tonnes, spread over the end-uses given in Figure 11.

The traditional end-use sectors of catalysts, glass, and metallurgy will remain important in the future but usage in magnets and advanced ceramics will show the fastest growth. By the year 2000 the market is expected to have increased to about 53 000 tonnes REO.

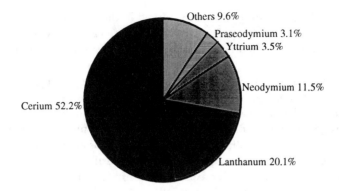

Figure 10   *World: Rare Earths Consumption by Product 1990* (%)
(from CRU International Ltd.)

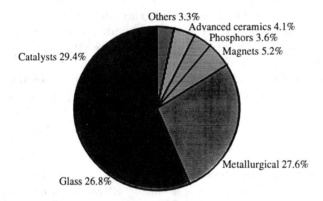

Figure 11   *World: Rare Earths Consumption by End-use 1990* (%)
(from CRU International Ltd.)

- **Sodium metal**, made by the electrolysis of molten sodium chloride (with chlorine as a by-product) is used as a reductant in the production of titanium metal, in chemical synthesis, and in the manufacture of tetraalkylleads. The market for the last has declined as leaded gasolines have been replaced by unleaded. Two end-uses show potential: sodium azide, used in air-bags in cars, can be made from sodium; and rechargeable sodium sulfur batteries are under study for electric vehicles.
- **Sodium cyanide** is made from hydrogen cyanide (HCN) and sodium carbonate or caustic soda. HCN is purpose-made (from ammonia and natural gas) or arises as a by-product of acrylonitrile production (ammonidation of propylene). Small quantities of HCN are made from formamide. In the USA about 20% of production capacity is by-product. World capacity is $450\,000\text{--}500\,000\,\text{t}\,\text{a}^{-1}$.

  Sodium cyanide can be considered as a portable form of HCN. Whilst HCN is moved in bulk, its extreme toxicity is a concern and sodium cyanide is a safer alternative whether in solid or liquid form. The market is driven by the state of the gold business where it is used in the leaching process. The high price of gold in the late 1980s led to a growth in capacity with new plants in, *e.g.* Australia and the western US, specifically to serve gold producers in the area.

It is also used in the selective flotation of sulfidic ores for the separation of metals like lead, zinc, copper, molybdenum, and iron; electroplating and the surface hardening of steel parts; and in organic synthesis.

- **Sodium hydrosulfide**, which is made by saturating a solution of caustic soda with hydrogen sulfide, is used in the synthesis of organic chemicals and dyes, in kraft pulping, and in the leather industry.
- **Sodium hydrosulfite (sodium dithionite)**, is made in the UK from sodium formate, caustic soda, and sulfur dioxide. It is used in pulp bleaching, textile dyeing, and clay bleaching.
- **Sodium sulfate** is made from salt and sulfuric acid using the Mannheim process but natural (mined) sodium sulfate and by-product material is also available. The main end-use is detergents where sodium sulfate acts as a processing aid and filler. The trend towards concentrated powders and superconcentrates is reducing demand. Sodium sulfate is also used in kraft pulping as an alternative to caustic soda.
- **Sodium sulfite**, which is made from sodium carbonate solution and sulfur dioxide, is of growing importance in the pulp and paper industry for the chemithermomechanical (CTMP) process. It is also used in the water treatment industry with dechlorination being an active segment. It also has uses in photography, textile bleaching, food preservation, ore flotation, and oil recovery.
- **Sulfur dioxide**, made by burning sulfur in air (the first stage of the sulfuric acid process) and also available in smelter off-gases, is used to make sulfites and hydrosulfites, in pulp and paper, water and waste treatment, in the manufacture of corn products such as high fructose corn syrup (HFCS), and in metal and ore refining. It has been suggested that sulfur dioxide has strong potential in the power generating industry in the removal of particulate matter from stack gases.
- **Zinc chemicals**, principally zinc oxide and zinc sulfate. The world market for zinc oxide is in the range $500\,000$–$600\,000\,\mathrm{t\,a^{-1}}$. Its main use is in rubber compounding with car tyres being of particular importance. It is used in latex and oil-based paints as a mildewcide. It is also used as a chemical intermediate, in ceramics, agriculture, in glass, and in photocopying paper. The market for zinc sulfate is in the region of $100\,000\,\mathrm{t\,a^{-1}}$. The principal market is agriculture where it is used in both plant and animal nutrition as a source of micronutrient zinc. It is also used as a source of zinc for zinc dithiocarbamate fungicide, in the manufacture of zinc stearate, in corrosion inhibitors, and in rayon manufacture.

Many hundreds, if not thousands, of inorganic chemicals have been omitted from this brief review. The descriptions which are given above will serve to show the diversity and importance of this sector of the chemical industry. It is hoped that they also show that the inorganic and organic sectors of the chemical industry are indissolubly linked. They depend on one another for markets in the case of the inorganics, and for raw materials in the case of the organics. Above all, inorganic chemicals reach into every facet of modern life. Without them life as we know it would not exist.

(The reader is referred to 'Speciality Inorganic Chemicals', RSC Special Publication No. 40, for detailed description of production and uses of the important lesser-tonnage chemicals.)

CHAPTER 2

Environmental Aspects of Inorganic Chemicals Production

A. K. BARBOUR

1 INTRODUCTION

Environmental considerations and the management of environmental effects are now top priority issues for the managements of industrial companies in all sectors and in the vast majority of countries. No sector has expended more time and capital expenditure on improving the environmental performance of its processes and products to meet present-day requirements and expectations than the chemical and metallurgical industries; there is no doubt that this level of effort will continue as regulatory criteria tighten and public expectation becomes ever more demanding. This is largely because of management responses to the real or perceived concerns of the general public, and some scientists, about issues which are directly concerned with the activities of industry and the technologies upon which it is based. Such concerns are rarely related to the consumption and disposal of products by the general public. Most of these issues relate to perceived health effects which are generally thought to be particularly marked for children and old people. Some are related to ecological effects. Particularly vociferous responses are generated by perceived visual effects or those which are detrimental to amenity. Paradoxically, occupational health effects generally arouse less publicity and hence, less apparent concern. Examples are given in Table 1.

This chapter summarizes some of the major issues which confront, and will continue to confront, managements in the inorganic chemicals industry, under the following headings:

- Legislative/Regulatory regimes with emphasis on the United Kingdom and EC countries;
- Scientific issues in standards setting;
- Some areas where new technology assists environmental protection; and
- Areas where new basic information and new technology is required.

2 LEGISLATIVE/REGULATORY REGIMES

In the United Kingdom, implementation of the Environmental Protection Act

Table 1

Issue	*Concern*
Emissions from chemical and nuclear installations	Health of near neighbours, particularly children
Exhaust fumes from cars and trucks	Health, particularly children, and nuisance
World-wide distribution of persistent pesticides, *etc.*, *e.g.* DDT, Lindane	Ecology and, ultimately, human health
Safety of nuclear plants and nuclear waste disposal	Major accident hazard, human health
Incinerators	Human health, amenity
Safety and emissions from chemical plants/installations	Major accident hazard, human health
Waste disposal areas	Nuisance, amenity
Marine, beach, and river pollution	Adverse visual and amenity effects on fishing, surfing, bathing, *etc.*
Global issues, *e.g.* damage to ozone layer, 'acid rain', global warming	Human health, ecology, irreversible damage to 'planet earth'
Mining and mineral extraction	Visual intrusion, amenity, health effects from dust and noise, ecological effect
Food additives	Health effects
Groundwater contamination	Health effects

1990 has altered substantially the regulatory basis under which chemical and metallurgical plants regarded as possessing major polluting potential are authorized to operate. Such processes are now called 'prescribed processes' essentially defined as processes emitting, or having the potential to emit 'prescribed substances'. Relevant examples of the latter include:

(i) *Emissions to atmosphere.* Particulate matter, metals and metalloids, oxides of carbon, oxides of sulfur, oxides of nitrogen, organic compounds, asbestos, glass and mineral fibres, halogens and their compounds, phosphorus, and compounds of phosphorus.

(ii) *Emissions to water.* Mercury and its compounds, cadmium and its compounds, and essentially all non-ferrous metals.

(iii) *Deposition on land.* Alkali and alkaline earth metals and their oxides, slag containing metals and metalloids, and contaminated plant equipment, *e.g.* refractories, electrodes, *etc.*

The major prescribed processes (Part A processes) are controlled by Her Majesty's Inspectorate of Pollution (HMIP) while those regarded as of lesser polluting potential (Part B processes) are supervised by Local Authority Environmental Health Officers, at present for emissions to air and 'nuisance' only.

Processes of each type require a formal Authorization to be granted before they can operate.

When fully applied, the new system will require prescribed processes to meet the rigorous criteria attainable by the use of Best Available Technique, efficiently operated. Thus, the UK tradition of point-source control for emissions to atmosphere will be maintained, though the health-based criteria provided by ambient standards may well be introduced more widely through future EU legislation.

Emissions from such processes are controlled by Her Majesty's Inspectorate of Pollution applying the principles of Integrated Pollution Control (IPC) and Best Available Technique Not Entailing Excessive Costs (BATNEEC) to minimize the production of all wastes, the irreducible minimum of which will be disposed according to the Best Practicable Environmental Option (BPEO). These principles apply to emissions to atmosphere and discharges to water and so imply, at least for processes under HMIP control, the acceptance of the uniform fixed emission standards provided by the application of Best Available Technology, efficiently operated.

However, the National Rivers Authority (NRA) continue to base their management of river and estuarial quality on the Quality Objectives/Quality Standards approach which, though valid scientifically, was essentially unacceptable to other EC countries for the most toxic pollutants. To date, no conflict has been apparent between the fixed emission approach now adopted by HMIP through the application of BAT and the achievement of Quality Objectives adopted by the NRA.

Purification of process streams to acceptable emission criteria inevitably generates solids which, where they cannot be recycled back to process, will require permanent disposal in a landfill or destruction by incineration in an HMIP-authorized plant. Since disposals to land are, by definition, difficult to control on the basis of BAT-based emission criteria, they are currently covered by more qualitative 'Duty of Care' regulations. It seems likely that new UK landfills will in future have to meet the stringent design criteria now imposed in other countries such as the USA and Holland. The other major policy issue relating to solid waste disposal is whether or not the UK will be able to retain its current policy of co-disposing many biodegradable industrial and commercial wastes with domestic waste in the context of strong opposition to this policy by both the USA and EU countries.

Several other important departures from previous UK practice are embodied in the new system which, overall, may be summarized as:

- Control of *all* emissions from new plant by the use of Best Available Technique Not Entailing Excessive Costs (BATNEEC). (The use of the word 'Technique' rather than 'Technology' permits HMIP to retain the management, design, and maintenance aspects which were such an important feature of the former 'Best Practicable Means' system.)
- Production of wastes requiring permanent disposal to be reduced to the absolute minimum whose disposal will be judged by a Best Practicable

Environmental Option (BPEO) analysis to minimize adverse environmental impact.

- Publication of authorizations and their review at regular intervals, say every three years.
- The upgrading of existing plant to meet current environmental criteria, based on BATNEEC, on a timetable to be agreed with the Inspectorate. Three to four years might be considered appropriate.
- Agreed protocols of performance monitoring and the publication of results, in terms comprehensible by the non-specialist, in HMIP reports.

The new system undoubtedly provides a rational and scientific basis for the point-source control of emissions and should be fully acceptable to all but the most antagonistic members of the general public.

Teething troubles have accompanied the introduction of the system, as would be expected in a pioneering endeavour being introduced at a time of a deep slump in industrial activity. The major issues include:

- Since an HMIP authorization currently requires detailed information about all aspects of a process from input raw materials and specifications through to end-products and specification (and all managerial stages in between), it is not surprising that there have been complaints regarding the time to prepare it and the time for the Authorization to be granted. The costs to the operator in obtaining an Authorization have also been criticized.
- Real problems exist in defining BATNEEC in many industries. It would probably be more appropriate for HMIP to concentrate on methods for attaining the emission levels provided by BAT rather than the basic BAT process technology being authorized.
- A full BPEO study of all emissions is both scientifically demanding and costly; a justifiable argument in many cases is that only those emissions having major environmental impact should be included in the BPEO study.
- The real costs of implementation are excessive and could lead to uncompetitiveness, particularly at a time of tight resources and when further additions to unemployment due to plant closure is very undesirable.

Nevertheless, the proposals provide a basis for the chemical and metallurgical industries to meet the highest practicable environmental standards for their processes, and it is noteworthy that many of the requirements of the IPC System are now embodied in BS7750 ('Environmental Management Systems') and in the EU Regulation 'Eco-Management and Auditing Systems'. Another major initiative is the 'Responsible Care' Programme developed by the Chemical Industries Association; all member companies have to accept and implement this programme as a condition of membership. This is an important development and industry-wide declaration of intent which would have been quite impracticable only a few years ago.

If the principles of IPC continue to be embodied in future EU Legislation, and the 'NEEC' provision is not interpreted too liberally, then the major processes in the chemical and metallurgical areas will indeed operate on a 'flat playing field'.

Moreover, if, as seems likely, UN-related organizations such as the World Bank and UNEP also adopt such an approach then there is at least an embryonic 'flat playing field' internationally.

A logical and, some would say, overdue corollary of the Integrated Pollution Control system is the formation of an Integrated Inspectorate to implement it. The UK Government has announced its intention to form such an Inspectorate by joining with HMIP, the pollution control functions of the NRA but the implementation of this decision remains for the indefinite future.

2.1 Plant and Process Safety

The potential for serious accidents at chemical and metallurgical plants probably causes the general public at least as much concern as the effects of process emissions. Such serious accidents brook no argument. Unlike most environmental issues, there is no debate that the accident has happened; only prolonged analysis to determine the reasons and prevent recurrence. Detailed analysis of the steps taken by the process industries and Government to improve standards and to minimize the likelihood of recurrence is outside the scope of this chapter. Regulatory steps include:

(i) *CIMAH Regulations* ('Control of Industrial Major Accident Hazards') – For plants storing and handling above certain 'trigger' values of toxic chemicals, there is introduced the very important analytical methodology of the Safety Case, together with designated emergency control centres, manning, and command. The neighbouring population is made fully aware of the activities taking place at CIMAH plants.

(ii) *COSHH Regulations* ('Control of Substances Hazardous to Health') – In due time these regulations, properly applied, should ensure that the exposure of plant workers to the products they make is kept to the absolute minimum practicable; that new exposure criteria are adopted without delay, including biochemical testing wherever practicable; and adequate health checks are made and recorded long-term, together with exposure data.

It is thus clear that Regulations, and, in principle, systems for implementing them, are now in place to:

- control emissions from production processes at the level prescribed by 'Best Available Technology';
- minimize waste production from processes;
- manage production processes by fully adopting the 'quality approach;
- audit performance on a regular basis to ensure both current and future compliance and the capital expenditures thus implied;
- design and operate plants in the safest practicable manner;
- provide the maximum practicable health protection for the workforce; and
- publish performance data in a comprehensible form and to advise the general public of all relevant safety aspects.

3 STANDARDS SETTING

Although, as noted above, IPC is a technology-based, point-source system for managing and minimizing process emissions to the environment, health-based systems will continue to play a vital part in setting criteria. This will remain true in the environmental area as the emphasis on ambient or environmental quality standards grows, as well as in the more traditional areas of occupational health and purity of food, drinking water, and some beverages.

The accurate setting of health-based criteria, balancing necessary safety margins for agreed populations with availability and cost, remains the most fundamental, and probably the most challenging, aspect of the environmental scene. Industry must face the fact that major pressures are inexorably pushing in the direction of ever-tighter standards. These pressures are generated by both sociological and scientific reasoning.

Sociologically, we all feel that Industry should be, and should be seen to be, safe both to the people who operate its plants and to those in the outside world, particularly near-neighbours. This is peculiarly true of the chemical and metallurgical process industries. But what is safe? We have long since passed the time when exposure of work-forces to the atmospheres of chemical plants caused acute, clinical poisoning. Clinical lead poisoning, for example, is now extremely rare. Increasingly, attention is being turned to the possible effects of chronic, long-term, low-level exposure which may ultimately lead to degenerative disease of various types. Some scientists take the position that exposure to a level of toxicant which gives rise to any detectable biochemical change, whether or not it is reflected by changes in a person's feeling of well-being, or clinical diagnosis, cannot be tolerated and that exposure standards should be set at levels which avoid these subtle biochemical changes. Those who take this view sometimes express it by saying that work-force exposure must not be greater than general population exposure – the obverse of the more general view that permitted workplace concentrations safeguard operator safety and hence general population safety, since the latter is generally several orders of magnitude lower.

Continuing developments in analytical chemistry in all its manifestations provide something of a scientific base for the above sociological concerns. Modern embodiments of gas–liquid chromatography, thin-layer chromatography, atomic absorption spectroscopy, mass spectroscopy, and so on permit the detection of a very wide range of chemicals in miniscule concentrations in a wide range of substrates. We can now talk confidently, even if we do not comprehend fully, of concentrations in the p.p.m. and even p.p.b. range for many pollutants – concentrations which would have been quite beyond the limits of detection until recently. There is no sign that this revolution in analytical capability has yet spent itself, nor the extension of its results to toxicological and ecological implications.

Complementing these developments in strictly chemical analysis are those in biochemical testing: the measurement of the effect of minute concentrations of chemicals on biochemical systems. The main gap in this area naturally continues to be the precise relevance of such '*in vitro*' testing to the human situation. Animal testing is most useful for determining short-term effects or counter-effects of the infective or irritant type. Tests on animals are much less accurate in identifying and evaluating long-term effects such as carcinogenicity, mutagenicity, and teratogenicity. This is because laboratory evaluations of such effects are frequently conducted on several types of animal at concentrations necessarily much higher than normal human exposure or intake. Additionally, vocal public pressure has built up in recent years against the principle of testing potentially biologically-active substances on animals and there is no doubt that other methods of evaluation should be used to the maximum extent possible, consistent with human safety.

For the present, animal testing is justified because:

- workers in plants producing chemicals have the right to the best health protection possible;
- consumers also have this right and must be assured that all possible steps have been taken to ensure that products, prepared and consumed reasonably, are safe; and
- regulatory requirements must be met.

Nevertheless, increasing effort is being devoted to epidemiological surveys, *i.e.* judging the effects of pollutants on the basis of statistical analyses of large populations (sometimes not particularly large) of persons exposed to above-average concentrations of the toxicant under study. Such epidemiological approaches have widely-appreciated defects; particularly the separation of the effects of one pollutant from another, personal living factors, synergistic effects, and so on. But in many cases the epidemiological approach is the only tool currently available to judge the effects of pollutants on human populations – and naturally, judgements tend to be weighted towards the 'safe' side. This is particularly the case where judgements have to be made on possible carcinogenic, mutagenic, and teratogenic effects of chemicals and metals.

A wide range of regulatory agencies is now involved in the establishment of environmental and health criteria. Probably correctly, the international agencies, often notionally advisory, are gaining in influence and status. Examples are the World Health Organization, the International Agency for Research on Cancer, the International Commission on Radiological Protection, and various sections of the Commission of the European Community. Nonetheless, the influence should not be minimized of national agencies such as the Environmental Protection Agency of the United States and the American Committee of Governmental Industrial Hygienists. Important criteria are not only related to the concentration of potential toxicants in air: human ingestion through water and food is probably even more important in the minds of the general public in judging 'safety'. This is illustrated by the growing popularity of relatively expensive 'organic' foodstuffs, bottled waters, and the pressure to fulfil, at great cost, EU-inspired regulations on the purity of drinking water. Such waters, *i.e.* those supplied at the tap, rather than those marketed as bottled beverages, have to meet criteria such as:

triazine herbicides: below $100 \, \text{ng} \, l^{-1}$ with total herbicides and pesticides below $500 \, \text{ng} \, l^{-1}$

nitrates   :   regulatory level below 50 p.p.m.: ultimate target level
                         25 p.p.m.
lead   :   below 50 p.p.m. probably reducing to 25 p.p.m.

The foregoing will have demonstrated the close inter-relationship between
toxicology and environmental protection (or ecology as it is sometimes termed).
The integration between these two scientific areas is often not as close as it should
be and all interested persons should do what they can to improve scientific
communication between scientists working in these areas, and with managements
too.

4 MODERN PROCESS TECHNOLOGY AND ENVIRONMENTAL PROTECTION

Other chapters in this book deal with modern technology for the production of
most of the major inorganic chemicals. This presents a brief review of some of the
problems encountered in non-ferrous metals production and how relatively
modern process technology substantially reduces environmental impact or makes
it much easier to control.

The production of non-ferrous metals, particularly zinc/lead/cadmium, copper,
nickel, and aluminium, starts of course with mining/extraction, often by open-pit
methods, almost invariably followed by enrichment of the ore by milling and
flotation. The first of the Royal Society of Chemistry 'Issues in Environmental
Science and Technology' (ed. R.E. Hester and R.M. Harrison, published 1994) is
entitled 'Mining and its Environmental Impact', so this aspect will not be covered
here.

The processing of non-ferrous concentrates requires the effective management
and control of several common environmental problems:

- control of particulate emissions to atmosphere;
- where sulfidic concentrates are processed (*e.g.* copper, nickel, zinc/lead/
  cadmium) devising efficient systems for fixing sulfur dioxide, usually by
  conversion to sulfuric acid;
- maintenance of safe in-plant working atmospheres;
- disposal of solid waste arisings either by marketing or by environmentally-
  acceptable deposition as waste; and
- limiting liquid effluents to acceptable concentrations of heavy metals and
  other contaminants.

4.1 The Sulfur Dioxide Problem

4.1.1 In Copper Production. This became a first priority issue for the non-ferrous
industries when, in the late 1960s, it became clear that reverberatory smelting
of copper could not meet the 90% capture, emission, and tight neighbourhood
ambient air criteria then being formulated by the US regulatory agencies.
Although as a percentage of total US emissions of sulfur dioxide, the
non-ferrous metals industries – particularly copper – represented only about
one-eighth, the effects of the concentrated emissions around copper plants were

both severe and highly visible. This is because most of the sulfur dioxide is evolved from reverberatory smelting in gas streams containing around 1% by volume and hence impracticable as a feed gas to a contact sulfuric acid plant; the balance from the second or converter stage is cyclic in composition (see also Chapter 4).

The environmental imperative has led to the elimination of reverberatory smelting in most developed countries and its replacement by processes such as flash smelting and, occasionally, electric smelting, which fortunately had been under development for reasons of economies of labour, power, capital cost, and/or purity of product. At about the same time, the development in Germany of the double-contact, double-absorption process for sulfuric acid manufacture permitted emissions from this plant to be reduced by a factor of four or five to below 500 p.p.m. compared with single-contact sulfuric acid plants.

The most widely used embodiment of the flash smelting system was developed in the 1950s and early 1960s by Outokumpo. A generally similar technology has also been developed for the smelting of sulfidic nickel concentrates by the International Nickel Company and its recent adoption, together with much improved converting stages, has resulted in large reductions in sulfur dioxide emissions from their nickel production operations at Sudbury, Ontario.

Flash smelting has been adopted world-wide and, in combination with modern equipment for converting the matte produced, achieves acceptable performance for all environmental parameters. In particular, the sulfur dioxide off-gases (10–14%) are eminently suitable for conversion into sulfuric acid by means of a conventional double-contact acid plant.

In the flash smelting of copper the three stages of conventional copper smelting are carried out continuously, *i.e.* roasting, smelting, and partial conversion. The flash smelting furnace comprises three sections: a reaction shaft, a settling zone, and an offtake. The top of the reaction shaft is provided with several concentrate burners, through which is fed dried concentrate, fluxes, and oxygen-enriched air (usually about 40% O_2).

The overall reaction with chalcopyrite may be summarized as:

$$2CuFeS_2 + 2\tfrac{1}{2}O_2 + SiO_2 \rightarrow \underset{\text{(matte)}}{Cu_2S.FeS} + \underset{\text{(slag)}}{FeOSiO_2} + 2SO_2$$

With enriched air, the process is autogenous and so highly energy-efficient.

Particle separation takes place in the settler part of the furnace with the matte passing through the slag phase to the base of the furnace. Matte compositions are usually in the range of 45–65% copper; the matte is then upgraded by conversion, usually in a Peirce–Smith converter, to blister copper containing around 99% copper. Converting takes place sequentially:

$$\text{the 'slag blow': } 2FeS + 3O_2 + SiO_2 \rightarrow \underset{\text{(slag)}}{2FeO.SiO_2} + 2SO_2$$
$$\text{the 'copper blow': } Cu_2S + O_2 \rightarrow 2Cu + SO_2$$

With adequate sealing of the converter mouths, sulfur dioxide concentration is in the range 4–6% and so is suitable for conversion into sulfuric acid.

Since the slag from both the matte-smelting and converting stages contains significant concentrations of copper, slag cleaning is usually practised either by processing in an electric furnace or by flotation.

4.1.2 In Primary Zinc, Lead, (and Co-product Cadmium) Production. World primary zinc is produced by both electrometallurgical and pyrometallurgical processes. About 80% of current production is made by the electrolytic route, with the Imperial Smelting blast-furnace process for the simultaneous smelting of zinc and lead concentrates the most important of the pyrometallurgical routes.

As with copper and nickel processing, success in the high efficiency conversion of smelter-derived sulfur dioxide into sulfuric acid depends firstly on the basic smelting technology delivering a gas of adequate, constant concentration, preferably in at least the 5–7% range, and then the provision of highly efficient cleaning/dedusting facilities. These may comprise combinations of wet scrubbers (packed towers, venturis, *etc.*), electrostatic precipitators, and bag-collection plants.

The electrolytic process for zinc production is invariably based on high-grade zinc concentrates which can thus be desulfurized by efficient fluid-bed roasting. The blast-furnace process is unique in that it treats simultaneously a mixture of zinc and lead concentrates. For this reason fluid-bed roasting is inapplicable since the lead content causes the charge to agglomerate and 'hang-up'. However, updraught sintering on a conventional moving grate is readily operated to provide a gas of the required strength. With both processes, conventional practice world-wide is to convert the cleaned sulfur dioxide produced to sulfuric acid using a double-contact acid plant. Where particularly onerous ambient air criteria exist for sulfur dioxide, it may be necessary to scrub the effluent gases from the double-contact acid plant. Such scrubbing is also required in some countries to eliminate the very small amounts of sulfur dioxide emitted from primary aluminium smelters (mainly from the carbon and pitch used in anode manufacture), following the main fluoride-scrubbing section (see Section 4.2).

To make such 'smelter' sulfuric acid fully competitive with acid derived from elemental sulfur ('brimstone' acid) it will usually be necessary to reduce mercury levels to below 1 p.p.m. using one of the several technologies which are available. Cadmium levels also have to be similarly low so that neither of these elements contaminate products such as fertilizers made using the acid. A further necessity from a marketing standpoint is to lighten the relatively dark colour of smelter acid 'as made'. Whilst all these additional steps add cost to the production of sulfuric acid from smelters, they are viable in the context of current environmental management thinking, compared with the cost and likely impracticability of disposing of the acid as 'waste' or at distress prices (see also Chapter 1).

Regrettably, the energy demands and costs of processes to convert large tonnages of sulfur dioxide into elemental sulfur seem now to rule out such technology which, some years ago, was under active development to the level of large-scale trials. Such technology would have been advantageous in that it is much easier and cheaper to store and transport elemental sulfur than sulfuric acid, thus extending greatly the geographic link between the smelter as a producer of acid and the market for that production.

4.2 The Control of Fluoride Emissions in Primary Aluminium Smelting

In contrast to zinc, lead, copper, and nickel, primary aluminium is produced from refined alumina rather than mainly sulfuric concentrates.

In the classical Hall–Herault process, still essentially the only primary smelting route, alumina is dissolved in a cryolite melt contained in carbon-lined cells or 'pots', prior to electrothermal reduction.

Hence, the most important emission to be dealt with from the reduction stage consists mainly of fluorides. These can be captured by either wet or dry scrubbing with the latter now dominant since it achieves higher recoveries and has no associated potential liquid effluent problem.

In the dry-scrubbing approach a proportion of the alumina feed to the cells is diverted through a specific design of bag-filter plant where it reacts with the off-gases from the cells, effectively neutralizing their fluoride content. The material thus collected is returned directly to the cells, thus providing an immediate recycle of both the alumina used and the fluoride captured. Naturally, great care has to be taken that the alumina used for scrubbing fluorides is carefully controlled with particular attention necessary to the design, fit, and maintenance of the cell hoods to minimize in-plant concentrations.

4.3 Solid Waste Disposal

In all refining processes, the feed materials contain elements other than those major production metals for which the operation was designed. Some of these elements have markets and, if isolated in marketable form, are correctly termed 'co-products'; some have no market, or are isolated in a non-marketable form and thus require permanent disposal in an environmentally-acceptable manner. The market and marketability for many trace and impurity metals varies greatly and environmental considerations are a relatively recent and important factor.

Iron is a common and major impurity in non-ferrous metal concentrates. As the red-mud by-product of producing alumina from bauxite by the Bayer process it is almost invariably a waste and the same is true where it is contained in the jarosite $[(NH_4,Na)Fe_3(SO_4)_2(OH)_6]$ by-product of electrolytic zinc production. However, as a constituent of pyrometallurgical slags it may find a market as a foundation material or it may have to be disposed permanently as a waste. This is one of several examples where the same material may either be marketable or a waste, depending on specific circumstances.

In the production of primary zinc by the electrolytic process, several metals, present in relatively minor amounts in concentrate feed, are isolated as chemical precipitates in the preparation of the high-purity zinc sulfate solution on which the process depends. Examples are lead, mercury, cadmium, nickel, cobalt, and manganese. In general, these metals are further processed by specialists and ultimately find a market, though environmental considerations relating to products are introducing quite severe restrictions on outlets for mercury, cadmium, and lead.

Feeds for the production of primary copper frequently contain minor amounts of arsenic which volatilizes during processing and has to be captured efficiently to maintain acceptable in-plant and external emission levels. Markets world-wide, nowadays restricted by regulation essentially to copper–chrome–arsenic timber preservatives, by no means balance arisings, even allowing that smelters naturally favour feeds with low arsenic contents. In these circumstances disposal of arsenic-containing waste products in an environmentally-acceptable manner – particularly historical arisings – is a formsaidable task.

The foregoing paragraphs indicate the complexity of categorizing accurately the by-products of non-ferrous metals production, many of which are the essential raw materials for the long-established secondary metals industry, now more popularly termed 'recycling'. The following discussion will outline means for the disposal of 'wastes' from these industries, the term being used in its true sense of materials which cannot forseeably be used as sources of economically marketable products in competition with primary sources. Other materials which may be reprocessable under one set of economic circumstances, but not under others, should never be categorized as wastes but should be kept in professionally designed containers and storages until it is practicable economically to reprocess and market them.

4.3.1 In Primary Zinc, Lead, (and By-product Cadmium) Production. By the electrolytic process. The electrolytic process consists of the following main stages:

- concentrate roasting, to give zinc oxide calcine, plus sulfur dioxide which is converted conventionally to sulfuric acid;
- calcine leaching with sulfuric acid to give an impure zinc sulfate solution;
- solution purification; and
- electrolysis of the solution to give zinc metal, together with sulfuric acid which is recirculated to the leaching stage.

The cathodes are then stripped, melted, adjusted for composition, and cast.

For many years the major environmental problem for the electrolytic process was disposal of iron residues from the leaching stage. Such residues arose because the normal types of zinc concentrates used in the electrolytic process conventionally contain 5–12% iron and, during roasting, most of this combines with zinc oxide to form zinc ferrite $(ZnO.Fe_2O_3)$ which is largely insoluble under normal leaching conditions. Thus ferrite removes significant quantities of zinc from the circuit, together with nearly all the lead contained in the original raw material. These leaching residues were unattractive to lead smelters and became unsightly dumps at most electrolytic plants. They also placed a significant economic penalty on the process because of their zinc content.

Three distinct approaches to iron removal have ameliorated this environmental problem. It is removed by precipitation in one of the following forms:

- jarosite, $Na(K,NH_4)Fe_3(SO_4)_2(OH)_6$ (33–35% Fe)
- goethite, $FeOOH$ (63% Fe)
- haematite, Fe_2O_3 (70% Fe)

The first process is most widely used. It depends on dissolving all the iron present by leaching at a higher temperature (80–95 °C) and with stronger acid (80–140 g l^{-1} initially) prior to precipitating the crystalline easily-filterable jarosite by the addition of base. By this means, a relatively iron-free solution is obtained leaving behind a residue which is a technically acceptable feed to a lead blast furnace. Jarosite is a chemical precipitate in which significant concentrations of heavy metals remain absorbed and it has to be stored carefully in sealed holding areas with provision to return run-off water to process. It is uneconomic to recycle because of its low iron content and the presence of heavy metals.

For this reason, and because an unacceptable level of metals-containing leachate has been found at some jarosite disposal areas, rather expensive processing to haematite or goethite is practiced at some operations. Such materials can be disposed to the production of ferro-concrete and, perhaps, to iron smelters.

In whatever form iron leaves the electrolytic zinc production circuit, it is clear that an acceptable waste disposal site must be made available, as an insurance against variations in market demand, and that it must be designed as a permanent disposal site to the highest possible engineering standards for total containment.

By the zinc–lead blast furnace (the Imperial Smelting process). This process can successfully treat relatively impure zinc concentrates – and secondary materials – which would be uneconomic to treat by the electrolytic process. It is also suitable for the economic treatment of zinc–lead feeds containing appreciable amounts of copper. It should therefore be considered as a combined zinc and lead smelter, a particularly important concept in that its environmental performance should always be compared with an electrolytic zinc plant plus a lead blast furnace, together producing the same output as an Imperial Smelting furnace.

The basic stages of the Imperial Smelting process comprise:

- sintering by the updraught method, in which the sulfidic concentrate feed is desulfurized and decadmiumized, to give hard sinter together with sulfur dioxide which is converted conventionally to sulfuric acid;
- blast furnace reduction, in which sinter is reduced by CO to zinc vapour, which passes out of the upper part of the furnace to the condenser, and molten slag with lead bullion which collect for tapping at the hearth; and
- condensation, in which the zinc vapour issuing from the shaft is absorbed in circulating lead thrown up into a shower of droplets for subsequent zinc separation by cooling in launders.

This shock-chilling of zinc vapour is a vital technical aspect of the process because it freezes the reversion reaction by which zinc vapour undergoes rapid oxidation.

In blast-furnace practice about 75% of the cadmium contained in concentrates (usually in the range 0.1–1.0%) is volatilized during the sintering stage, the remainder passing to the furnace in sinter. The sinter gas (approximately 6.5%

sulfur dioxide) is cleaned by either wet or dry processing and the solubilized cadmium recovered, often by ion-exchange. Any mercury present is also recovered chemically.

In tonnage terms, much the major solid waste from the Imperial Smelting process is the furnace slag which is granulated immediately after removal from the furnace. It contains varying amounts of zinc, depending on process operating conditions, and small amounts of lead, both as oxides. Due to its siliceous, glass-like nature, it is inert and innocuous and is harmless environmentally. In some countries it has been used as a grit-blasting medium and as a landfill material. This is another example where the same material has a market in some circumstances but, in others, it is a waste requiring permanent disposal.

4.3.2   In Primary Copper Production. Slags from the flash smelting stage usually contain about 1% copper in the Outokumpo process and somewhat less using INCO technology. The former are usually cleaned by settlement in an electric furnace. Slags from the converter stage may contain from 2 to 12% copper, depending on operating conditions; it may be present as oxide, matte, or metallic copper. It may be cleaned electrically, alone or together with flash smelting slag, or it may be comminuted and cleaned by flotation.

After cleaning, it may find a market as fill or alternatively it may be permanently disposed of as waste; at this stage it is innocuous environmentally.

4.3.3   In Primary Aluminium Production. The major solids disposal aspect of primary aluminium production relates to the carbon cell or 'pot' linings. After a service life which may be in the order of eighteen to twenty-four months, these become hard and anthracitic in nature but they may well have a content of cryolite in the range 40–50%.

They are usually disposed to specific landfills as special waste, though in some EU countries disposal to deep sea remains permissible since no detectable environmental damage appears to result. Whilst not a major environmental problem, a practicable method of recovering the cryolite would be advantageous, both environmentally and from the economic point of view. There is an extensive literature on the subject but apparently no commercial-scale practice.

5   PRODUCT ASPECTS AND NEW TECHNOLOGY

The cycle of minerals extraction, metal refining, fabrication, use in products, and recycling is an overall cycle which is not replicated in products based on organic chemicals. The latter, including plastics, can be degraded to simpler building-blocks by thermal, oxidative, bacterial, or radiation-induced reactions. It is thus true that metals use is metals relocation and that the management of the adverse environmental impact of the metals *per se* has to be based on:

- management of *environmental concentrations* to levels which cause no significant harm to human health or to ecological systems; and
- major emphasis, embracing logistics, technology, and economics, to *raise the present level of recovery and re-use.*

As noted earlier in the section on arsenic arisings from copper production, where market restriction for environmental regulatory reasons forbids disposal of toxic metal by-products in minute concentrations throughout large areas of the environment, the only alternative at present is to seal the materials in their most chemically stable form into total containment waste disposal areas. Such areas will require very long-term monitoring because their contents, if accessible, are likely to be highly toxic and metals/metalloids do not 'disappear' from the environment. What has already happened with arsenic seems likely to occur also with mercury and cadmium where many outlets in products have been restricted severely by environmental legislation. Similar developments are likely for lead and tin, though direct primary outputs are rather easier to match with market demands than metals which are produced as relatively small by- or co-products from other major metals. Thus, the environmental impact of products as well as processes requires very careful evaluation and, in particular, all environmental disposal options need close analysis. Perhaps life-cycle analysis will provide a means of doing this, always with the proviso that appropriate toxicological and ecotoxicological data exist and are used.

5.1 Metal Arisings from the Manufacture and Use of Products

Such materials can theoretically be regarded as candidate raw materials for recovery by the secondary metals industry which has operated for many decades under similar environmental regulatory control regimes as are applied to the primary metals industries. Many operations, in practice, utilize a combination of raw materials, the blend of which depends on availability, price, and processability.

Of the material currently recycled, it is useful to distinguish between NEW scrap (*i.e.* material generated in the fabricating process which is generally recycled within the process or that of the feed metal supplier) and OLD scrap which is material returned from its end-use when time-expired in the market. New scrap naturally achieves a high level of recycling whereas, as noted below, considerable variation exists in the level of recovery of old scrap. This is a major area requiring attention if recycle levels generally are to be raised.

Examples of such materials are:

    (i)   *From manufacturing processes*: swarf, off-grade material, off-cuts, *etc. recycled* direct to process (*new scrap*).

    (ii)   *Packaging*: steel drums cleaned and *disposed as waste* when time-expired.

    (iii)   *Time-expired fabricated products (old scrap)*: recovery depends mainly upon the size of the item, its value and hence the *economics and logistics of collection, e.g.*

        ● *Beverage Cans.* Following successful initiatives in the USA, the European primary aluminium industry is achieving considerable success in collecting and reprocessing used beverage cans. In addition to the direct environmental benefits to the community, the process for recovering metal from such secondary sources requires only about 10%

of the power required to manufacture the equivalent tonnage of primary metal.

- *Car Hulks* are collected. Batteries and other non-ferrous components and car bodies are largely *recycled* through the relevant secondary industries.
- *White Goods* such as refrigerators, washing machines, dishwashers, *etc.*, are collected and *recycled* to a much lower extent than car hulks though this may possibly improve with the need to recover CFCs from refrigeration and air-conditioning equipment.
- *Constructional and Electrical Fabrications* in aluminium, copper, and lead are largely *recycled*, though with a considerable time delay since such fabrications have a useful life of many years.

As noted earlier, metals never disappear from the earth. However, to recover them from extremely dilute sources would require large supplies of very cheap power. Thus, for practical reasons, metals recycling will never closely approach 100%. Some metal uses are **dissipative**, *e.g.* in paints, plated, or galvanized protective coatings, plastic stabilizers, *etc.* and so are ineligible for recovery. Some EU countries have now enacted legislation for 'forced' recycling of some items, *e.g.* batteries, including some with very low metal contents; and, in Germany, plastic packaging, and, probably in the near term, consumer durables of most types.

5.2   Factors Restraining Increases in the Recycling Level

5.2.1   Regulatory Definitions. Materials scheduled for recovery/recycling in fully authorized operations must not be subjected to similar restrictions on movement, transfrontier shipment, definition, and insurance as are applied to wastes requiring permanent disposal. This has been difficult to implement because:

- the *same material* may be termed a *waste* from a primary process but is the *raw material* for a secondary (recycling) process and
- *economics* may dictate the designation. What may be a *waste* (*i.e.* not worth recycling) when primary prices are *low* may well become economically attractive to recycle (*i.e. a raw material*) when primary prices are *high*. These price variations may either be *short-term* (*e.g.* derivative from the business cycle) or *long-term* (*e.g.* derived from substitution or restrictive environmental regulation of use).

5.2.2   Logistics and Technologies for Collection/Separation. The separation of 'scrap' into its constituent metals and the development of systems for valuing and processing the separated metals has been developed to a considerable degree of sophistication by the secondary metals industry. Noteworthy examples are the collection and processing of secondary steel from car bodies, *etc.*; the recycling of lead-acid automobile batteries and other lead-containing products such as sheet; and the recycling of aluminium beverage cans noted earlier. However, embryonic schemes to separate scrap more efficiently at the domestic level seem to have made relatively little progress. This provides some justification for recent legislation in some EU countries, notably Germany and France, to 'force' such collection and separation. Positive encouragement should follow to design consumer durable components for ease of separation and processing when they become time-expired. These developments aimed at influencing the long-established market systems for dealing with 'scrap' and 'secondaries' will be watched with interest for they must succeed if the recycling level is to be raised significantly and permanently.

Some changes in legislation will also be required (*e.g.* lengthening the storage periods for 'wastes') to accommodate the variations in price which always accompany free markets.

5.2.3 Process Technology. Improved process technology to convert a wider range of secondary feeds into metals will doubtless be required but will only be developed when an assured supply of processable raw material is seen to be available under relatively predictable economics.

In the interim, the versatility and adaptability of the substantial secondary raw-materials industry should not be underestimated. The earlier sections on primary processing technology show that the metallurgical processing industry is well used to dealing effectively with both variations and impurities in feed materials and substantial variations in economics. However, the primary sector is used to having relatively well assured supplies of feed materials containing economically viable concentrations of the desired metals. Such conditions will be necessary if increasing proportions of secondary feed materials are to be processed into metals fully competitive on specification, cost, and reliability with metals obtained from primary sources.

6 SOME AREAS WHERE NEW TECHNOLOGY IS DESIRABLE

As noted earlier, the non-ferrous metals industries span mining, processing, refining, fabrication, and recycling – widely diverse technologies practised in almost all areas of the world with equally divergent economies, environments, and legislation/regulatory requirements. It is quite impracticable, therefore, to enumerate exhaustively all the aspects of these industries where improved technologies would improve economic and/or environment, health, and safety performance. Even a selective list must necessarily be subjective, as outlined below:

6.1 Resources/Extraction/Concentration

Thinking inspired by the 'Club of Rome' in the early 1970s that world resources of many metals would prove to be inadequate to meet future needs in the short term (20–30 years) have demonstrably been proved incorrect. A combination of factors accounts for the erroneous conclusions drawn by some from the 'Club of Rome' report. These factors include:

- a failure fully to comprehend that few mining companies explore for resources more than 20–30 years 'ahead';
- whilst world population growth has continued as predicted, world prosperity has not, and so demand growth has been relatively modest;

- substitution by plastics and composite materials, together with the miniaturization of much electrical and electronic equipment, has also restricted anticipated growth; and
- recycling is having some limited impact on the demand for primary raw materials and this impact will increase if the measures to stimulate recycling mentioned earlier show positive success.

Even though 'metals use is metals relocation' it is clearly impracticable, for the reasons already given, to recover and re-use anything approaching the total tonnages of primary metals extracted. Hence, the practicability of primary resource recovery is a current issue just as is the practicability of increasing the level of recovery from the idealized 'mine above the ground', which is the term often used to describe used metals not permanently disposed as waste and hence eligible to be recycled. For primary materials, total tonnage is not an issue (the outer mile of the earth's crust contains all the tonnage required for the forseeable future); the main considerations relate much more to the concentrations available both of desired and impurity metals and the costs, particularly power costs, of extraction and conversion to usable metals. Similar general considerations apply to the increased recovery in reusable form of metals from the 'mine above the ground'. Whilst it is true that much energy has already been expended on these materials, there remain considerable energy requirements in collection, separation, and reprocessing secondary metals.

Such thinking thus points the need for:

- improved exploration/extraction techniques, particularly at increased depths and
- improved concentration technologies.

Even with present technologies and operations the power involved in extraction, transportation, comminution, and separation is enormous and it is clear that any reduction in power costs would be highly beneficial. Such a reduction is almost certainly a practical impossibility in the real world for the foreseeable future.

For these and for environmental reasons underground solution mining might conceivably provide an answer to some of these problems provided that the following formidable technological developments succeeded:

- *efficient extractants* which must be economic, as selective as possible and environmentally acceptable;
- *above-ground purification systems* in which ion-exchange, solvent extraction, and membrane technology might play a part; and
- *guaranteed environmental protection* both below and above-ground.

6.2 Processing and Refining

Whilst present processing and refining technologies are generally adequate to provide refined metals at acceptable costs and, at the same time, to meet increasingly stringent environmental criteria, there is no doubt that continually increasing power costs will be a major issue for all sectors of the metals industry.

This is particularly the case as environmental pressures continue to mount.

Thus, any technologies which can economically raise concentrations of metals are greatly beneficial both to production costs and in the practicability and reliability of meeting environmental criteria. The preceding section stresses this point in relation to both primary and secondary metals extraction systems but it is equally important in processing/refining and in environmental protection/recycling. Examples include:

- The sulfur dioxide issue already discussed in the context of the manufacture of co-product sulfuric acid from smelters. A robust and economic means for concentrating sulfur dioxide gases to say 80–90% would make liquid sulfur dioxide a more generally practicable and more economically transportable option. Such material might also be a practicable basis for reviving technologies for the economic production of elemental sulfur by reduction.
- The precipitation of heavy metals in aqueous effluents by liming usually produces an alkaline sludge containing minor amounts only of the metal. Consequently, in general, such sludges are uneconomic to reprocess for their metal values and have to be permanently disposed as toxic wastes. An economic concentration technology, applicable to a range of metals and tonnages, would increase the recycling level somewhat and would deter the highly undesirable creation of further toxic waste tips. Such technology would also be very advantageous in the decontamination of metal-contaminated land.
- Increasing the recycling level of metals would undoubtedly be aided by better separation of time-expired metal containing household and other items. However, such education will only be partially successful and, in any case, takes no account of time-expired automobiles, white goods, and electronic equipment. Improved dissolution, concentration, and separation methods will be necessary if these items are to avoid disposal in permanent waste areas and instead provide raw materials for increased recycle.

6.3 Toxicology and Eco-toxicology

Accurate toxicological data, relevant to human beings and all other appropriate species, should be the basis for all environmental and health criteria, other than the global issues which are at present unquantifiable. That such data are not always available when environmental and health criteria are being established is partly due to lack of resource and partly because of the need to develop improved test methods, particularly for long-term, low-level exposures. Consequently, increased reliance has to be placed on epidemiological studies and on thinking by analogy with those toxicological data which do exist. Paucity of relevant data is even more apparent when considering eco-toxicological effects. Whilst it is likely that toxicological evaluations will always be based to some extent on animal testing, public distaste for this method is real and all efforts should be made to use alternative methods consistent with not compromising the accuracy of the evaluation. Improved methods of test and interpretation are particularly necessary for the accurate evaluation of the carcinogenic, teratogenic, and mutagenic potential for chemicals, including non-ferrous metals.

In the absence of the development of more, and more reliable and appropriate, evaluation methods for toxicological and eco-toxicological properties, criteria will continue to be established on a non-scientific, even political basis, leading to the misallocation of priorities and much consequential expenditure and effort which is largely wasted.

CHAPTER 3

Chemical Engineering for the Industrial Chemist

D. C. FRESHWATER

1 INTRODUCTION

In a book intended to be read amongst others by young chemists who envisage a career in industry, it would seem inappropriate to enlarge at length on the curriculum which they should follow in obtaining their chemistry qualification. This in any case is so well established that there is little that a single chapter in this book could do. What therefore will be done in this chapter is to highlight some of the differences between the world of the chemical industry and that of the college chemistry department. This will serve to show the importance in industry of the technology of chemical engineering. This is an essential study that is required in the design and operation of large-scale chemical plant and it thus behoves the tyro industrial chemist not merely to be aware of it but to have some working knowledge of the subject, albeit in an elementary way. It will help them to understand the problems to be solved on an industrial chemical operation which are so different, and usually additional, to those encountered in performing a chemical reaction in the laboratory. It will also give them an appreciation of the task of the chemical engineers with whom they will undoubtedly have to work.

No-one who has visited, let alone worked at, a large plant engaged in the production of chemicals can help but be struck by the lack of any resemblance between its operations and appearance and those of the typical chemistry laboratory. This alone should suffice to show that for the chemist who intends to work in industry, some background into the ways in which operations are performed in the chemical industry is not merely desirable but essential. In the laboratory, if we wish to carry out a reaction to make some specified product we are able, usually, to handle this in equipment made of a near-ideal material, glass. The flask or vessel is charged by hand from bottles of reagents taken from the shelf or weighed out on a convenient balance in amounts small enough to be transported with ease. Heat is supplied by a gas flame or electric mantle with an energy output far greater than is needed for most reactions, whilst cooling is provided through a virtually unlimited supply of tap water. At the end the products are removed from the reactor, filtered or decanted, and tested for purity all in easy hand-controlled operations. Generally we are more interested in getting the product that we want than in its yield or in the products of any side reactions. Nothing could be further from the situation that obtains when we want to perform the same reaction on an industrial scale. The quantities to be handled are such that mechanical means of transport must be used. The scale of operations is such that limitations in heat supply are a serious consideration, as are the requirements for cooling. The materials from which the vessels and other containments are made have to be strong enough to take the mechanical strains and resistant to the chemicals so as not to corrode away.

Subsequent operations such as separation assume a considerable importance, both because of their engineering demands or because they are often very energy intensive. Finally, we have to be concerned above all about the yield and the selectivity of the reactions that we are carrying out. All these differences may be summed up by saying that in the laboratory, the chemist is the master of his environment but in the plant he is its slave.

We come now to one other general point before any detailed exposition is attempted. It relates to another important difference between the usual operating conditions on chemical plants compared with those in the laboratory. In the latter, the processes that are carried out in a chemical preparation or synthesis are sequential and time-variant. That is to say, each step follows a logical sequence and is not normally carried out until the previous step has been completed. This is effectively 'Batch Processing' in industrial terminology. Although batch operations are conducted in certain circumstances, in industry, the vast bulk of chemical operations are carried out continuously. Thus raw materials are fed in at one end of the process in a continuous stream and, at as constant a rate as can be managed, products are also removed continuously. The reasons for this are technical and economic. It is easier to ensure a standard quality of product if operating conditions are maintained at a fixed level and this requires flow rates, temperatures, pressures, and concentrations to be kept as constant as possible at each point in the process. Furthermore, to operate in a continuous manner enables the plant to be used to its fullest capacity. A batch operation requires the plant to undergo time-consuming start-up and shut-down operations for each batch of material that is produced. Such periods, being non-productive, mean that this type of operation is not used unless the product is especially valuable or if the nature of the process is such that continuous operation is not possible. For example, in the production of some polymers the reaction time is significantly long and it is necessary to control conditions of temperature and pressure in a complex sequence which depends upon the degree of polymerization that is occurring. However such situations are encountered much less commonly than large scale continuous operations. Indeed, it is this type of processing which is typical of the modern chemical industry and especially of the inorganic side of the business. It means amongst other things that the plant which is capital-intensive is fully utilized, 24 hours a day, seven days a week unlike that in many other manufacturing industries. However, apart from this significant economic aspect, continuous processing requires a high degree of control using automatic process instrumentation. More will be written about this later.

There is one other important feature of the chemical industry to which attention must be given, and that is safety; or to give it a more descriptive and more modern title, 'Loss Prevention'. It is a topic which unfortunately receives little consideration in most undergraduate chemistry courses except for a rudimentary treatment of safety in the laboratory; most of which takes the form of descriptions of proscribed behaviour. In the chemical industry the matter of safe operation has to be taken very seriously. This is not because the industry *per se* is especially unsafe. Indeed, it has an excellent record in terms of numbers of accidents per employee. Rather does this concern stem from the potential hazards posed by large-scale chemical processing. Although fires, explosions, and toxic releases are rather rare, the potential damage that could result from an extensive breakdown on a chemical plant could be very large. This perception of hazard is important in two ways. First, it makes the chemical engineer and others concerned with the management and operation of chemical plants very aware of the need to follow the best practice in safety. Second, since there is a general awareness by the public of the potential hazard it ensures that management at all levels up to and including the Board are very safety conscious and do all that they can to promote safe practice. A subsequent section will cover this in more detail.

This necessarily brief outline of the chemical industry from the point of view of a chemical engineer is presented as the background to the next sections which will explain further the nature of chemical engineering and the work of the chemical engineer with both of which, the chemist entering industry should have some familiarity. In the next section, some of the topics touched on above will be developed in order to illustrate how the chemist may better be prepared to work in the chemical industry.

2 CHEMICAL PROCESS DESIGN

Whilst the chemist will not be expected to produce a process design, he ought to be familiar with the procedures used not merely from the point of background knowledge but, more importantly, because he may well find himself to be a member of the design team.

The design process starts from the idea that there exists the need or market demand for an alternative supply of a known material; or, more rarely, for the production of a completely new product.

The primary market information will have been prepared by those in the company skilled in this area, sometimes assisted by outside experts. The design team has then to produce a preliminary design on which costings can be made to enable a more firm assessment to be made of the market prospects. If this looks promising, then the design team will move on to prepare a detailed design.

The preliminary design begins with the determination of a process route and to decide this a number of seemingly simple questions must be answered. These may be summarized as follows:

(i)   What is the feed?
(ii) What is the product?

(iii)   How much must be produced?

(iv)   Are any of the materials in the feed, intermediates, or product(s) especially hazardous?

(v)   Does the proposed process route involve extremes of pressure and/or temperature?

(vi)   Where is the plant to be located?

The significance of these questions may best be illustrated by a simple example. Consider the design of a plant to produce fresh water from sea water.

Two extremes may be supposed: (i) to produce water for a town of 300 000 people on the coast of the Red Sea and (ii) to produce water for a lifeboat that has to provide for 30 shipwrecked people in the N. Atlantic. The two feeds differ in salinity by a factor of perhaps three and the quantities required will vary from five million to as little as eight gallons per day.

The tolerance of salinity in the product (*i.e.* its purity) will also vary between two extremes, that being tolerated by the survivors of a sinking boat being much greater than the sophisticated taste of town dwellers. Provided the process to be chosen does not involve electrolysis and subsequent production of hydrogen as well as chlorine, then there are no significant hazards. However, the nature of the process used for large-scale production is obviously going to be very different from that which is feasible in the lifeboat. We do need to make sure that the proposed process is feasible, *i.e.* it does not contradict known chemical and physical laws nor involve high risk reactions.

Having settled on a process for a known feed and a given product in quality and quantity, the next step is to draw up a flow sheet of the proposed process. This is an especially creative operation. There are many ways of achieving the desired condition of various stages of the process and one may often have a choice as to the order of the steps. For example, heating and cooling may be accomplished in a number of different ways and, likewise, the operation of mixing when necessary may either precede, be contiguous with, or follow a heating step. The ordering of the various steps is something of an art despite progress that has been made in recent years in analysing this procedure. Also, there are many alternative ways of ordering these steps all of which can achieve the same end-product. However, they will not all necessarily have the same efficiency or process elegance. Someone once said, 'Good prose is the right words in their right order, Poetry is the best words in their best order' and a good design illustrates the difference between one that is merely adequate and another that is demonstrably much better. A fuller discussion of the process of choosing and ordering the process steps is given by Rudd and Watson.[1] Let it be sufficient to suppose that this has been done and that all the steps and their sequence have been determined. The process may now be set out in diagrammatic form in a flow sheet or flow diagram. An example of this for the relatively simple process of making caustic soda from soda ash is shown in Figure 1. The next step is to calculate the material and energy balance for the process.

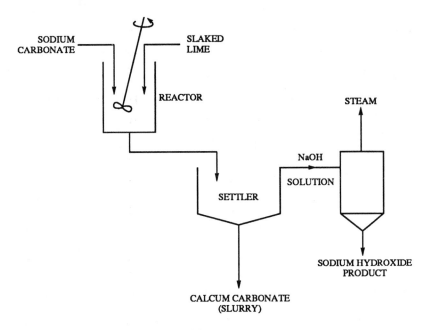

Figure 1 *Simplified Diagram of Sodium Hydroxide Production*

3 BALANCES

The preparation of material and energy balances comprises in essence, the application of the first (and occasionally the second) law of thermodynamics. We will suppose for simplicity that the operation is continuous and steady-state, *i.e.* flows, temperatures, pressures, and compositions do not change with time at a particular point in the system. The starting material and product(s) are known, as is the process route together with the expected yield and the reaction energies. Thus an overall balance of materials can be found, as can an estimate of the energy that has to be provided or removed. Since a preliminary design has already fixed the units in the process, each one can now be considered individually and its own material and energy balance calculated. Quite often it will be necessary to perform some of the chemical engineering unit operations procedures to solve the balances for a particular unit, but for the present purpose this need not be examined. It may be useful to continue with the simple example of the causticizing unit. In fact, this process is deceptively simple and some of its finer points will illustrate the concerns of the chemical engineer as different from those of the pure chemist. The reaction may be written as:

$$Ca(OH)_2 + Na_2CO_3 \rightarrow 2NaOH + CaCO_3$$

Applying simple reaction laws we see that the reaction constant and hence yield is proportional to the reactant concentration and inversely proportional to the square of the product concentration. Although the reaction rate is dependent upon temperature, the heat effects in this reaction are small and can be shown to have little effect on the rate. We are thus left with the concentration at which we should operate and also the matter of how much slaked lime should be used. Clearly, an excess of the latter would result in a more complete reaction (and greater yield). In general when one of the reagents is especially cheap (in this case lime) we use an excess so as to ensure that the more expensive reagent is fully consumed. However, it turns out that too much lime hinders the settling time of $CaCO_3$ to such an extent that the economic benefits of using an excess are lost. Thus we shall assume that the reagents used are in their stoichiometric amounts. Next we have to decide the concentration of sodium carbonate solution that is to be used. The conversion decays exponentially from near 100% at very low concentrations of sodium carbonate to about 86% for an 18% solution. Detailed costings (which are beyond the scope of this chapter) show that the optimum sodium carbonate concentration is about 13%. This is based on reaction time and settling time, both of which affect the capital cost and also on the need to concentrate the NaOH to 30% which involves considerable energy costs.

We are now able to draw up a flow diagram for the system and this is shown in Figure 2. Lime, sodium carbonate, water, and recycled sodium carbonate are mixed and then reacted in a series of three agitated tanks, the final mixture of NaOH solution, unreacted Na_2CO_3 solution and $CaCO_3$ going to a thickener or settling tank. The overflow from the thickener, consisting mainly of NaOH solution, goes to the evaporator system whilst the sludge settling out at the base goes to a series of three washers. This sludge contains a significant amount of NaOH adhering to the solid particles which must be recovered. That is done by washing the solid particles in three stages using as washwater condensate from the evaporators. The hot water has a reduced viscosity and so settling is more rapid. The wash liquor from the first washer goes to the initial reactor, thus recycling the NaOH that it contains. To complete the system, the evaporators (three in series) remove water so as to produce a 30% solution of NaOH. The hot liquor from the evaporators is cooled thereby precipitating unconsumed sodium carbonate. The latter goes back to the first reactor whilst the NaOH solution leaves as the final product.

This process has been described in some detail because it exhibits a number of features emphasizing the peculiar approach of the chemical engineer. First notice how great use is made of the recycling of materials so as to prevent unnecessary loss. Second, observe that several factors other than purely chemical ones enter into the decision as to the concentrations to be used. Third, observe how apparently small factors such as the use of hot water for wash in the thickeners can be incorporated to increase the cost-efficiency of the process.

Having developed this flow diagram we are now able to determine the material balances for the system. The calculation basis will be $100 \, kg \, h^{-1}$ of sodium carbonate producing a 30% solution of sodium hydroxide. Taking the equation quantities this requires $69.7 \, kg \, h^{-1}$ of $Ca(OH)_2$ and produces $75.5 \, kg \, h^{-1}$ of NaOH and $94.3 \, kg \, h^{-1}$ of $CaCO_2$. However since the sodium carbonate concentration is 13%, conversion is found to be 93.5% hence we see that

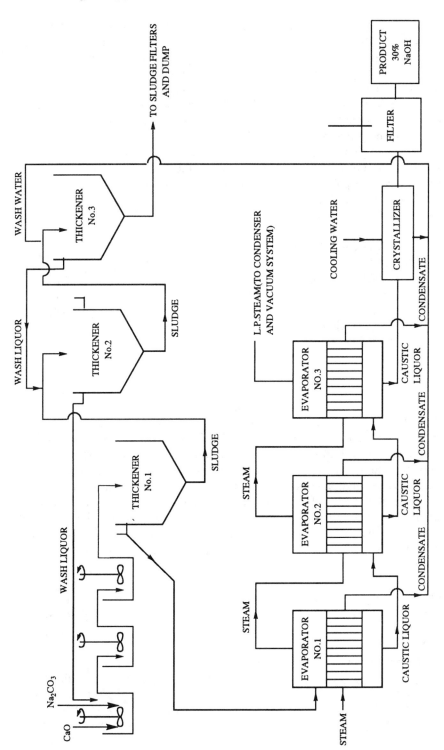

Figure 2

$670 \, kg \, h^{-1}$ of water are required and the actual production of NaOH is $70.5 \, kg \, h^{-1}$. Lime is added rather than slaked lime so an additional $16.9 \, kg \, h^{-1}$ of water are added together with $52.8 \, kg \, h^{-1}$ of pure lime. These sums assume that both CaO and Na_2CO_3 are pure. In fact both contain impurities but since the lime is in excess (due to incomplete reaction) and the impurities are insoluble and so will leave in the thickeners, these can be neglected. Also, the impurities in Na_2CO_3 amount to less than 0.5% and can likewise be neglected. A word of caution here. When you know that you are not dealing with pure chemicals although the impurities may be neglected in a material balance of this kind it is important to know what they are and if they are likely to affect the process due to such actions as corrosion, catalyst poisoning, or hazardous waste. In this example, fortunately, these factors can be dismissed. Thus completing the material balance on the agitator/reactors we have:

		$kg \, h^{-1}$
(i)	Insoluble solids leaving	
	$CaCO_3$	88.1
	$Ca(OH)_2$ unconverted	4.53
	Insol. Residue (from Na_2CO_3)	0.032
	Total	92.662
(ii)	Solution leaving Agitators	
	NaOH	70.5
	Na_2CO_3 (unconverted)	6.5
	H_2O	670.149
	Total	747.149

These figures are presented in two sets because it is necessary to know the mass of insoluble solids per unit mass of solution in calculating the balances on the thickeners. The calculation of these is complicated by the presence of recycle and so only the final results will be given here. Suffice it to say that: (i) a series of fairly straightforward simultaneous equations have to be solved and (ii) in order to set up the equations, some assumptions have to be made regarding the amount of solution that adheres to the solid at the exit of each thickener. For the sake of simplicity we shall report only the composition of the overflow leaving the first thickener (and going to the evaporators) and that of the sludge leaving the last thickener.

	$kg \, h^{-1}$
Overflow(solution) from thickener 1	
H_2O	587.087
NaOH	70.2
Na_2CO_3	6.423
Undertow(sludge) from thickener 3	
H_2O	92.508
NaOH	0.160
Na_2CO_3	0.015
$Ca(OH)_2$	4.533
$CaCO_3$	88.1

As a matter of interest, the amount of NaOH in the sludge from thickener 1 is 9.85 kg h^{-1} and these figures show the efficacy of washing.

We will not deal here with the balance around the crystallizer whose function it is to recover most of the sodium carbonate contained in the overflow from thickener 1. The amounts concerned are small and need not seriously affect the remainder of the balance which is concerned with the evaporators. However, at this stage we shall note that all the calculations so far have been based on a feed rate of 100 kg h^{-1} of sodium carbonate. This produces 70.2 kg h^{-1} of NaOH, a rate of production far too small to be of commercial interest. But knowing what rate we need, the figures in the balance can be calculated by simple proportion. Suppose, for example that we need 16 500 kg h^{-1} of a 30% solution then we simply multiply the quantities by $16\,500/(70.2*30) = 69.8$. In considering the evaporation stage it is convenient to work in more realistic quantities and so we shall take the figure of 16 500 kg h^{-1} of a 30% solution as our target.

Thus the total feed to the evaporators is going to be 47 114 kg h^{-1} containing 4980 kg h^{-1} of NaOH so the concentration is 10.5%. Since the final concentration is to be 30% the water to be removed is 30 280 kg h^{-1}. If this were all done in one evaporator it would require roughly that amount of steam. However, by dividing the load between several evaporators in series and making the water evaporated from the first act as the heating medium for the second, and so on, this steam load can be reduced considerably. This is known as the 'multiple-effect' principle and is effected by operating the series of evaporators at successively reduced pressures. The extreme limit of this technique is five evaporators. We shall consider three or 'triple-effect' evaporation. The actual calculations will not be reproduced here but rather a table is given (Table 1) which summarizes the results for single- (*i.e.* 1 stage), double-, and triple-effect evaporation.

From this it is seen that single-effect evaporation requires 2.65 times as much steam as triple-effect. However, the size of the evaporators for triple-effect is about three times that of a single-effect machine. When total operating and fixed charges are calculated it turns out that the triple-effect yearly cost is less than half that of the single-effect. This shows, in an abbreviated way, the kind of technical and economic calculations that have to be made at various stages in a process design.

In the meantime we are left with a wet slurry of calcium carbonate which contains about 50% water by weight and is too wet to handle other than as a slurry, *i.e.* it has to be piped and held in tanks. This is regarded as a waste or non-useful product, something that has to be discarded from the process.

Table 1   *Evaporator Calculations*

	Single-effect	Double-effect	Triple-effect
Area in each, m^2	530	539	540
Total area, m^2	530	1 075	1 621
Steam required, kg h^{-1}	31 300	17 000	11 800

Fortunately, the material is non-toxic and not hazardous but before it can be disposed of its bulk has to be reduced. This is done by filtering the slurry and thus producing a filter cake that can be handled like a solid. In some processes this is recycled by burning it to produce fresh lime but this is not generally economic.

Finally we complete this example with a summary of the material and energy balances for the process (Table 2, opposite).

4   UNIT OPERATIONS

The classical order in which the Unit Operations of Chemical Engineering are generally considered is as follows:

  (i)   Fluid Flow.
  (ii)   Heat Transfer.
  (iii) Evaporation.
  (iv)   Absorption.
  (v)   Distillation.
  (vi)   Extraction.
  (vii) Drying.
  (viii) Size Reduction.
  (ix)   Sedimentation.
  (x)   Filtration.

This is a somewhat old-fashioned list and modern treatments lump all the first eight together under the generic heading of 'Transport Processes' before elaborating on the detail of any of them. However, this treatment is beyond the scope of the present work. More importantly, it masks the differences between the various operations that are highlighted by the descriptive names given to them in the above list. We shall now proceed to outline the important principles involved in these unit operations. The illustrative example given above does not include all of these by any means, but it does provide a useful introduction to the application of the chemical engineering approach. We shall now consider the individual unit operations and give examples of their application.

4.1   Fluid Flow

In the chemical industry fluid flow is of prime importance. Not only is its application obvious in situations like the transport of liquids and gases, but is also of prime importance in the heating and cooling of fluids, in mixing, absorption, and so on. Wherever possible, materials are processed in the fluid form because of ease of handling. For example, solids are dissolved as a first step. Where this is not possible then the solid material is ground to a powder which is then mixed with a gas or liquid to make it behave as if it were a fluid. Examples of this are fluidized beds and slurries.

The flow of a fluid in a closed pipe or vessel is controlled by the energy changes in the system. The energy may be subdivided into four terms. First there is that due to difference in level (potential energy). Then that due to differences in pressure

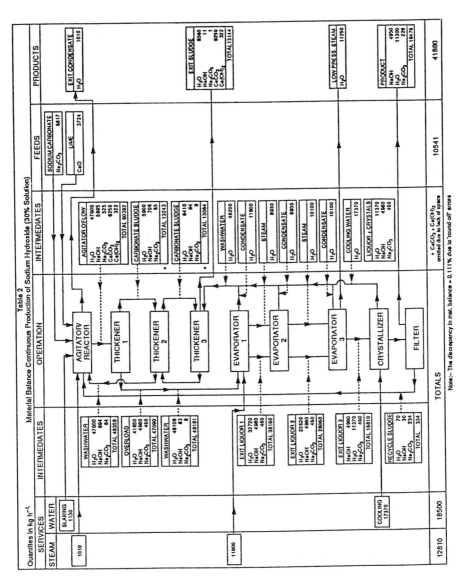

Table 2

Material Balance Continuous Production of Sodium Hydroxide (30% Solution)

between one point and another (pressure energy). Third is the energy possessed by any mass in motion (kinetic energy); and finally the energy lost due to friction.

A fluid will flow in a pipe or confined apparatus from point A to point B only if there is a driving force available to overcome these energy differences. This force is usually apparent as a pressure drop or difference between two reference points. The calculation of this is necessary to determine the energy requirements for transport. Determination of the first three (potential energy or 'head', pressure energy, and kinetic energy) is a simple matter of elementary physics. To find the frictional resistance recourse has to be made to semi-empirical methods which relate the frictional resistance to the physical properties of the material and to the regime of flow. The latter is characterized by computing the Reynolds' number which is the ratio of drag forces to inertial forces in the system. Provided that this dimensionless number is greater than 3000 then the flow is described as turbulent and experimental values of so-called friction factors may be used. These apply to any Newtonian fluid, be it liquid or gas, and the Reynolds' number is found simply from the expression Dgv/μ where D is the characteristic dimension of the apparatus (diameter in the case of a pipe), v is the velocity, g is the density, and μ is the viscosity, all expressed in consistent units. Having calculated the value of Re (Reynolds' number) one then makes use of charts in which a friction factor f is plotted against Re. Such charts cover a range of Reynolds' number commonly from 2000 to 500 000 where a single line or small family of lines covers all cases. This shows the significance of this flow criteria. Having found the friction factor, the energy required to overcome friction is found by substituting in an equation such as the Fanning:

$$H_f \text{ (head loss due to friction)} = 4flv^2/2gD$$

where, for pipe flow, l is the length of pipe, D is the diameter, v is the flow velocity, and g is the gravitational constant.

From such calculations the pumping energy required can be found. This way of proceeding is not confined to flow in pipes but can be extended to flow in other shapes and also to flow of a moving solid body, *e.g.* a particle, through a stationary fluid. It was using this technique that settling times for the thickeners were calculated in the NaOH example.

It was said earlier that energy is required to move a fluid from point A to point B and that this was made up of potential energy (level), pressure energy (head), and kinetic energy (motion). Since energy must be conserved, the energy at point B must equal that at point A. This fact is the basis of most forms of measurement of fluid flow. Consider a short section of horizontal pipe through which a fluid is flowing. Potential energy at A equals that at B (the pipe is horizontal) and for a short section, frictional loss may be ignored. So the sum of the pressure energy and kinetic energy at point A must be the same as that at point B. By putting an artificial restriction of known diameter in the line between point A and B and measuring the difference in pressure between point A and that at the restriction, the difference in kinetic energy can be calculated since the velocity has to increase through the restriction (continuity of flow). Hence the velocity through this restriction can be calculated and the flow rate found. This is the basis of the orifice meter, perhaps the most common flow measuring device. This is illustrated in Figure 3.

Some practical considerations about the flow of fluids will now follow.

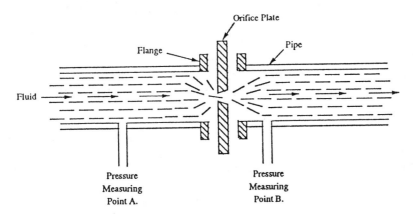

Potential Head at A = Potential at B (Pipe Horizontal)

$$\text{Kinetic Head at A} = \frac{M_A V_A^2}{2g}$$

where M_A is mass flow rate of fluid at A and V_A is its velocity.

$$\text{Kinetic Head at B} = \frac{M_B V_B^2}{2g}$$

For continuous steady state flow

$$M_A = M_B$$

Hence,

$$\text{Pressure head at A} = \frac{P_A}{\rho_A}$$

where ρ_A is fluid density.

$$\text{Pressure head at B} = \frac{P_B}{\rho_B}$$

$\rho_A = \rho_B$ for non-compressible fluid (liquid)

Hence,

$$P_A - P_B = \frac{V_A^2 - V_B^2}{2g}$$

Since the ratio of pipe diameter and orifice diameter is known V_B can be expressed in terms of V_A. Hence V_A can be found by measuring P_A and P_B.

Figure 3 *Orifice Plate*

Fluids are transported around the process through pipes. These pipes have to be made of a material that will withstand the pressure exerted on the fluid and also any chemical attack that may occur due to the nature of the fluid. Pipes for the transport of services (steam, water, and air) are usually of steel whilst those for chemicals may be of steel, special alloys, polymers, or even specially resistant metals such as vanadium. Pipes come in standard lengths, usually about 5 m and so must be joined together. Water pipes and steam pipes at low pressures usually employ simple screwed fittings or unions such as are seen on domestic supplies. At higher pressures flanged joints are needed and process pipes always use these as a matter of safety in operation. The flow has to be controlled and this is accomplished by valves fixed in the pipe at appropriate points. For service lines, simple valves may be used that operate essentially in the same way as the domestic water tap. Where the control is to be by instrument, then specially designed valves are necessary. The control of flow on process lines needs to be very positive (safety again) and is provided by a combination of isolating cocks (these are larger versions of the glass taps on burettes) and specially-designed control valves. In recent years ball valves have become quite widely used for process lines.

The energy required to tranport fluids is provided by pumps which come in an enormous variety of designs. They may conveniently be classified as follows.

4.1.1 Pumps for Liquids. (i) *Centrifugal* – These are the most common type. They operate by imparting a high velocity to the liquid with a rotating device and then converting this kinetic energy into pressure energy in a specially shaped casing. Single-stage pumps are capable of pressures up to about 3 atm. They can be operated in series (multistaged) to give much higher pressures. They give a continuous flow at a constant pressure.

(ii) *Piston type pumps* operate by displacing liquid in a cylinder with a piston. Can give very high pressures but unless they are multicylinder they give a pulsating flow in both quantity and pressure.

(iii) *Other types.* There are numerous other types of liquid pumps such as metering pumps, gear pumps, and liquid jet pumps but for these the reader is referred to a specialist text.[2]

4.1.2 Pumps for Gases. Two of the main types are the same in principle as centrifugal and piston type pumps above but traditionally have different names which is confusing.

(i) *Fans (centrifugal).* These operate in principle the same as above. Because of the low density of gases the pressure attainable is very much less in a single-stage fan (typically 0.3 atm. maximum). They can be multistaged to get higher pressures but this is not common.

(ii) *Compressors.* These are piston operated machines which work rather like a bicycle pump. Because of the compressibility of gases the gas being compressed gets hot and has to be cooled. With several stages they can reach high pressures, *e.g.* in ammonia synthesis plants pressures of several hundred atmospheres are common.

(iii) *Centrifugal compressors.* They operate on the same principles as the turbine.

They are used for very large installations at moderately high pressure, *e.g.* LPG liquefaction prior to distillation. They are very expensive and somewhat inflexible machines.

(iv) *Blowers.* These are positive rotary machines which deliver a constant flow of gas at a pressure intermediate between that obtained by a fan and that of a compressor. The usual limit is about two atmospheres.

(v) *Other types.* Again there is a large variety of devices for specialized purposes. The only one worth special mention here is the jet ejector, common for vacuum operation. It works by transferring momentum from a high velocity fluid jet (usually steam) to the surrounding gas.

This completes this brief survey of fluid flow; for further details the reader should consult the bibliography at the end of this chapter.

4.2 Heat Transfer

After transport, perhaps the most important operation is the heating or cooling of process materials. When this is a fluid, as is generally the case, such heating has to be by indirect means. Thus the fluid is contained in a vessel or a pipe through which it is flowing and heat is applied to the outer wall of the container. Very often the medium conveying this heat is another fluid, *e.g.* steam or, in the case of cooling, water. Heat is tranferred by conduction, convection, and radiation. Except in special cases, *e.g.* high-temperature furnaces, radiation is not a principal mode and most chemical process operations rely on the other two mechanisms. The laws governing conduction are relatively simple, there being a linear relationship between the rate of heat flow and the temperature difference; the proportionality constant being the thermal conductivity for the material. Conduction is usually confined to solids since as soon as one part of a fluid is heated, density differences cause it to move, and so convection comes into play. Thus conduction is significant in so far as the walls of the pipe or container are solid. To maintain good rates of heat transfer, vessels are made of metal wherever possible since these have high thermal conductivities. However, since fluids are involved usually on both sides of the vessel it is necessary to study convective heat transfer. The convection may be natural, *i.e.* caused by density difference; or forced, *i.e.* caused because the fluid is in turbulent motion. The latter is by far and more common and indeed one aims in fluid heat transfer always to have turbulent flow. So it is forced convection which we shall consider. It is not surprising then that the Reynolds' number plays a role in determining heat transfer rates in forced convection. Emphasizing once again that the engineer is interested above all in the rate at which things happen, what we need to calculate is the rate of heat transfer for a given situation. This depends on the geometry of the apparatus and the properties of the fluids being heated and cooled. For the usual continuous flow, the heat needed to be transferred in a given time is known. If this is not met by the design proposed, then the design must be modified.

Let us see how this rate is determined for a simple but common case, that of a liquid being cooled by another liquid in a jacketed pipe, *i.e.* one pipe concentrically enclosed within another. A general equation for this situation has been developed which shows that the rate of heat transfer multiplied by the pipe diameter and divided by the thermal conductivity of the fluid is proportional to the Reynolds' number raised to the 0.8 power multiplied by another dimensionless group, the Prandtl number which is raised to the 0.33 power. The first group, hD/k, is known as the Nusselt number and is the ratio of convective heat transfer in the fluid to that predicted by pure condution. The Prandtl number is given by the expression $C_p\mu/k$ where C_p is the specific heat at constant pressure, μ is the viscosity, and k the thermal conductivity of the fluid. It is the ratio of kinematic viscosity to thermal diffusivity of the fluid. The geometry of the system is included in the Reynolds' number. The proportionality constant has been found from hundreds of experiments on different fluids and the equation holds for all Newtonian fluids in turbulent flow, which testifies to its remarkable generality. This relationship with some modifications can be used for a very large number of heat-transfer situations encountered in the process industry.

As a simple example, consider the heat transfer in a waste-heat boiler generating steam from the hot gases leaving a furnace. The gases pass through tubes immersed in the boiling water. Knowing the composition of the gases we can calculate their density. Their viscosity and thermal conductivity can be found in standard tables. Suppose there to be 100 tubes each 2.5 cm diameter and 1.7 m long. The hot gas flow rate is $1500\,\text{kg h}^{-1}$. The gases flow in parallel through all the tubes.

Using the data given and substituting in the equation for forced convection we find that the Nusselt number is equal to $0.023(8320)*(0.79)$. Note that the Reynolds' number is sufficiently high to justify using this equation. Since D in the Nusselt number is 0.025 m and k is 0.016 then h the coefficient of heat transfer is equal to $1.5\,\text{w m}^{-2}\text{h}^{-1}\,°\text{C}^{-1}$.

Since the heat-transfer coefficient for boiling water is of the order of several thousand, it follows that the main resistance to heat transfer is on the gas side and this will determine the overall rate of heat transfer. Knowing the surface area of the tubes we could now easily calculate both the total heat transferred and also the exit gas temperature.

Similar types of calculations are performed to determine the required size of common heat-exchangers like reboilers and condensers. This section has concentrated on forced convection heat transfer because that is the most important in the process industry. However, pure conduction is also significant in some heavy chemical operations where processes or reactions involve solid materials. Often such processes are batchwise and this introduces additional problems in their determination. For such kinds of operations the reader is referred to the text by Beek.[3]

Next it is proposed to look at the kinds of equipment used for process heat transfer. Probably the most important and certainly the most widely used is the 'Shell and Tube' heat exchanger. The jacketed pipe has already been mentioned and is occasionally seen being used for small applications. However, this is too expensive as well as unwieldy for any but the very smallest heating problems. Nevertheless the shell and tube exchanger is a direct development of this in that instead of each individual tube having its own jacket a bundle of tubes is mounted in one big jacket or shell. The diagram (Figure 4) makes this clear. The tubes are fixed at each end into a tube plate which in turn has a cover and an inlet pipe. Since each tube has two ends there are two tube plates with a shell mounted between them surrounding the tubes. The shell has branches for the inlet and egress of one fluid whilst the other enters and leaves through the respective tube plates. Flow is usually arranged to be roughly countercurrent, *i.e.* the hot fluid enters one end and leaves the other, whilst that being cooled flows in the opposite direction. This type of construction ensures a large area of heat-transfer surface in a construction that is compact, strong, and less expensive in metal than the double-pipe arrangement. Within this basic pattern many different modifications are found, most of them relating to arrangements for causing the tube-side fluid to pass through the shell several times using dividing sections on the tube cover. Any standard text on heat transfer will show these details but the basic principle is that just described.[4]

4.3 Evaporation

Evaporation is the removal of water from a solution by means of heat which causes the water to vaporize but leaves the solute behind. Thus it is assumed that the solute is non-volatile at the temperatures encountered. An example of this has already been given in the causticizing process. Essentially, evaporation is a special case of boiling heat transfer. It is special because of several factors. The first of these is the presence in the medium of a non-volatile solute. This may well come out of solution during evaporation and will especially tend to occur on the heat transfer surfaces because the evaporation here is more severe. Such solid depositions greatly impair the heat transfer performance and evaporators are often designed to avoid or minimize this. One way is to have rapid forced circulation past the heating surfaces. Another is to pass the solution once only over the surface. A

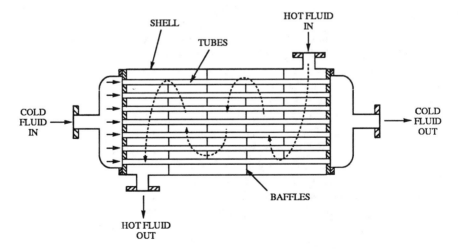

Figure 4 *Cross Section Through Simple Shell-and-tube Heat Exchanger*

second problem is that as the solution becomes more concentrated, so the boiling temperature increases due to the elevation of boiling point effect. This is one of the factors that limits the number of effects that may be operated in series. Again the viscosity of the solution increases with concentration and this may in some cases cause difficulties leading to the adoption of externally powered stirrers. The biggest factor, however, affecting the design and operation of evaporators is the cost of energy. The multiple-effect principle already described was introduced and is used to overcome this problem. Other schemes that are used take the water vapour coming from the top of the evaporator and compress it thus raising its temperature so that it can be re-used as the heating medium. Special problems that may be encountered especially in the inorganic chemical industry are the possible corrosive nature of the material being evaporated. This makes it necessary to construct the heating surface of a material chosen not so much for its good heat transfer properties as for its resistance to corrosion. For example, the thermal conductivity of stainless steel is only about one third of that for copper, but there are situations where the much greater resistance to corrosion of the former makes its use in an evaporator preferable to the latter.

A facility that is almost always required with multiple-effect evaporation is a means of producing vacuum. The necessity to operate each successive stage at a lower pressure than the one before it means that the final stage is usually boiling a liquid at a pressure well below atmospheric. This means that a vacuum producing device is required, attached to the condenser on the last stage. A common method is to use a jet ejector type of pump, but rotary pumps are often used nowadays because of their greater energy efficiency and the lessening of possible contamination from the discharge.

4.4 Mass Transfer Operations: Absorption and Distillation

We now come to a whole series of unit operations which all belong to the same class in that their principal feature is the transfer of mass across a phase boundary. Examples of this are absorption, distillation, liquid extraction, and drying but the basic phenomenon occurs in others too. What is of concern here is the separation of a mixture by removal of one component preferentially to the others that are present. Absorption is a good example to begin with. Here we are taking a gas mixture and removing one component by dissolving it in a liquid in which the other gases in the original mixture are not soluble. The component being separated may be valuable, *e.g.* removing ammonia from synthesis gas to separate it from the unconverted nitrogen and hydrogen, or it may be an undesirable material such as H_2S from crude oil pretreatment. In any case the mechanisms are the same, as is the technique of solution.

Absorption then makes use of the degree of solubility of a gas in a given liquid and so it is necessary to know the solubility relationships in terms of the amount of the material in solution that is in equilibrium with a given concentration in the gas phase. The data are generally given as partial pressure, p, of the solute gas in equilibrium with its concentration in solution, c, and the familiar relationship known as Henry's law: $p = Kc$ expresses it mathematically. Since this represents conditions at equilibrium this state cannot be attained in a real time-limited continuous operation, but rather represents limiting conditions. The difference between the existing concentration at a particular partial pressure and the equilibrium concentration represents the driving force available to cause the solute gas to move out of the gas phase and into the liquid. Since it is desired to keep this driving force at a maximum, absorption apparatus is arranged so that the incoming gas which contains the highest solute concentration meets the outgoing liquid with the highest concentration of dissolved solute. Likewise at the other end of the apparatus, the outgoing gas being weakest in the solute meets the incoming liquid which contains little or no solute in it. This is countercurrent operation and, as with most operations in processing, is assumed to be continuous. The equilibrium can be changed in the direction of increasing the solubility by increasing the pressure and this effect is often used in industry. For example, CO_2 is removed from synthesis gas by dissolving it in water at a high pressure. Where corrosive materials are being handled it is often not possible to use this effect because many corrosion resistant materials have low mechanical strengths, *e.g.* ceramics and glass.

In gas absorption as in all mass transfer operations, the transfer occurs across a boundary between two phases, in this case a liquid and a gas. In the countercurrent continuous-flow apparatus that has been supposed, the bulk liquid and the bulk gas bodies are both in turbulent flow but at the interface there is a stagnant layer or 'boundary film'. In the bulk, forced convection assures an even distribution of the differing component at any particular cross-section, but transfer is limited by molecular diffusion at the film on either side of the boundary. This situation is illustrated in Figure 5 which shows the changes which are believed to happen in this system. The driving force is represented by the difference in bulk concentrations in each phase. Equilibrium is supposed to exist at the interface itself

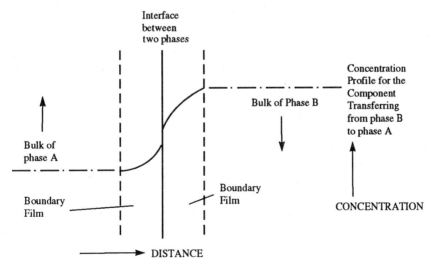

Figure 5 *Representation of Mass Transfer Between Phases*

where there is a step change in concentration, whilst there is a gradual change from these compositions to those in the main body of each fluid. Again, as engineers, we want to know the rate of mass transfer per unit surface area for a given driving force. This is often expressed as $\mathcal{N}_a = K(\Delta C)$ where $\mathcal{N}_a$ is the number of moles of component A transferring across a unit area per unit time, ΔC is the concentration difference or driving force, and K is a proportionality constant usually termed the 'Mass Transfer Coefficient'. Obviously, to maximize $\mathcal{N}_a$ we must maximize both K and ΔC. The latter is achieved by countercurrent operation which maintains the greatest driving force through the absorber. The former is a function of both the system and the geometry of the process. The system properties are fixed for a given solute and solvent, and so K can be increased only by changes in the flow regimes. Thus one wishes to maintain turbulent flow conditions so reducing the thickness of the film each side of the interface. Since transfer takes place through this film only by molecular diffusion, it is desirable to have as thin a film as feasible. Much of the skill of the designer of not only absorption towers but other mass-transfer apparatus turns on producing turbulent mixing conditions for the two phases inside the equipment.

Absorption takes place in vertical towers in which the liquid (solvent) enters at the top and flows down by gravity, and the gas containing the soluble component enters at the base and leaves, stripped of most of its solute at the top. Intimate contact between the gas and the liquid is achieved by causing the liquid to flow over grids or packing inside the tower which have a large surface area per unit volume, thus exposing a large area of liquid to the uprising gas. It is necessary that this packing also has a large voidage so that the loss of energy (pressure drop) experienced by the gas in passing up the tower is not excessive. This is typical of the conflicting desiderata that the chemical engineer encounters in design. He is always having to compromise between two or more ideal conditions neither of which is satisfactory if pursued to its ultimate. The gas needs to have energy put into it by a pump to force it through the tower, and since this is expensive the pressure drop through the apparatus must be kept low. Often instead of packing or grids the absorption tower is fitted with special plates which occupy the cross-section of the cylindrical part at intervals of perhaps 50 cm. These plates have pipes (downcomers) to carry the liquid from one stage to the next and the rising gas is forced to pass through holes in the plate over which the liquid is flowing. Intimate contact is achieved by the gas bubbling through a depth of liquid. There are many industrial types of plates, the two most common being the perforated plate described above and the bubble cap plate. For a description of the types of apparatus and design techniques see reference 5.

It is not intended to go into the design of absorption systems here. Suffice it to say that the diameter of the tower is determined by the required throughput whilst its height is fixed by the degree of recovery that is desired.

Distillation is the next operation to be considered. This is a most widely used technique for separating mixtures of liquids. So useful is it that mixtures which are commonly in a gaseous state, *e.g.* air or light hydrocarbons, may be liquefied for ease of separation. The whole technique depends upon the simple physico-chemical phenomenon that when a mixture of two or more liquids is partially vaporized, the vapour has a different composition from that of the liquid from which it arose. Generally it is richer in the more volatile component or components. If one now takes that vapour and condenses it, then repeats the partial vaporization process, one will obtain a vapour even richer in the more volatile or 'lighter' components. Successive repetitions will enable one to end with a nearly pure material of the most volatile component. However, because each vaporization must be partial (clearly total vaporization would result in no change in composition) the process described above would produce less and less material at each stage and so would not be commercially viable. This snag is overcome by making the process of successive vaporization and condensation continuous and countercurrent in a tower type apparatus. This then becomes what is known as fractional distillation. The vapour is produced in a heat exhanger, called a reboiler, at the base of the tower and passes up the tower contacting successive pools of liquid on the way. At the top of the tower, some of the vapour is condensed and returned to the tower to provide these pools of liquid. This is called 'reflux'. The rest of the vapour which is the refined product is taken off and condensed separately. In passing down the volumn (by gravity) the liquid is contacted with the rising vapour in a series of stages. At each stage vapour is condensed and in giving up its heat generates more vapour from the liquid with which it has come into contact. This new vapour is richer in the more volatile component and when it passes on to the next stage and is itself condensed it generates more vapour even richer in the light component. The process operates continuously so that there is a continuous flow of light material out of the top of the apparatus, whilst the liquid which must leave the base of the tower to maintain continuity is denuded of light material and is rich in the less volatile or heavy components.

This arrangement is shown diagrammatically in Figure 6.

The key to the process is the difference in composition between the liquid and the vapour in equilibrium with it, and the chemical engineer has to use this as his basic data. These equilibria can be calculated from vapour pressure data for ideal mixtures such as benzene/toluene, but usually the data has to be found from experimental work. Fortunately most systems encountered have already been studied and very large compilations of data are available (see Ghemling and Onken, Achema data books on Vapour Liquid Equilibria[5]). Knowing the equilibria relationships, the number of contacting stages required for a given degree of separation can be determined. Note that this fixes the number in an ideal sense independent of the quantity to be processed. The actual size of the apparatus, in particular the diameter of the distillation tower, depends upon the throughput.

The separation performance of a distillation tower depends critically upon the ratio between the amount of liquid fed back to the top of the tower and that taken off as product. This is known as the reflux ratio, R. The significance of this can be seen from a simple construction based upon the equilibrium diagram which is a plot on a molar basis of the composition of the vapour that is in equilibrium with a particular liquid composition. Such a construction is shown in Figure 7. The feed composition is represented by the point x_f and the product composition by the point x_d. If we assume that all the vapour is condensed at the top of the tower, then clearly x_d must equal y_d, the composition of the vapour at the top of the tower. It

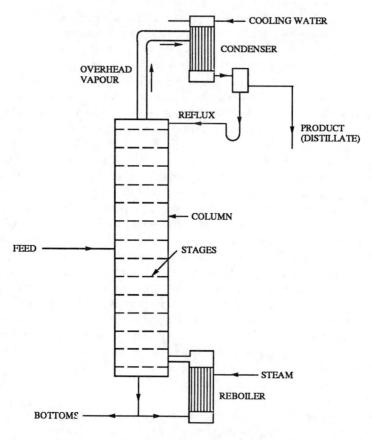

Figure 6   *Simple Continuous Distillation Unit*

can be shown that for a simple near-ideal system the interstage compositions lie on a straight line passing through the point x_d, y_d and having a slope equal to the ratio $R/(R + 1)$.[6] This is called the 'Operating Line'. The number of stages required for the separation can be found simply by stepping off right-angled steps between the operating line and the equilibrium curve until the feed composition is reached. This is shown in Figure 8 for several different cases. The first [case (a)] is where the operating line is coincidental with the x,y diagonal. This represents the best possible separation, *i.e.* with the fewest steps. Since the slope of the line $R/(R + 1)$ is 1 then R must be infinitely large. This condition is known as 'total reflux' and is one in which the totality of vapour produced at the top of the tower is returned to the tower, *i.e.* there is no product. Case (b) shows another extreme. Here the operating line intersects the equilibrium curve just above the feed composition. Clearly it is impossible to reach the feed composition no matter how many steps are drawn. This condition is known as 'minimum reflux' and, although useful analytically, is not a realizable condition in practice unlike the total reflux one which can be achieved. So industrial distillation towers must operate somewhere between total and minimum reflux, *e.g.* case (c). Just where this is depends upon the designer,

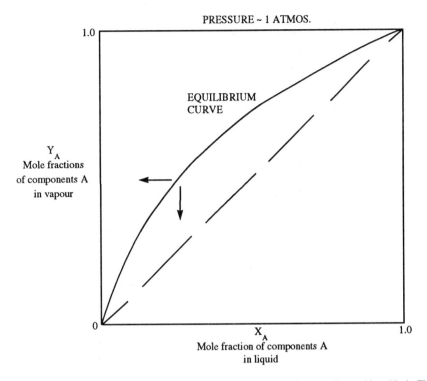

Figure 7 *Equilibrium Curve Showing Typical Equilibria Relations for a Near-ideal Two Component System*

who weighs the economics of capital cost represented by the number of stages against operating cost which depends on the reflux ratio. Since one has to work with the system that is given, and with the apparatus provided by the designer, the only variable available to the operator for control of quality is the reflux ratio. This has been stressed because it is important. Distillation is not only the most common method of separating liquid mixtures, it is also the only one which stands alone, *i.e.* it does not have to be supplemented by another operation as do absorption and extraction.

The apparatus used for distillation is similar to that for absorption. The towers are vertical cylinders containing either grids and packing or trays at spaced intervals. A typical tray is shown diagrammatically in Figure 9. A typical simple system will have feed entering the tower part way up, product or distillate leaving the condenser at the top, and the lower volatile material or 'bottoms' leaving the base of the tower at the reboiler. The whole will be supplied with instruments to control the flow of feed and of reflux and to control the heat supplied to the reboiler. This is the basic minimum of control for steady-state operation. Usually a distillation unit will comprise a train of towers each one separating out a different component of the original mixture. A simplified diagram of such a train is shown in Figure 10.

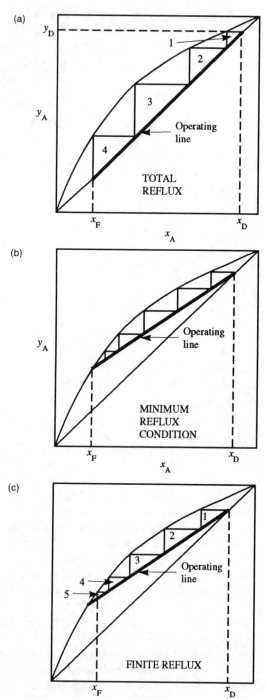

Figure 8 *(a) Total Reflux ~ Minimum No. of Stages to go From* x_F *to* x_D*; (b) Minimum Reflux ~ Going From* x_D *to* x_F *Requires an* **Infinite** *No. of Stages; (c) Finite Reflux ~ Showing No. of Stages to go From* x_D *to* x_F

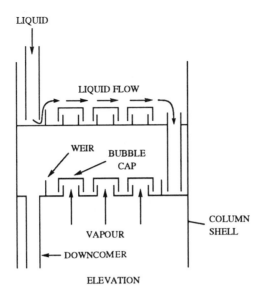

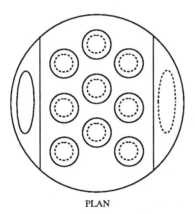

Figure 9 *Outline of Bubble-cap Plate*

4.5 Extraction

The third method of separation to be considered is liquid extraction. This process makes use of the fact that in a mixture of two liquids, one component may be soluble in a third liquid with which the other is immiscible. This process is widely used in the refining of lubricating oils to remove waxy components, and is also a primary step in the recovery of uranium from its ore. Essentially, a homogeneous liquid mixture is contacted in a suitable apparatus with a second liquid (termed the solvent) in which one of the components of the original mixture is preferentially soluble. It is essential that the final mixture is heterogeneous, *i.e.*

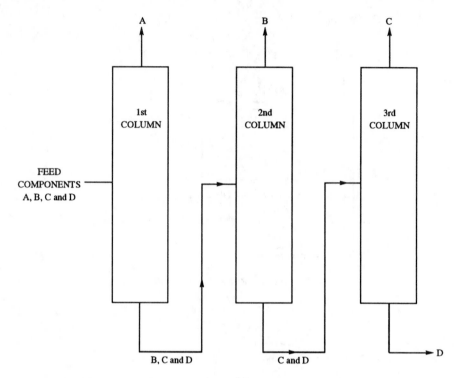

Figure 10   *Multi-column Arrangement for Separating a Four-component Mixture into Four Products*

forms two liquid layers, as that rich in solvent containing the solute can then be separated physically from the non-solvent layer containing the non-soluble material. The two layers are termed respectively, 'extract' and 'raffinate' and the process is shown in Figure 11.

Again the basic information for the determination of extraction is the equilibrium data for the system. A typical representation of this for a common type of three component mixture is shown in the triangular phase diagram Figure 12.

The simple system shown in Figure 11 can be represented on such a diagram as shown in Figure 13. Point F represents the original two-liquid mixture or feed and point C the solvent. Note that the feed contains no solvent or third component whilst the solvent is pure. Adding feed to solvent results in a mixture whose composition is given by point M which lies on the straight line joining F and S. Not only does this lie on this line but its position is dictated by the relative amounts of F and S. Thus if these are equal then M will lie equidistant from F and M. Since M lies in the heterogeneous region, such a mixture is not stable and will separate into two phases. The compositions of these will lie at the ends of the tie-line passing through M; these tie-lines representing the equilibria compositions of extract and raffinate. Since there are now two liquid phases these may be separated by decanting the top layer and thus achieving two new liquid mixtures, one much richer in solute than the original. Notice that complete immiscibility between solvent and the third component does not exist and that the raffinate also contains

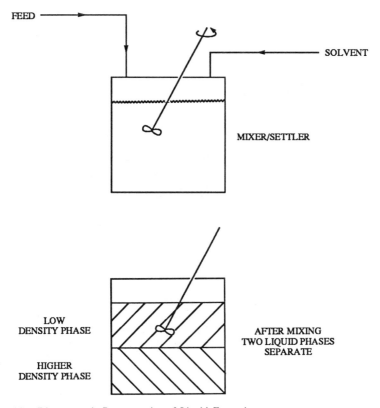

FEED

SOLVENT

MIXER/SETTLER

LOW
DENSITY PHASE

HIGHER
DENSITY PHASE

AFTER MIXING
TWO LIQUID PHASES
SEPARATE

Figure 11 *Diagrammatic Representation of Liquid Extraction*

some solvent. This means that solvent recovery, which is essential for economic operation, has to be practised on both the raffinate and the extract phases. This in turn requires distillation stages to be added to complete the process.

Liquid extraction is used where distillation is difficult because of a very small difference in the volatilities of the two components, or where the temperatures associated with distillation might damage one of the materials present.

The apparatus used for extraction nowadays is generally of the tower type containing contacting stages and employing countercurrent continuous flow.

Perhaps the best detailed treatment of liquid extraction is the book with this title by Treybal[7] to which the reader is referred.

4.6 Drying

This is an important operation, especially in the inorganic chemical industry where many of the products are solids which have to be dried before they can be stored, packed, or distributed. Drying is generally understood to mean the removal of unwanted water from a solid, although it is sometimes applied to the drying of organic liquids for special purposes, *e.g.* transformer oils. All solids contain water which is absorbed from the atmosphere but many processes of

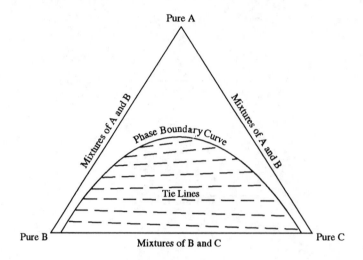

Figure 12   *Typical Phase Diagram for Type I Three-component System. Any point inside the triangle represents a mixture of all three-components. Any mixture whose composition is represented by a point inside the phase boundary curve will separate into two-phases. The composition of each phase will lie at the ends of a tie-line passing through the mixture point*

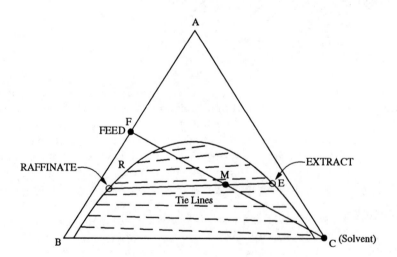

Figure 13   *Simple Single-stage Extraction Process. Feed (a mixture of A and B) is mixed with solvent C. The **overall** mixture composition (A, B, and C) is represented by point M. At equilibrium this separates into Raffinate, R and Extract, E*

manufacture involve precipitation and subsequent filtration of a solid product. This will leave it with a water content of perhaps 20 to 30% most of which it is necessary to remove. This is accomplished by heating the wet material in a relatively dry gas stream (usually air). The gas, having a low partial pressure of water vapour, will pick up moisture from the wet solid since the partial pressure of water at the solid surface will be the vapour pressure of water at the surface temperature. Drying is usually thought of as a two-stage process. In the first, the solid is assumed to be thoroughly wet on the surface and the drying is simply evaporation of water into a stream of gas. As the drying proceeds the surface of the solid develops dry areas since the water held in the interstices of the solid cannot diffuse to the surface at as fast a rate as it is being removed by the stream of warm dry gas. During the first stage, the drying takes place at a constant rate which depends upon the temperature and gas velocity. When the second stage is reached, the rate decreases steadily with time. Eventually a stage is reached where no further drying occurs (the so-called 'equilibrium moisture content' is reached). The drying is then complete for industrial purposes.

Drying is often performed as a batch process unless the quantities involved are rather large, *e.g.* fertilizer manufacture. Continuous driers are large tubular rotating drums set at a slight angle to the horizontal. The wet material enters at the top and the hot gases at the bottom, thus achieving countercurrent flow. The slow rotation of the drum causes the material to shower down through the hot gases during its passage through the machine, thus exposing a large surface area to the drying action.

Drying is amongst the most empirical of all chemical engineering mass-transfer operations, and all driers of any significant size have to be designed by experiment upon the material to be dried using a small scale version of the type of drier that is intended to be used.

4.7 Size Reduction

Most reactions, whether chemical or physical, occur on surfaces and rates are nearly always expressed in relationship to surface areas. Thus when solids are processed the first step is usually to crush or grind them to a powder so that the surface area per unit volume is very much increased. This then accelerates the rate at which the desired reaction will take place, be it chemical such as catalysis or physical such as dissolution. To make a powder from a solid by mechanical means requires rather large amounts of energy and, as might be expected, this energy is proportional to the amount of new surface that is produced. Crushing and grinding are unfortunately very inefficient processes in energy terms. Theoretically the energy requirement ought not to be more than that represented by the surface energy of the material multiplied by its new area. If this were taken as the efficiency measure, then size reduction processes would have an efficiency of less than 1%. The underlying reason for this low efficiency is the lack of understanding of the exact nature of the phenomena that occur when a solid is broken. This in turn leads to there being a very large variety of machines for size reduction which have developed largely by empirical evolution.

Crushing is generally used to signify the reduction of material from very large pieces, such as may originate in large scale mining or quarrying operation, to fairly coarse material perhaps 5 to 10 cm in size. Grinding generally implies taking a feed of an already crushed material and producing a powder which may vary from particles as small as a few microns to ones as large as 50 mm. The machines for crushing tend to be very large pieces of heavy engineering equipment, whilst those used for fine grinding often are pieces of precision engineering. This is a specialist area which we shall not expand on here, but instead will mention some of the problems associated with the processing of powdered materials. It was said earlier that solids were made into powders and then suspended in a fluid for ease of handling. However, such mixtures possess quite different properties in many ways from regular fluids and require special consideration and treatment. First consider dry powders. These are characterized by their particle-size, or more properly by the particle-size distribution since no powder is made up of particles of one uniform size. Often the distribution is ignored and an average size used although this can lead to problems in handling. To illustrate some of these problems, consider the simple act of pouring sugar through a small glass funnel. Ordinary household granulated sugar is one of the purest chemicals found in the home. No problems will be experienced in getting this to flow through the funnel. Now try to pour finely powdered sugar (castor sugar for example) through the same funnel, and immediately flow problems are seen. This is exactly the same material chemically, only its granular size has changed but this size change makes a profound difference to its handling properties. Quite serious problems were experienced some years ago in the design and operation of large vertical storage bunkers for granular materials because of flow factors, and much effort in recent years has gone into the study of the design of these. So not only are the size and size distribution of a powder important, so also are its flow properties. When admixed with a fluid these flow problems are eased but not removed entirely, because the mixture no longer behaves like a regular or Newtonian fluid. In particular, its viscosity is not constant irrespective of stress. This leads to special problems in pumping and control but these, whilst not completely understood, are well known by chemical engineers who have adopted methods for dealing with them. Despite these problems it still remains the objective of the engineer to transform solids into powders for processing purposes.

4.8 Sedimentation

Many products are obtained as precipitates, the calcium carbonate in the early design example for instance, and these have to be settled-out before further treatment. It would be uneconomic to try to filter the whole reaction mass because of the very large volume of liquid that would have to pass through the filter, and so the mixture is concentrated with respect to its solids content. This was done in the example quoted by allowing the solid particles to fall to the bottom of a holding tank by gravity. The operation is known as 'sedimentation'. The size of the tank is determined by the size and concentration of the solid particles in suspension, which in turn fix the length of time required for them to settle in a given depth. Knowing the particle size and concentration, this time can be calculated by the application of simple hydrodynamics (Stokes' Law). In practice, precipitates do not have a common particle size but rather a range of sizes so that some cohorts of the population settle at a slower rate than others. Since the faster moving particles have to overtake the slower ones, interference occurs and a phenomenon known as 'hindered settling' is usually present. Because of this, settlers or sedimentation tanks have to be designed on the basis of laboratory tests on the material to be settled. The methods for doing this are well tried and tested, and settling tanks of very large sizes (30 to 100 m diameter) are in operation.

The particles which settle to the bottom form a sludge or slurry containing perhaps 50 to 80% water. This is fairly easily pumped out of the bottom of the tank and goes to a filter system as a concentrated slurry. Sometimes gravity settling is not adequate, either because it is too slow or because the solid particles are so small that Brownian motion is significant compared to the gravitational force on them. In such cases the settling rate may be artificially enhanced by causing the suspension to enter a centrifugal field. Here the force on the particles may be many times greater than that of gravity, and rapid and effective settling is obtained although at a cost since the machines that do this are expensive.

A special case of sedimentation occurs when essentially the same technique is used to sort particles of the same material into different size ranges. Although this is sometimes done with the particles in a liquid suspension it is much more common for the particles to be separated this way in a stream of air. There are many proprietary devices for this operation which is termed 'Classification'. It is widely used in many inorganic powder processing operations, either to sort out oversize material or undersize product or both.

4.9 Filtration

This follows naturally on from the previous discussion. In filtration the wet material, usually containing more than 50% water or other liquid is fed to a membrane in which the holes or pores are small enough for the water to pass through but too small for the solid particles to pass. Filters vary enormously in type, and range from simple mesh belts moving in a continuous ribbon over rollers onto which the slurry is fed at one end and the drained material scraped off the top of the belt at the other, to highly sophisticated automated machines operating at fairly high pressures and sequencing a series of complex operations to simulate continuous production. It is well beyond the scope of this chapter to describe either the latter, or even several of the variety of machines that are commonly in use. Instead some brief ideas of the underlying principles will be presented.

Figure 14(a) and (b) represent a filter membrane and it will be seen that once a layer of material starts to build up on the septum, any liquid that is being removed has to pass through this layer as well as through the septum. To do this a driving force must be applied. In the first filter described above, the driving force is simply gravity but this is satisfactory only with the coarsest materials and where good liquid removal is not so important. In general, a pressure has to be applied by a pump. Usually this pressure is applied through the slurry but an important class of filters uses atmospheric pressure by creating a vacuum on the clear liquor side of the septum. The major interest in practice is again the rate of flow of filtrate. Knowing the concentration of slurry and the throughput of material, a simple

(a)

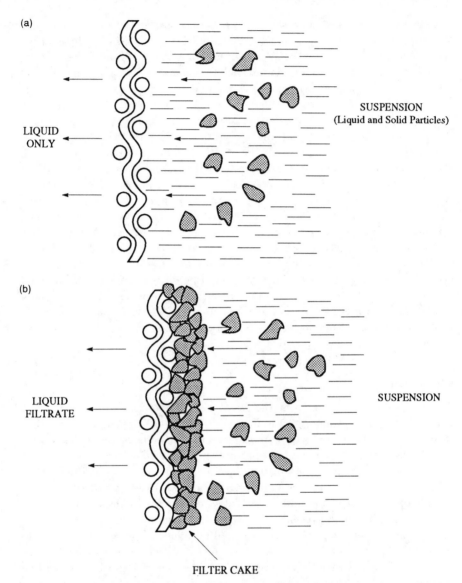

SUSPENSION
(Liquid and Solid Particles)

LIQUID
ONLY

(b)

LIQUID
FILTRATE

SUSPENSION

FILTER CAKE

Figure 14   *(a) Magnified Cross-section of Filter Septum (Woven Cloth Type) at Start-up; (b) The Same After Filtration has Progressed*

balance enables the required rate of flow of filtrate to be found. Since the rate of flow is expressed in terms of unit area, when this basic rate is calculated it is easy to find the filter area that is needed.

In filtration as opposed to most operations, the fluid is assumed to be in viscous rather than in turbulent flow and the rate equations are based essentially on the Poiseuille equation. The main factors are thus the viscosity of the fluid, the depth of filter cake, and its porosity. Ideally we wish to keep the depth or thickness small, the porosity high, and the viscosity low. In practice we have to compromise between a small depth, which means frequent breaks in operation to remove the cake, and a greater depth with greater resistance and hence longer filtration time. The porosity is largely a function of the particle sizes of the solid. A solid with many small particles is going to have a much poorer porosity than a coarse granular material. The conditions of precipitation ought to be controlled carefully to try to obtain the best particle size for filtration. Finally, the viscosity is a factor that can be reduced by taking advantage of the fact that liquid viscosities are greatly reduced as the temperature is increased.

Filtration is a somewhat Cinderella operation, being regarded often as a necessary evil. But much can be done by the application of modern techniques to reduce the nuisance that used to be associated with this separation.

Here we come to the end of this very brief review of the main unit operations of chemical engineering. What this has tried to do is to give an oversight of these with a glimpse of the underlying principles; not to enable the chemist to become a chemical engineer, but rather to alert him or her to the problems that the engineer has in transforming chemical processes into commercial production.

5   CONTROL AND INSTRUMENTATION

Stress has been laid repeatedly upon the continuous nature of industrial chemical operations. There are two main reasons for this continuous operation. The first is the need to make the fullest possible use of expensive plant. Large scale chemical plant is very costly to build and install, and even when it is not working it costs money due to depreciation, obsolescence, and so on. The second reason is that continuous operation makes it much easier to maintain a required quality of product.

Continuous operation was not practiced until well into the twentieth century and the reason for this was that it depends for its successful application on the continuous measurement and control of process conditions. Flows, temperatures, and pressures have to be monitored and maintained at fixed predetermined values if continuous steady-state operation is to be achieved. Other variables may also be required to be monitored, *e.g.* pH, but the most common and important are the three mentioned already. The measurement of *flow* has been touched upon earlier in the section on fluid flow. Almost all flow measuring devices make use of some application of the energy balance and its variation at different flow rates. The measurement of *temperature* on an industrial level is usually by means of thermocouple or resistance thermometers in which some electrical effect linked to temperature is utilized. The measurement of *pressure* is performed in a variety of ways. One of the simplest is a gauge in which the pressure inside a pipe or vessel is transmitted to an element which distorts under the force exerted. This distortion is then displayed visually and used as an indication of pressure. This by itself is only an indicator and is not useful as a control device. What we need for this is a device which converts pressure into some easily transmitted signal, preferably electrical which can then be amplified and used for control. Such a device is the piezo-electric cell in which a material whose electrical conductivity is sensitive to pressure is used. The distinction is made here between 'measurement' and 'control'. Without measurement there can be no control, but measurement by itself is not enough. The sensing device which detects a change in operating conditions must be such that this change can be used to actuate a control mechanism. This has to be independent of the operator since he cannot be expected to observe every variable on a complex unit for every minute of his shift. Moreover, there may well be systems where the response time of the operator is too slow.

Thus there is a need for automatic control. Most automatic control operates on what is known as 'Feed-Back'. The sensor detects a change in the measured variable. It transmits this change to the controller which compares the value with the desired or 'set' value. According to whether the deviation from the desired value is negative or positive, the controller then actuates a valve or other device which changes some input to the system in a direction calculated to restore the measured variable to its proper value. This sequence is known as the 'Control Loop'. A simple example will help to make this clearer. Figure 15 shows a heat-exchanger system in which a process liquid is being heated by steam. The flow of process liquid is controlled at a fixed predetermined level, but its input temperature may be subject to some variations caused by a previous stage. It is desirable to maintain a constant outflow temperature. This means that the supply of steam to the heat-exchanger must be varied to meet the variations in the incoming temperature of the process stream. This is achieved by (a) measuring the temperature of the outlet stream; (b) transmitting this temperature to the controller; (c) comparing, in the controller, this temperature with the desired value; and (d) sending a signal to the valve on the steam line to open or close as the case may be. This is the classic control system which generally works very well.

It may be noticed that the control is applied after the variation has worked through the system to the outlet. In recent years there have been attempts to overcome this by the use of so-called 'Feed Forward' control. In this the temperature variations of the incoming stream are monitored and, by using a model built into the control system, the steam supply is varied in accordance with

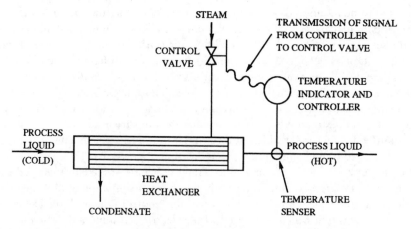

Figure 15 *Diagrammatic Representation of Conventional 'Closed-loop' Control Scheme*

this incoming variation rather than waiting for it to show in the outlet. This is a more complex way of working and although much favoured by the theoreticians has found little application in practice. What is most common nowadays is the use of computer monitoring of process variables. The control room, which is the nerve centre of the modern process plant, will have many monitors, with their keyboards, on which the operators can call up the value of any significant process variable. However, computer control as such is still the exception rather than the rule and the individual control loop as described above is usual.

This then is the key to the successful working of large scale continuous chemical processing and it needs to be appreciated as such by anyone working in the industry.

6 SAFETY AND LOSS PREVENTION

We come now to the final but not the least important part of this brief survey of the work and problems of the chemical engineer: the matter of safety and loss prevention.

Any spillage or escape of material represents a loss of material, and hence a reduction in the economy of a process. However, in the chemical industry any spillage or leak may have far more serious consequences because of the hazards associated with chemically-reactive material. These hazards are, in order of importance: fire, explosion, and toxic release. They are ordered in this way because although toxic release is potentially more dangerous to life than explosion, which in turn might injure more people more severely than fire, the facts are that fire is the most frequently occurring hazard and that explosion certainly and toxic release possibly are usually consequences of serious fires.

The severity of any incident is measured in different ways in different countries, but in Europe it is usually rated in terms of the number of fatal injuries. There are two reasons for this, the first being that death is final and its relative seriousness cannot be argued. The second is that the severity of incidents bears a distinct relationship to the number of incidents of this type. Although the relationship is not exact it is well enough known that an incident serious enough to cause death is only about one hundredth as likely as one which causes bodily injury and that, in turn, this lesser incident is only about a fiftieth as likely as one which causes minor injury. The moral of this is that by paying attention to minor incidents and reducing their frequency one will reduce the statistical chance of major injury or death occurring. Thus the fatal accident rate, FAR, is commonly used for determining the actual and/or probable safety of a process. This number is strictly defined and it is the number of deaths that occur in a plant or industry per thousand workers expressed as a proportion of the total hours worked. A typical figure for European chemical industry is 4.0, which means that for every thousand workers that are employed in the industry there will be four deaths during a working lifetime, say 10^8 hours. This figure is significantly lower than that in many industries; for example in agriculture it is about 8, in mining nearer 12, and in some hazardous jobs such as steel erecting, 20 or more. This gives a target figure to work towards or even to get below. Since half the fatal accidents may be attributed to accidents which have little or nothing to do with the process, *e.g.* falling off a ladder or being run over, then we must try to ensure that the actual process risks give a FAR of 2.0 or less. The risks and hence the FAR for either a particular process or a complete plant may be calculated (although the latter is very time-consuming) and this is a part of the work these days of the chemical engineer. Indeed in the UK it is essential to do this for any new plant before permission to build and operate it is given. The actual methods of doing so are beyond the scope of this survey, but are well set out in several books the most important of which are listed at the end of this chapter. Essentially they develop from (a) a knowledge of the failure rate of components of a unit and, in particular, sensitive components like valves and instruments allied to an analysis of the plant function which relate the overall risk of failure to the failure of an individual component or components. If this risk is unacceptably high, then the design must be modified.

A simple example might make this clearer. Suppose a process requires a reactor to operate at a pressure well above atmospheric. Then there is a risk of the vessel bursting due to the pressure rising well above the design limit. The risk of the vessel bursting at the design pressure is well known and is so small as to be negligible. However, the design pressure may be exceeded either by maloperation by the operator or by the failure of an automatic valve. This risk is guarded against by installing a safety valve which releases the pressure to a vent line if it rises say 10% above design pressure. However, safety valves do not always work and this is another hazard. The situation that may occur is shown in Figure 16. The numbers for failure rates are illustrative only but of the right order. This shows that the risk for the simple system described is such that an operator might be killed once in a hundred years. If this is unacceptable, then the design must be changed, perhaps by installing another safety valve in parallel or by more frequent testing of the safety valve to ensure that it is in working order. The techniques outlined are known as 'fault and event tree analysis' and are part of a new armoury of weapons wielded by the engineer to make his plants more reliable and safer.

In addition to the concern of the engineer with the safety of personnel working on the plant, he must also consider the general public who might be adversely affected by a large scale fire, explosion, or toxic release. This kind of study reaches its peak in the analysis of risks considered when it is proposed to build a nuclear power station. It is also something that has to be done when a large new chemical complex is to be built in this country, where there are no longer any truly remote areas. For an example of the kind of detailed study that is possible the reader is referred to the report on a proposed petrochemical complex to be built next to the town of Riijmond in Holland (see bibliography).

This concludes an all too brief survey of the significance of chemical engineering in the modern chemical industry. The material in this chapter cannot give more than an overview, but the chemist who works in the industry will undoubtedly come across many of the problems discussed herein. It is hoped that what has been written will provide a useful introduction and when it becomes necessary (as it undoubtedly will) for he or she to look further into some of these topics, the bibliography will provide a useful starting point.

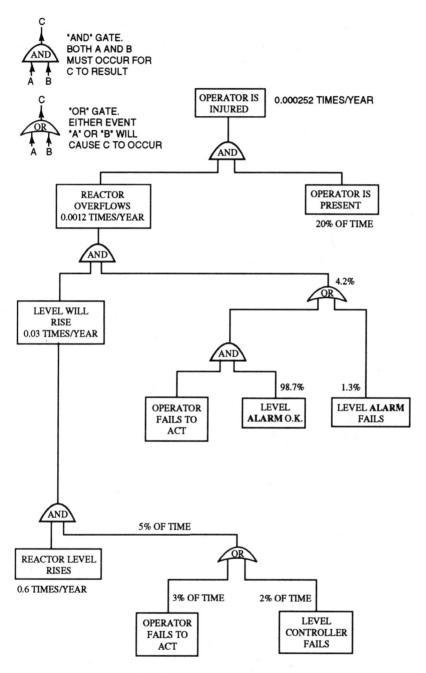

Figure 16   *Simple Fault-tree Showing How Risks to Operator May be Analysed and Assessed*

7   BIBLIOGRAPHY

This bibliography contains a number of books additional to those shown in the references throughout the text so that the reader may follow up particular topics in more detail.

Part I   Introduction

1. D.F. Rudd and J. Watson, 'The Strategy of Process Engineering', John Wiley, New York, 1968. This was the first book to systematize the design process and it revolutionized both teaching and practice in this area. It is somewhat advanced and a simpler text by Rudd and Siirola entitled 'Process Synthesis' (Prentice Hall, Netherlands, 1973) is also recommended.

Part II   Unit Operations

One of the best comprehensive texts at a fairly elementary level is W.L. McCabe, J.L. Smith, and P. Harriot, 'Unit Operations of Chemical Engineering', 3rd Edn., McGraw Hill, New York, 1985.

2. 'Fluid Movers, Pumps, Compressors, Fans and Blowers', produced by the editors of 'Chemical Engineering', McGraw Hill, New York, 1979.
3. W.J. Beek and K.M.K. Muttzall, 'Transport Phenomena, John Wiley, New York, 1975.
4. D.Q. Kern, 'Process Heat Transfer', McGraw Hill, New York, 1990. Although first published in 1955 this is so well-regarded a book as to have a new printing made in 1990 to satisfy renewed demand. The best book on heat transfer equipment yet written.
5. T.K. Sherwood, R.L. Pigford, and C.R. Wilke, 'Mass Transfer', McGraw Hill, New York, 1975. This is a classic text by three of the most able academics of their time.
6. V. Gmehling and U. Onken, 'Tabulations of equilibria data', Dechema, Frankfurt, 1977–90. An ongoing series of volumes containing the best information available in the open literature on phase equilibria in systems.
7. C.S. Robinson and E.R. Gilliland, 'Elements of fractional distillation', 4th Edn, McGraw Hill, New York, 1950. Although most distillation calculations are performed by computer nowadays, the computer program is only as good as the users understanding of the process. This, despite its age, remains the best book ever written on distillation.
8. R. Treybal, 'Liquid Extraction', McGraw Hill, New York, 1980. A specialist book by an expert.
9. C.A. Smith and A.B. Corripio, 'Principles and Practice of Automatic Control', John Wiley, New York, 1985. After a useful introduction to the principles this quickly gets into theory at a fairly senior level. However, it contains a valuable and up-to-date Appendix on types of measuring devices, control valves, and the like.

Safety, being a relatively new subject in its present-day quantitative and predictive form and being of such importance in Europe, deserves a section to itself.

The most important book is undoubtedly F.P. Lees, 'Loss Prevention in the Process Industries'. This is a massive two volume work that should be in the reference library of every process company, since there is scarcely a situation that may arise which is not covered by the methodology in here. It is not a book for the individual unless he aims to become a specialist.

Two books specially recommended to the tyro are (i) 'What Went Wrong', a survey of dozens of cases of disasters or near disasters with simple explanations of their causes. Morals are included showing how they and similar incidents can be prevented. This book is written by T. Kletz and published by Gulf Publishing Co. Also by the same author is 'Hazop and Hazan' a simple relatively non-mathematical treatment of the important task of Hazard Analysis with which modern chemists will undoubtedly become involved. This is published by the Institution of Chemical Engineers, Rugby.

There are now several textbooks for students use of which probably the most useful is that by D. Crowl and S. Louvar entitled 'Chemical Process Safety'. However these books are really for student and/or practicing engineers.

Finally, to complete this section reference is made to the Report to the City of Riijmond in Holland on the siting of a proposed petrochemical complex. This report publishes the detailed calculations and conclusions of two consulting organizations who were commissioned separately to report on the possible hazards of this installation within the city limits. It is both comprehensive and readable and is a marvellours example of the extent to which safety studies can be taken.

11. A Report to the Riijmond Public Authority on the Risks of Six Potentially Hazardous Industrial Plants, Reidel Publishing Co., Dordrecht, 1981.

CHAPTER 4

Production of Sulfuric Acid and Other Sulfur Products

A. A. TRICKETT

1 INTRODUCTION

Sulfuric acid is the largest volume commodity chemical produced. In the past the acid production of a country was a reasonable measure of its gross national product (GNP), but with the significant changes that have occurred in both technology and world economics this is no longer the case. However, sulfuric acid is still a major chemical building block of the economies of many developed nations, and it will continue to be an important inorganic chemical into the next century.

The uses of sulfuric acid are too numerous to be covered in detail in this chapter, but a breakdown of major uses is given in Table 1. By far the main use of sulfuric acid continues to be for the production of phosphatic fertilizers. Other major uses such as pigment manufacture, uranium recovery, and steel pickling have declined significantly over the last two decades, due mainly to changes in the size of the industry as well as the adoption of new technology. This trend is expected to continue, with the only real growth in world acid production being as a result of increased fertilizer manufacture.

In the early 1970s typical capacities for larger acid plants were in the range 500–1000 ton day^{-1} of 100% H_2SO_4. This situation has changed, and there are now single stream units with capacities as high as 3500 ton day^{-1}.

The other shift within the industry has been the increase in the number of new plants which do not use elemental sulfur as a feedstock. The prime reasons for this change have been the increased pressure from governments to reduce SO_2 emissions from non-ferrous metallurgical operations and for chemical companies to cease the dumping of weak, contaminated sulfuric acid into the environment. This has resulted in the construction of new metallurgical sulfuric acid plants and sulfuric acid decomposition plants, both of which will be described in detail in the next section (see also Chapter 2).

While sulfuric acid is the largest volume inorganic chemical produced, it is not the only sulfur product of industrial importance. The world production of liquid sulfur dioxide is by comparison two orders of magnitude lower than that of sulfuric

Table 1   *Uses of Sulfuric Acid*

End use	%
Fertilizers	68
Petroleum Refining	8
Mining, Metallurgy	5
Organic Chemicals	3
Inorganic Chemicals	3
Rubber, Plastic	2
Pulp and Paper	2
Water Treatment	2
Rayon, Cellophane	1
Miscellaneous	6

Reproduced from Chemical Products Synopsis, September 1992, Copyright Mannsville Chemical Products Corp., with Permission

acid production, but it is a high-value chemical and its uses are growing. The production and uses of liquid sulfur dioxide and liquid sulfur trioxide will be covered later in this chapter.

2   SULFURIC ACID MANUFACTURE

2.1   Types of Sulfuric Acid Plants

There are three types of sulfuric acid plants which can be classified based on the feedstock to the plant. The best known is the sulfur-burning acid plant which uses elemental sulfur or brimstone as a feedstock. This type of plant is a major energy producer and is frequently a supplier of acid for captive use. A flowsheet of a double absorption sulfur-burning acid plant is shown in Figure 1.

The other two primary feedstocks for sulfuric acid plants are sulfur dioxide-containing gases from non-ferrous metallurgical operations, including the roasting of pyrites, and spent sulfuric acid. The acid plants required to process these feedstocks are very similar in design, as both require cleaning of the gases prior to the manufacture of acid in the catalysis and absorption sections of the plant. They are sometimes referred to as 'cold' gas plants.

In the production of non-ferrous metals such as copper, lead, zinc, nickel, and molybdenum, the mined ore is first processed to produce a sulfide-rich concentrate. This concentrate then enters a pyrometallurgical process, such as a roaster or smelter, where it is subjected to conditions which liberate the sulfur atom in the metal sulfide as sulfur dioxide gas. This gas, which contains dust, volatile metals, and sulfur trioxide generated in the pyrometallurgical process must be cleaned before the sulfur dioxide present can be converted into sulfuric acid. A flowsheet for a typical metallurgical acid plant is shown in Figure 2. Typical requirements for the gas cleaning section of a metallurgical acid plant are as follows.

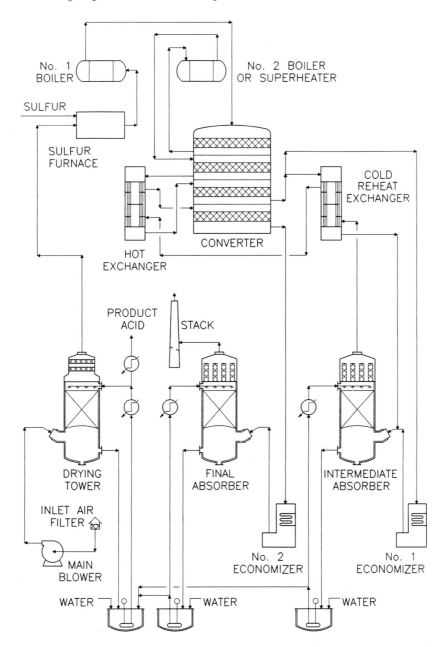

Figure 1 *Flowsheet of a Typical Sulfur Burning Acid Plant*

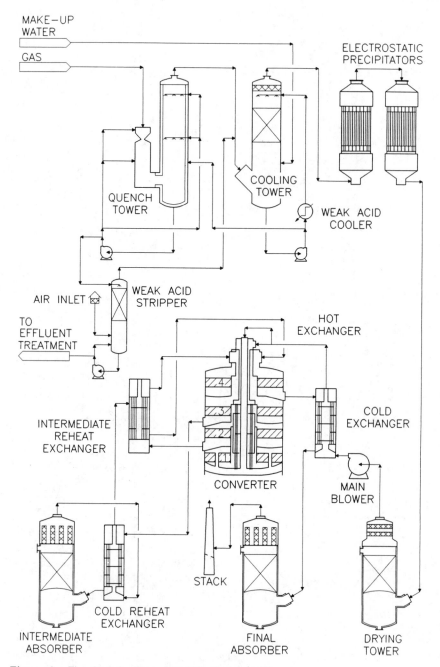

Figure 2 *Flowsheet of a Typical Metallurgical Acid Plant*

- Solid contaminants must be removed, giving a final gas quality of $1-2\ mg\ Nm^{-3}$.
- Sulfur trioxide content of the gas must be reduced to the level of $15-25\ mg\ Nm^{-3}$
- Halides, typically fluorides and chlorides, must be removed to prevent damage to downstream brickwork and converter catalyst.
- The gas must be cooled to a temperature of 35–40 °C to satisfy the acid plant water balance.
- As a result of the build-up of acid and impurities in the scrubber loop, it will be necessary to have a purge from the system.

In many of the uses of acid the H_2SO_4 molecule is not incorporated into the final product; instead it is used as a catalyst or as a means of removing water. In the process it becomes contaminated with organic compounds and water. The level of contamination varies depending on the application. In some instances the level of contamination is low enough that the acid may be re-used in a second application without it being necessary to reconcentrate the acid or remove the organic contaminants.

When the organic contamination is high, such as alkylation spent acid from petroleum refinery operations, or sulfonation spent acid from the manufacture of detergents, the only effective means of purifying the acid is by thermal decomposition. This is achieved by atomizing the acid in a furnace maintained at 1000–1100 °C. The gases generated in the furnace, containing sulfur dioxide, sulfur trioxide, water, carbon dioxide, and nitrogen, are then cleaned and processed in a similar manner as metallurgical gases. Plants using spent acid as a feedstock are referred to as Sulfuric Acid Regeneration (SAR) plants. A flowsheet of the front end of a typical SAR plant is shown in Figure 3.

2.2 Process Description

The building blocks of the three different types of acid plants are the same, particularly in the dry-gas and strong-acid systems. For the purposes of clarity, process descriptions for each of the three types of plant will be given; however, where systems or equipment are the same, reference will be made to the earlier description. With tightening environmental legislation it is very unusual for new sulfuric acid plants to be built as single absorption units, so these process descriptions will be based on the double absorption process.

2.2.1 Sulfur-burning Acid Plant. (i) *Gas System.* Liquid sulfur from a storage tank or sulfur pit is atomized in a horizontal brick-lined furnace, as shown in Figure 1. Atmospheric air is drawn in through an inlet air filter by the Main Blower where it is compressed to 3600–4500 mm of H_2O. The air then passes to the Dry Tower where it is dried to a dewpoint of below −40 °C by counter-current contact with a stream of H_2SO_4 which has a strength typically in the 98% range. The dried air then flows to the furnace where the sulfur is burned at a temperature of 1000–1150 °C to generate an exit gas stream containing 10–12% SO_2. The hot gas

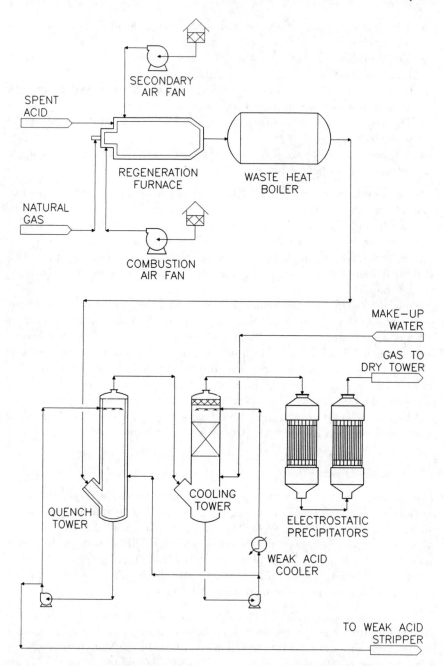

Figure 3 *Front End of a Typical Sulfuric Acid Recovery Plant*

is then cooled to about 420 °C in either a firetube or watertube boiler, the latter being found in larger plants with high pressure (60–65 bar) steam systems.

The cooled gas then enters the Converter, which is a vertical, cylindrical vessel divided into four separate layers or beds of vanadium pentoxide catalyst. The gas exits the first bed at a temperature of 615–635 °C with approximately 60–65% of the SO_2 now converted to SO_3. The gas is then cooled to 435–450 °C in either a superheater or a second boiler, before entering the second catalyst bed where further reaction of SO_2 to SO_3 occurs, increasing the conversion of SO_2 to SO_3 to 85–90% and the temperature to 520 °C.

After leaving the second catalyst bed, the gas is further cooled in a Hot Gas Exchanger reducing the temperature to 435–455 °C by exchange with cold gases from the Intermediate Absorber. The gas then flows to the third catalyst bed, where the percentage of SO_3 converted is increased to 95% and the temperature to about 470 °C. Cooling of the gas prior to entering the Intermediate Absorber is done in the Cold Reheat Gas Exchanger and No. 1 Economizer which are situated in parallel between the Converter and the Intermediate Absorber to reduce gas pressure drop. The gas entering the Intermediate Absorber is typically at a temperature of 150–160 °C.

In the Intermediate Absorber, the gas contacts a counter-current stream of 98% H_2SO_4, the gas is cooled, and SO_3 absorption takes place. Prior to leaving the tower the gas passes through high-efficiency fibre bed mist eliminator candles situated at the top of the tower. This reduces the acid mist content of the gas, and it is then reheated in two stages in the Cold Reheat and Hot Gas Exchanger, before passing to the fourth bed of the converter at a temperature of 415–420 °C.

In the fourth catalyst bed further reaction of SO_2 occurs, increasing the overall conversion of SO_2 to SO_3 to greater than 99.7%. The gas at a temperature of 440–455 °C then flows via the No. 2 Economizer where the gas is cooled to 110–120 °C before entering the Final Tower. The remaining SO_3 is absorbed in a counter-current stream of 98% H_2SO_4. The gas then exits the tower through high efficiency, fibre bed mist eliminator candles and enters the Stack for discharge to the atmosphere.

(ii) *Acid System.* In its simplest form, the acid system of any H_2SO_4 plant can be thought of as two circulating loops, one which is getting weaker as a result of gas drying, the other which is getting stronger as a result of SO_3 absorption. In order to control the strengths in both loops, it is necessary to have a cross-flow of acid from one loop to the other, and to add additional water to the system to satisfy the material balance.

The main components of any acid loop are a pump tank, acid pump, and an acid cooler. The strength of the acid in both of the absorber loops is 98–99% H_2SO_4, as this is the optimal strength for effective SO_3 absorption. The drying acid system, however, can be as low as 93% or as high as 98–99% H_2SO_4, dependent on the gas strength to the plant and the acid product strength required.

The acid system shown in Figure 1 is a simple system with three dedicated loops, each with its own pump tank, pump, and cooler. In the Dry System the acid at a strength of 98% enters the Dry Tower at a temperature of 70 °C where it contacts the incoming wet gas. The acid decreases slightly in strength and increases to a temperature of about 95–100 °C as it leaves the tower. In the pump tank further mixing of acid occurs by virtue of the crossflow from the absorber system. The acid, now at a temperature of 100–105 °C, flows to the cooler where it is cooled to 70 °C by heat exchange with cooling water in a shell and tube acid cooler.

The two absorber systems are exactly the same in concept as the Dry System, only the temperatures are slightly different. The typical temperature into both towers on new plants is 75–80 °C. Some older plants, particularly in Europe, operate the absorption towers at lower temperatures.

With an inlet temperature of 75–80 °C and an inlet concentration of 98.5% H_2SO_4, the exit temperature from the tower will be approximately 105–110 °C and the concentration approximately 99%. Water of dilution and the crossflow from the Dry System are added to the pump tank to control the acid strength. This further increases the temperature of the acid to 110–115 °C, the heat being removed in the acid cooler.

(iii) *Steam System.* The steam system operates on treated water, with softened water being used for steam systems at lower pressures, and demineralized water being required for high pressure systems. The treated water flows to a deaerator to remove residual oxygen and is then pumped by the boiler feedwater pumps through the two Economizers, which are usually located in series, and then on to the Boiler. The steam generated in the boiler leaves the system via the steam drum, is superheated and then can be exported for use outside of the acid plant battery limit, used for driving the Main Blower via a turbine drive, or sent to a turbo-generator to produce electric power. These options will be discussed in more detail in the section on energy recovery.

2.2.2 Metallurgical Acid Plant. (i) *Gas System.* The gas from the metallurgical operation will typically be cooled in a boiler and go through primary solids removal devices such as cyclones and hot-gas precipitators before entering the acid plant battery limit at a temperature of 300–350 °C. The partially cleaned gas then passes to a Quench Tower in which the gas is cooled to its adiabatic saturation point, typically 65–75 °C, by contact with a stream of weak sulfuric acid. In the process some of the solids and SO_3 are removed.

The gas, which is now saturated with water vapour, passes to a Gas Cooling Tower where it is cooled by contact with a stream of cold weak acid and further cleaning takes place. The gas, now at a temperature of 35–40 °C, passes to two stages of Wet Electrostatic Precipitators in series where the acid mist is removed from the gas before flowing to the Dry Tower for water removal. The gas, after drying, passes to the Main Blower which increases the pressure to approximately 3500–4000 mm of H_2O. The dry gas, now at a temperature of 100–120 °C, is heated up to converter temperature by passing through the Cold and Hot Heat Exchangers.

As can be seen from a comparison of Figures 1 and 2, the main differences between the two flowsheets downstream of the Dry Tower are the location of the Main Blower and the fact that the gases in a metallurgical acid plant have to be heated up to converter temperature by contact with hot converter gases. This means that there is less heat available for recovery from the gas system, and it is thus uncommon to have steam-raising equipment on a metallurgical acid plant.

Once the gases have been heated up to converter temperature the flowsheet is basically the same as for the sulfur-burning flowsheet, with some minor temperature differences dependent on the SO_2 strength from the metallurgical operation.

(ii) *Weak Acid System.* The weak acid system as depicted in Figure 2 consists of two circulating liquor systems, one of which is adiabatic. The Cooling Tower circuit is the weakest and cleanest. Make-up water is added to this circuit, and there is a subsequent purge from the Quench System in order to maintain steady-state conditions. Weak acid in the Cooling Tower circuit is circulated by weak acid pumps via a plate cooler before being sprayed into the tower at a temperature of 25–30 °C.

The purge stream from the Quench Tower circuit can vary between 1–20% H_2SO_4 and also contains the solid impurities and condensed volatiles which have been removed from the gas stream. In many operations this effluent is neutralized. There is, however, increasing pressure to clean this effluent stream before discharging it to the environment (Chapter 2).

2.2.3 Sulfuric Acid Regeneration Plant. Gas System. As shown in Figure 3, sulfuric acid is atomized in a brick-lined furnace where it is decomposed at a temperature of 1000–1100 °C. The furnace temperature is maintained by burning a supplementary fuel in a pre-heated air stream. In some cases, such as on-site plants at refineries, the supplementary fuel used is H_2S, which oxidizes to SO_2 and H_2O. This increases the SO_2 gas strength to the acid plant.

The air is supplied to the furnace by the Air Blower. The furnace is usually maintained under a slight negative pressure generated by the Main Blower. The gas from the furnace is then cooled in a boiler to a temperature of approximately 300 °C. From this point on, an SAR plant is basically the same as a metallurgical acid plant, but the gas cleaning requirements are not as demanding.

2.3 Recent Developments in Equipment Design

Over the last two decades there have been no fundamental changes to the sulfuric acid manufacturing process, other than optimization to improve energy efficiency and the introduction of improved V_2O_5 catalysts. The developments that have taken place have been in the areas of improved mechanical design, and the use of improved materials of construction.

The two materials of construction that are now used increasingly in new sulfuric acid plant designs are fibre-reinforced plastics and stainless steels. While it will not be possible to cover all the developments in this text, some details will be given on the changes that have occurred on the major equipment items in each section of the plant.

2.3.1 Gas Systems. The major change that has taken place within gas systems is the use of stainless steel instead of carbon steel for the construction of gas exchangers and the converter vessel.[1] This innovation has been led by Chemetics International who introduced an all-welded stainless steel converter design to the industry in 1979–80.

---

[1] 'Converter design for SO_2 oxidation', *Sulphur*, March–April 1992, **219**, p. 26–39.

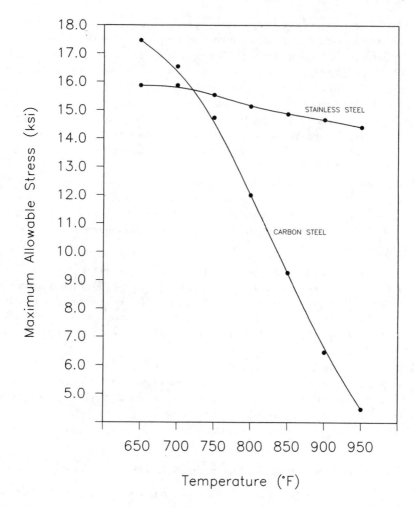

Figure 4   *ASME Allowable Stresses for Carbon Steel and Stainless Steel.* Reproduced from 1992
Edition of Boiler and Pressure Vessel Code, Section VIII, Div. 1; Table
UHA-23 and Table UCS-23, Copyright the American Society of Mechanical
Engineers, With Permission

The structural advantages that the use of stainless steel brings are illustrated in
Figure 4 which compares the ASME allowable stresses for stainless steel and
carbon steel over the operating temperature range encountered in the dry gas
section of a sulfuric acid plant. Stainless steels also have the advantage that they do
not have to be protected with surface coatings or refractory lining, as is the case
with carbon steels, in order to make them resistant to high temperature oxidation.

Traditionally acid plant converters were carbon steel vessels with cast iron posts
and grids to support and separate the catalyst beds. To protect the carbon steel
shell from high temperature oxidation, either metallizing or refractory brick lining
was used. These units suffered from a variety of problems, which have been

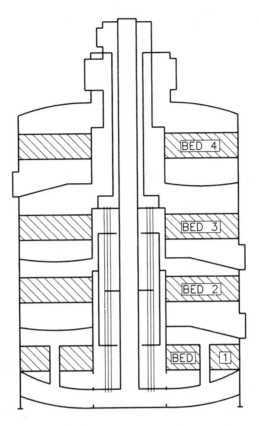

Figure 5 *Typical Chemetics Converter and Hot Heat Exchanger*

addressed in new stainless steel designs. The Chemetics patented converter design which incorporates an internal hot heat exchanger is illustrated in Figure 5. The benefits of this design are outlined below:

- gas bypassing between beds is eliminated, thus increasing overall conversion efficiency;
- elimination of the hot gas duct by internal location of the hot gas exchanger;
- improved gas distribution due to radial entrances to all beds;
- all duct connections are circular thus eliminating highly stressed areas in the shell;
- shorter heat-up time due to the lower metal mass in the stainless vessel compared to the standard post and grid design; and
- location of the first bed at the bottom which simplfies catalyst screening.

Stainless steel is also finding increasing application in gas exchangers handling gases above 470 °C. This is due not only to its resistance to high temperature oxidation, but also to its resistance to acid corrosion where carry-over is a problem downstream of absorption towers.

The changes that have occurred in strong acid tower design have all been in the

Table 2 *Nominal Composition of SARAMET®*

Component	%
Carbon	0.015 (max.)
Nickel	17.5
Chrome	17.5
Silicon	5.3

area of improved mechanical design to reduce maintenance requirements.[2] In most modern plants all three towers are constructed from brick-lined carbon steel, with the tower bottoms typically being dished. All-metallic towers have been installed by a number of acid plant contractors, using high-silicon austenitic stainless steel, which will be discussed in more detail in the section on acid systems. This approach is however considered to have limited application, and will likely only be used in smaller plants where an all-metallic design is more cost-effective, or where replacement of a tower is required during a short turnaround, and insufficient time is available to install a bricked design.

The rest of the major components within the gas system have not undergone any significant change over the last 15 years, other than improvements to lower the gas pressure drop through the system, or improve mechanical reliability.

2.3.2 Acid Systems. Prior to 1970, strong acid cooling was done using cast iron sections with cooling water cascading over the outside. These units were very inefficient from a heat-transfer standpoint, and acid cooler repair was the greatest cause of downtime on a sulfuric acid plant. In the early 1970s Chemetics developed an anodically-protected stainless steel shell and tube design, which proved to be extremely reliable. The coolers have now become the standard for the industry, with the number of units in service approaching 1000. Extensive use is now being made of this proven design to recover energy from acid systems by heating boiler feed water, or by transferring low grade acid heat to other systems. Chemetics coolers have also been used for the generation of low-pressure steam.

Ten years ago practically all strong acid piping systems were fabricated from cast iron. In the early 1980s Chemetics introduced a high-silicon austenitic stainless steel, known as SARAMET®, to the industry. This development, which is protected by patents, allows the use of smaller piping, and significantly reduces the use of flanges giving a higher integrity system. The material is practically resistant to corrosion at normal strong acid circuit temperatures and concentrations. The nominal composition of SARAMET® is given in Table 2.

The reliability of strong acid pumps has continued to improve as a result of advances in both materials and mechanical design. The vertical submerged pump design offered by Lewis Pumps has become the industry standard for strong sulfuric acid. A photograph of a Lewis Pump is shown in Figure 6.

These pumps are available in flow capacities up to $1150 \, m^3 \, h^{-1}$, and total heads of 10–45 m. They offer exceptional corrosion resistance achieved by the use of proprietary alloys which were specifically developed for this application. Lewis

[2] 'More reliable and efficient towers', *Sulphur*, November–December 1992, **223**, p. 31–44.

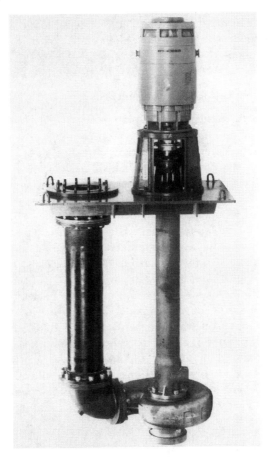

Figure 6 *Lewis Pump 10 MTH, H₂SO₄ Tower Pump.* Photograph Courtesy of Lewis Pumps

Pumps are also available for high-temperature service, greater than 120 °C in 98% H_2SO_4, for use in energy recovery applications.

2.3.3 Gas Cleaning Systems. Within gas cleaning systems extensive use is now being made of fibre-reinforced plastic (FRP). Once the gas has been quenched to its adiabatic saturation temperature, FRP is an acceptable material of construction, and it is finding application for both vessels such as gas cooling towers and retention vessels, as well as ducting. Perhaps the most significant innovation has been the use of FRP in wet electrostatic precipitators. Lurgi now offer a wet electrostatic precipitator design which has a plastic shell as well as plastic tubes. Other manufacturers, such as Joy, offer a plastic shell, with lead tubes.

Increased use is also being made of metallics in gas cleaning systems. For instance, non-metallic Karbate coolers are being replaced to a large extent by metallic plate heat exchangers for weak acid cooling. Typical materials being used are Alloy C-276 and 254 SMO. These materials have also been used for metallic wet gas condensers, which are an alternative to gas cooling towers with cooled weak acid loops.

2.4 Catalysts Used in Acid Manufacture

In the early days of the contact process, platinum was used as a catalyst. For more than fifty years, vanadium pentoxide has been the active ingredient in all commercial catalysts. Vanadium pentoxide contents of most catalysts are in the 6–9% range, with smaller amounts of alkali metal sulfates being added as promoters or modifiers. The support material on which the active ingredients are absorbed is porous silica.[3]

The major change that has taken place over the last 10–15 years is the introduction of ring-type catalysts. These catalysts were first developed by Haldor Topsoe, and they have now become the standard for new acid plants. The prime benefit that ring catalyst provides is reduced pressure drop. For an equivalent catalyst loading, ring catalyst gives approximately 50% of the pressure drop of pellet catalyst. With increasing energy costs and more efficient plant designs, the use of ring catalyst has become essential.

Most vanadium pentoxide catalysts become active at 400–420 °C, but by the addition of activity promoting agents like caesium, the 'strike temperature' of the catalyst can be reduced by 20–30 °C. These catalysts are finding increasing application in higher SO_2 strength designs, as reducing the inlet temperature to the first bed of the converter increases the potential for temperature rise over the first bed, while staying within the maximum outlet limitation of 625–635 °C.[4]

2.5 Product Quality

Most acid plants produce a product with a strength in the range of 93–98% H_2SO_4. Impurities in the product acid can arise from two sources: contaminants in the feedstock, and those resulting from corrosion of plant equipment due to the particularly aggressive environment within the sulfuric acid manufacturing process. As the purity of sulfur fed to most sulfur burning acid plants is high, the acid product from these plants is usually better than the product from either metallurgical acid or SAR plants.

In the past, the major impurity in the acid arising from corrosion within the plant was iron. The upgrading of materials of construction in the strong acid circuits has however led to a significant reduction in iron levels, with figures in the range of 5–10 p.p.m. being quite common. Most plants will have no difficulty in meeting the commercial grade acid specification shown in Table 3. As stated earlier, the highest-quality acid is produced by sulfur-burning plants. The only deviation from specification that may be encountered is high nitrate levels. This is caused by high local temperatures in the sulfur furnace causing oxidation of atmospheric nitrogen in the sulfur flame.

In metallurgical acid plants and SAR plants, the quality of product acid is

[3] 'Sulfuric acid plant catalysts –Shaping up for the future', *Sulphur*, March–April 1985, **177**, p. 33–40.
[4] 'Sulfuric Acid Catalysts', *Fertilizer Int.*, August 1990, **228**, p. 41–46.

Table 3 *Commercial Grade 93% Acid Specification*

Parameter	Specification
Strength	93.0% (min.)
Non-volatile Matter	200 p.p.m.
Oxides of Nitrogen (NO_3)	10 p.p.m.
Sulfur Dioxide	40 p.p.m.
Iron	40 p.p.m.

dictated by the performance of the gas cleaning system. If contaminants in the incoming gas stream are effectively removed, there is no reason why the acid quality should not approach that of a sulfur-burning plant. Common problems experienced in these plants are higher nitrate levels and the presence of trace amounts of organics in the product acid arising from the feed gas, which imparts a slight coloration to the acid.

Both metallurgical acid plants and SAR plants produce acid with higher levels of sulfur dioxide than sulfur-burning plants. This is due to the fact that the acid drying circuits of 'cold' gas plants handle gases which contain significant amounts of sulfur dioxide, resulting in much higher equilibrium levels of SO_2 in the crossflow acid stream. Product acid conforming to the specification in Table 3 can be drawn off from the final absorption system, or a product stripper can be used to air strip SO_2 from the acid.

2.6 Energy Recovery from the Sulfuric Acid Process

Within the sulfuric acid manufacturing process significant amounts of energy are generated as a result of the oxidation of sulfur to sulfur dioxide, and the subsequent conversion into sulfur trioxide. In addition, heat of absorption results when water and sulfur trioxide are absorbed in the strong-acid circuits.

Sulfur burning acid plants obviously have the greatest potential for energy generation, due to the feedstock as well as the fact that the gases entering the process are already hot, and heat from the converter does not have to be used to heat-up incoming gases as is the case for metallurgical acid plants and SAR plants. This section will therefore focus on energy recovery from a sulfur burning acid plant.

A modern sulfur-burning acid plant is capable of producing between 1.3–1.4 t of steam per tonne of 100% H_2SO_4 produced. This number varies based on the pressure and temperature of the product steam. The majority of steam produced in most plants is due to heat recovered from the gas phase.

Over the past decade, sulfur-burning acid plant designs have concentrated on maximizing energy recovery and the following changes have occurred in the design approach in order to achieve this:

- plants are designed with low pressure drops to minimize energy consumption of the main air blower;
- the blower is situated downstream of the Dry Tower to enable the heat of compression to be recovered as steam;
- all equipment drives on the plant are electric and the steam produced is then

let down to condensing conditions in a high efficiency turbo-generator, thus eliminating the inefficiency of multiple smaller steam drives; and

• gases are cooled as close to the dew-point as possible prior to entering the absorbers, thus minimizing the heat lost to the acid system.

Until the advent of a reliable acid cooler design, all the heat from the acid system was lost to the atmosphere via the cooling water system. When the Chemetics cooler was developed and proven, it became possible to recover some of the heat in the acid system. This was achieved by preheating boiler feedwater in either direct or indirect energy recovery loops. There are now a significant number of Chemetics coolers in operation world-wide, recovering low-grade heat from the acid system, and improving the overall energy efficiency of the process as a result.

During the mid-eighties, Monsanto Envirochem developed a heat-recovery system which was able to recover low-grade heat from the acid system, and convert this into low pressure steam. This system, known as HRS (Heat Recovery System) uses a boiler in place of the normal acid cooler and is capable of producing an extra 0.5–0.6 t of low pressure steam (300–700 kPa) per tonne of 100% H_2SO_4 produced.

As standard steels are used for this system, control of acid strength is crucial. There have been a number of failures in the early systems, but as experience improves these systems are becoming more common within the industry. Heat recovery systems are also offered by other contractors, using different approaches to achieve the same end. Within the scope of this chapter it will not be possible to deal with this subject in full, but reference can be made to numerous papers presented on this subject.[5–7]

2.7 Emissions from Modern Acid Plants

As stated earlier, practically all new plants constructed are double absorption units. The allowable sulfur dioxide emission from a double absorption plant is typically 2 kg of SO_2 per 1000 kg of 100% H_2SO_4 produced. This corresponds to an SO_2 conversion efficiency of approximately 99.7% for a typical sulfur burning plant. Modern catalysts are easily capable of meeting this limit, and as will be explained in the section on future trends within the industry, many new plants are being designed for even higher conversion efficiencies.

The other types of emissions to the atmosphere from a sulfuric acid plant are sulfur trioxide and sulfuric acid mist contained in the gases exiting from the final absorption tower. Not all the SO_3 in the gas stream is absorbed, so there will always be some SO_3 present in acid plant stack gases. In addition, the gases leaving the packed section of the absorption tower will contain acid mist droplets, which are effectively removed by high-efficiency mist eliminators. There will however be some residual acid mist left in the gas. The typical allowable emission limit for

[5] J.D. Carnes and J.R. Shafer, 'IMC Fertilizer Sulfuric Acid Plant Heat Recovery System', Central Florida and Peninsular Florida AIChE Convention, Clearwater, Florida, May 1992.

[6] P. Ludtke and R.T. Kreuser, 'The Lurgi Energy Recovery System in Sulfuric Acid Plants', Central Florida and Peninsular Florida AIChE Convention, Clearwater, Florida, May 1992.

[7] E.G. Lyne, G.A. Locke, and A.A. Trickett, 'A Secure Acid Heat Recovery System – Phosphate Complex Applications', AIChE Spring National Meeting, New Orleans, Louisiana, March–April 1992.

sulfur trioxide and sulfuric acid mist in a double absorption plant is 0.075 kg of SO_3 and H_2SO_4 combined (expressed as H_2SO_4) per 1000 kg of 100% H_2SO_4 produced.

2.8 Future Trends within the Industry

As with any technology, change will continue to occur as a result of the needs of the industry and the pressures of more stringent environmental regulations. Within the sulfuric acid manufacturing process, the likely changes that will occur over the next decade are in the areas of improved energy efficiency and reduced sulfur dioxide emissions.

2.8.1 Energy Efficiency. As explained in Section 2.6, significant advances have been made in recovering the energy from the strong acid systems of sulfuric acid plants. The conversion of this low-grade heat into low pressure steam is finding greater application, primarily within the phosphatic fertilizer industry. Further development is likely to occur, resulting in heat recovery systems with improved operability and reliability.

Within the non-ferrous metallurgical industry, changes in smelting technology have resulted in stronger SO_2 gases being available for acid manufacture. Also, as regulatory pressures (see Chapter 2) increase to further reduce sulfur dioxide emissions, regenerative SO_2 scrubbing technology will be implemented on a larger scale. This technology produces a high-strength SO_2 gas, which must be diluted significantly to allow the production of sulfuric acid using current technology. Both of these trends will result in the need for acid plants to process gases containing SO_2 levels above 12%, resulting in energy and capital savings.

2.8.2 Reduced Sulfur Dioxide Emissions. Most new acid plants are designed with a four-bed converter vessel, based on the SO_2 emission standard for a double absorption unit given in Section 2.7. To an increasing extent authorities in some countries are forcing new plants to meet even higher SO_2 conversion efficiencies. This is resulting in designs with higher than normal catalyst loadings, or in some cases, the need for an extra bed in the converter. The trend is expected to continue, and technical developments are anticipated in this area to meet this new requirement.

3 SULFURIC ACID CONTAINMENT

3.1 Acid Storage

Being a commodity chemical, sulfuric acid is stored and transported in large volumes. The most common grades of acid which are sold commercially are 93%, 96%, and 98%. The freezing points for these three grades are shown in Table 4.

Sulfuric acid is generally stored in carbon steel storage tanks. Tanks containing 93% acid are usually uninsulated, while higher-strength acids are stored in either insulated, or heated and insulated tanks dependent on ambient conditions.

Corrosion of storage vessels is a function of temperature and acid concentration. This is illustrated in Figure 7, where it can be seen that the lower strength commercial grade acids are more aggressive to carbon steel than 98% acid. The

Table 4   *Acid Freezing Points*

Acid strength / %	Freezing point / °C
93.0	−27.0
96.0	−14.7
98.0	−1.0

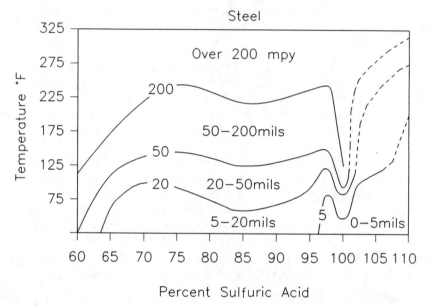

Figure 7   *Corrosion Curves for Sulfuric Acid.* Reproduced from M.G. Fontana/N.D. Greene, 'Corrosion Engineering', Copyright 1978, 1967 by McGraw-Hill Inc., With Permission

other factor to be taken into account is acid purity. The lower the level of iron in the acid, the higher the corrosivity. This effect is shown in Figure 8. As a result of the increased use of stainless steels in modern plants, the product acid is very low in iron, typically less than 5 p.p.m. Storage tanks can therefore be expected to corrode at higher rates than in the past.

Anodic protection systems for storage tanks have been in use for 15–20 years. With the reduction in iron levels in product acid, tank protection systems are now finding even greater application within the industry for tanks larger than 1000 tons. The principle of anodic protection as applied to storage tanks is the same as for Chemetics' coolers, whereby a passive film is formed on the metal in the presence of an electrolyte (H_2SO_4), by the application of a positive potential to the metal. Anodic protection systems do not eliminate corrosion, but they do reduce it significantly.

It is not generally cost effective to construct large acid storage tanks from stainless steel. It does however find wide application in small tanks for high-quality acid products, and is used extensively for vessels transporting sulfuric acid.

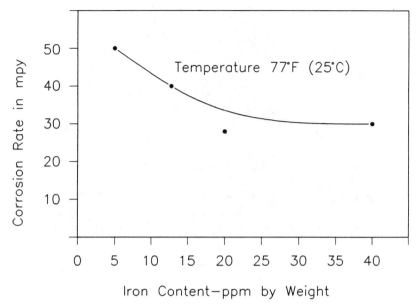

Figure 8 *Effect of Iron on Acid Corrosivity.* Reproduced from 'Carbon Steel Sulfuric Acid Storage Tank – Inspection Guidelines', Copyright 1986 by Marsulex Inc., With Permission

3.2 Transportation of Sulfuric Acid

Acid is transported in barges on inland waterways and in cargo ships on the high sea, and many barges still in use are made from carbon steel, although most large carriers have stainless steel tanks. This is done for two reasons: to reduce corrosion, as well as providing a general-purpose carrier that can transport a variety of different commodities.

The transportation of commercial acid by road is being done more and more in stainless steel tank trucks; the rationale behind this being the same as stated earlier for ships carrying acid. Baked phenolic linings are also used in tank trucks carrying speciality acids, such as electrolytic grade or low-iron acids. These linings are effective in maintaining the product quality as well as protecting the carbon steel shell of the tank truck. Linings do however have a finite life, and are subject to variations in performance dependent on the quality of the materials, workmanship, and the operating procedures adopted.

In North America, and to an increasing extent South America, acid from remote metallurgical operations is being moved to market, or to port for export, by rail.[8] Specialized tank cars, which are carbon steel with baked phenolic linings, are linked together in large-unit trains, containing up to 5000 tons in one shipment. Lining of these cars has proven to be essential due to the severe corrosion effects

---

[8] A.A. Trickett and D.W. Young, 'Transportation and Handling of Sulfuric Acid – A North American Experience', British Sulfur's 9th International Conference – Sulfur 85, London, England, November 1985.

Figure 9   *Hydrogen Grooving of the Shell of a Vertical Tank.* Reproduced from 'Carbon Steel
Sulfuric Acid Storage Tank – Inspection Guidelines', Copyright 1986 by
Marsulex Inc., With Permission

Table 5   *Properties of Liquid SO_2*

Boiling Point @ 1 atm	$-10\,^{\circ}\mathrm{C}$
Freezing Point @ 1 atm	$-75.9\,^{\circ}\mathrm{C}$
Latent Heat Vaporization @ 21.1 $^{\circ}$C (70 $^{\circ}$F)	$361.5\,\mathrm{kJ\,kg^{-1}}$
Specific Gravity of Liquid @ 0 $^{\circ}$C	1.436
Specific Heat of Gas @ 25 $^{\circ}$C and 1 atm	$C_p\ 0.622\,\mathrm{kJ\,kg^{-1}\,{}^{\circ}C^{-1}}$ $C_v\ 0.485\,\mathrm{kJ\,kg^{-1}\,{}^{\circ}C^{-1}}$

that became evident over the service life of a tank car.

When acid corrodes carbon steel, the electrochemical reaction produces
hydrogen as a by-product. These hydrogen bubbles result in a phenomenon
known as 'hydrogen grooving', which can occur on the shell of vertical carbon steel
storage tanks, as well as in the upper halves of shell manholes and horizontal tanks.
An example of hydrogen grooving is shown in Figure 9.

In summary, sulfuric acid can be reliably stored and safely transported, but
attention must be paid to the corrosive nature of this product. Tanks and
transportation vessels must be carefully specified, closely inspected during
fabrication, and well maintained while in service to ensure incident free operation.

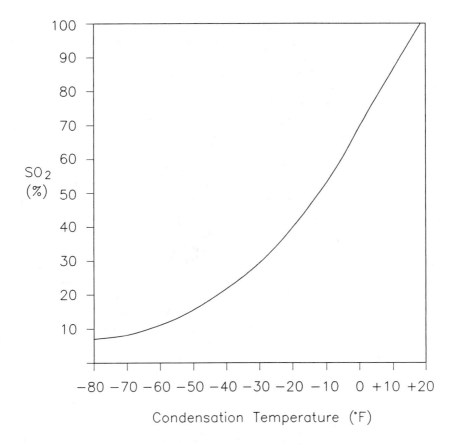

Figure 10 *SO₂ Condensation at Atmospheric Pressure vs. % SO₂*

4 LIQUID SULFUR DIOXIDE

4.1 Introduction

Liquid sulfur dioxide is a liquefied gas, which at atmospheric pressure boils at minus 10 °C. Some of its physical properties are given in Table 5. Liquid SO_2 is non-corrosive to carbon steel and can be readily transported by road and rail.

Figure 10 shows the condensation temperature of liquid SO_2 at atmospheric pressure as a function of gas strength. To stay within the range of most commercial refrigerants, liquefaction of SO_2 at atmospheric pressure would therefore be restricted to gas strengths above 12% SO_2.

4.2 Production of Liquid Sulfur Dioxide

Sulfur dioxide gases containing greater than 12% SO_2 can be generated directly by burning elemental sulfur, by flash-smelting processes handling sulfide ores, and by chemical means. Once generated, the processing required to produce liquid product is the same in principle, but varies with the characteristics of the feed gas.

For instance, the gases produced by a sulfur burner are both clean and dry, but contain small amounts of SO_3 due to the equilibrium achieved in the burner. To make commercial-grade liquid SO_2 will require removal of this contaminant, which can be achieved by passing the gases through an acid tower in which the SO_3 is absorbed prior to liquefaction of the gases.

By contrast, metallurgical gases leave the smelter containing a variety of contaminants as explained in Section 2.1. These contaminants must be removed in a gas cleaning system, and the gas dried before the SO_2 can be condensed. The drying of the gas is normally achieved by counter-current contact with sulfuric acid in a standard drying tower. This operation does, however, produce a weaker acid stream containing high levels of SO_2 which must be disposed of. This can be easily achieved by integrating the SO_2 plant drying circuit with the main acid plant system. Strong 98% acid is fed to the SO_2 plant to dry the gas, and the weaker 93% acid laden with SO_2 is then returned to the main plant dry acid circuit.

For sites where the potential does not exist for integrating the SO_2 drying system with an adjacent acid plant, the producer is faced with the problem of purchasing acid for drying the SO_2 gas, and then selling the weaker acid produced as a by-product. In this instance there is a big incentive to minimize the amount of acid both purchased and sold, by using the maximum possible acid strength for make-up to the circuit. For small plants alternative means of drying can be considered.

In liquefying sulfur dioxide from gas streams containing incondensibles, it is also necessary to have a means of dealing with the plant tail gases. This can be done using a scrubbing system to remove residual sulfur dioxide, or in the case where there is an acid plant on the site, by further processing of the tail gases in the acid plant thus producing acid from the remaining SO_2.

4.3 Processes for the Production of Liquid Sulfur Dioxide

Having produced a clean, dry gas, liquefaction can then be achieved by compression, refrigeration, or a combination of both. The choice depends on gas strength, cooling water temperatures, and the degree of liquefaction desired. Both compression and refrigeration flowsheets will be described, and these are illustrated in Figures 11 and 12 respectively.

4.3.1 Compression Flowsheet. For the purposes of this description, it has been assumed that the feed gas is from a flash smelter producing a gas containing in excess of 70% SO_2, and that the gases have been cleaned and dried prior to entering the SO_2 plant.

Dry gas is drawn into the plant at a temperature of 35–45 °C, and is then compressed in two stages to a pressure of 620–690 kPa (g) by a primary compressor. Between stages 1 and 2 the gas is cooled in an intercooler. After leaving the second stage of the primary compressor, the gas is cooled in a primary liquefier and the

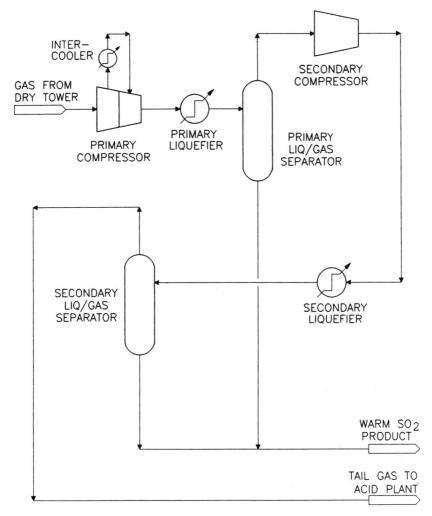

Figure 11 *Liquid SO₂ Production, Compression Flowsheet*

liquid is separated in a vapour separator prior to the gas entering the secondary compressor. At this point 60–70% of the incoming SO_2 has been removed.

In the secondary compressor, the pressure of the gas stream is further increased to 1900–2000 kPa(g) before passing to the secondary liquefier and separator where approximately 20% more of the incoming SO_2 is removed. The gas leaving the separator still contains appreciable amounts of SO_2, and cannot be discharged to the atmosphere. As shown in Figure 11, the gas would then be let down to the acid plant for further processing.

The liquid from the primary and secondary separators is collected and transferred to storage. Liquid sulfur dioxide is normally stored in carbon steel vessels under pressure.

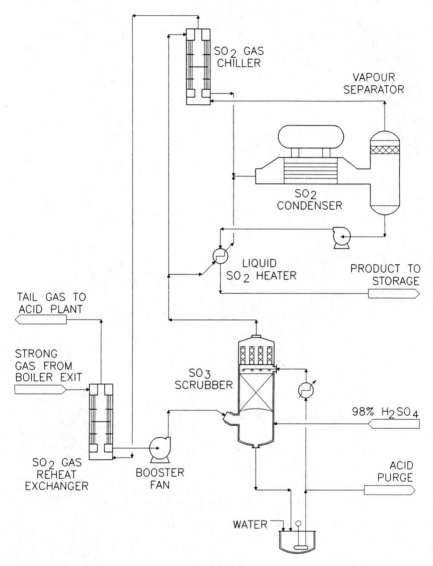

Figure 12 *Liquid SO₂ Production, Refrigeration Flowsheet*

4.3.2 Refrigeration Flowsheet. For the refrigeration flowsheet shown in Figure 12, it has been assumed that the feed gas is from a high strength (greater than 12%) sulfur burner feeding an acid plant. The gas enters the SO₂ plant after being cooled to approximately 500 °C in a firetube boiler. It is then cooled further in the SO₂ Gas Reheat Exchanger by interchange with the tail gases from the SO₂ plant.

The gases, now at a temperature of approximately 170 °C, are boosted to 15–25 kPa above the acid plant operating pressure by the Booster Fan. The gas then enters the SO₃ Scrubber for removal of sulfur trioxide, which if left in the gas would result in unacceptable acidity levels in the final product.

In the Scrubber the gas contacts a counter-current stream of 98.5% H_2SO_4 which removes the SO_3. The gas exits the tower via high-efficiency mist eliminators and splits into two streams, one entering the SO_2 Gas Chiller where it is cooled further and the other heating up the cold liquid product from the vapour separator, prior to entering the SO_2 Condenser. The Condenser, which is a stainless shell and tube unit, is cooled by a refrigerant which boils in the $-40 - -50\,°C$ range. From the Condenser, the gas flows to the Vapour Separator where the liquid is removed, with the gas flowing via the SO_2 Gas Chiller and SO_2 Gas Reheat Exchanger back to the inlet to the acid plant for further processing. The liquid from the Vapor Separator is heated and then transferred to storage.

4.3.3 Production of SO_2 by Chemical Means. Liquid sulfur dioxide can also be produced by the reaction of sulfur trioxide with sulfur to form sulfur dioxide gas which can then be liquefied by either compression or refrigeration as explained earlier. The reaction which is shown below is mildly exothermic:

$$S + 2SO_3 \rightarrow 3SO_2$$

This process was first developed in Germany, and involved feeding liquid sulfur and preheated oleum to a reactor containing oleum at $110\,°C$. Modifications made to this process by Schweizerhall and Stauffer Chemical Company involve the use of liquid and gaseous sulfur trioxide, respectively, instead of oleum.

4.4 Uses of Liquid Sulfur Dioxide

Sulfur dioxide is used in a variety of industrial applications. In the large volume uses SO_2 is typically not incorporated into the end product. Because uses are quite extensive, only the major industrial applications will be covered.

(i) *Pulp and Paper.* SO_2 was at one time used extensively in the pulp and paper industry, but due to changes in technology this has declined. The industry is however still a major user of SO_2 as a reductant in the production of chlorine dioxide, as a raw material for the production of sodium dithionite which is used for the bleaching of pulps, and as a pulping chemical in sulfite mills.

(ii) *Mining.* This is a growing market for liquid sulfur dioxide. It is used as a reducing agent for the separation of vanadium from uranium, as a precipitant for the recovery of copper cyanide from leach solutions, as a reductant in the recovery of bromine and iodine from seawater, and as a depressant in mineral flotation circuits.

(iii) *Water and Wastewater Treatment.* Sulfur dioxide is used to remove excess chlorine from water and wastewater after it has been chlorinated, to destroy cyanide in the tailings from gold leaching effluents, and in the treatment of effluents containing chrome.

(iv) *Wine Production.* SO_2 is used for sterilization, decolorization, and stabilization.

(v) *Petroleum.* SO_2 is used to remove aromatic and high-sulfur components from paraffins and naphthenic hydrocarbons. It is also used as a solvent in sulfonation

Table 6 *Properties of Liquid SO₃*

Boiling Point	44.8 °C
Melting Point (γ)	16.8 °C
Heat of Vaporization @ NBP	0.538 kJ kg^{-1}
Specific Gravity @ 30 °C	1.88
Specific Heat	3.22 kJ kg^{-1} °C^{-1}

reactions due to its capability to remove heat by vaporization.

(vi) *Electric Power Generation.* Sulfur dioxide is used in this application as a raw material for the production of SO₃ for the conditioning of flue gases to improve particulate removal.

(vii) *Sweeteners.* In the manufacture of corn syrups SO₂ is used for the removal of colour, and as an enzyme stabilizer and buffer for high-fructose corn syrup. It is also used for the clarification of cane and beet juices, and to control fermentation and facilitate softening in corn wet-milling.

(viii) *Chemical Manufacture.* Sulfur dioxide is used in the manufacture of dithionites, sulfolane, polysulfones, sulfites, bisulfites, resins, plastics, and miscellaneous sulfur–organic compounds.

5 LIQUID SULFUR TRIOXIDE

5.1 Introduction

Sulfur trioxide is the anhydride of sulfuric acid. It is a strong sulfonating agent for organics and also finds use in dehydrating applications. It has distinct advantages over other sulfonating agents such as sulfuric acid or oleum, as it does not generate a spent-acid stream which requires disposal. It is, however, a hazardous chemical and for this reason its use is being restricted increasingly to applications involving on-site generation.

5.2 Production of Liquid Sulfur Trioxide

As can be seen from the summary of physical properties shown in Table 6, liquid sulfur trioxide has both a high freezing point and a relatively low boiling point. This makes it a difficult chemical to handle, which is further exacerbated by its high reactivity, its explosive reaction with water, and its tendency to polymerize in the presence of trace amounts of water or sulfuric acid. When polymerization occurs, melting can result in explosions due to the high vapour pressures which are generated before melting occurs. All of these characteristics encourage the maintenance of small inventories, and on-site generation which allows immediate use in downstream processes.

5.2.1 Process Description. Liquid sulfur trioxide is produced by the distillation of oleum to produce a pure liquid SO₃ stream which is then condensed to form a liquid product. A process flowsheet for a typical liquid sulfur trioxide plant is

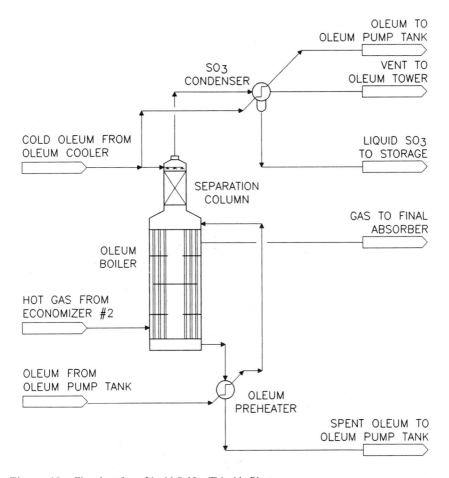

Figure 13 *Flowsheet for a Liquid Sulfur Trioxide Plant*

shown in Figure 13. Before describing the process some information on oleum will be provided.

Oleum, or fuming sulfuric acid, is a solution of 100% H_2SO_4 containing absorbed SO_3. It is produced in the absorption system of an acid plant by contacting a rich SO_3 stream in a packed tower with a circulating stream of oleum. On most acid plants oleum strengths between 20–37% are made directly in a single-absorption step. For strengths higher than this, distillation of part of the product is required to provide a pure SO_3 stream which can then be contacted with the product oleum to further fortify it.

As can be seen from Figure 7, the rate of corrosion of carbon steel increases significantly in the high acid/low oleum range. For this reason, when oleum is distilled the residual oleum strength should not be reduced below 17%. To maintain a margin of safety, many plants distill to 20% oleum. Because of the need to refortify the depleted oleum stream, liquid sulfur trioxide production normally takes place in conjunction with sulfuric acid manufacture.

As previously stated, liquid SO_3 is produced by vaporizing SO_3 from strong oleum and condensing the resulting gas. High strength oleum (typically 30–40%) is pumped from the oleum circulation loop on the acid plant to the Oleum Preheater, where it is heated by the depleted oleum stream exiting the Oleum Boiler as shown in Figure 13. The preheated oleum then enters the top of the Boiler and flows down the inside of the tubes in a thin film. Sulfur trioxide gas is boiled off from the oleum by transferring heat from hot process gas from the acid plant at temperatures in excess of 300 °C. The gas from the Boiler is returned to the acid plant absorption system.

The depleted oleum at a strength of 20–25% is returned to the oleum pump tank on the acid plant for fortification. The hot SO_3 gases now at a temperature of 140–145 °C leave the top of the Boiler and flow directly to the Separation Column. The purpose of the Separation Column is to remove any entrained mist or oleum droplets, and to condense any H_2SO_4 vapour present. In the Column, the hot SO_3 gas contacts a counter current stream of cold oleum taken from the delivery side of the cooler of the oleum loop on the acid plant.

The SO_3 gas is cooled to approximately 70 °C before it enters the SO_3 Condenser, where it is liquefied by transferring heat to a cold oleum stream from the acid plant oleum loop. The liquid SO_3 exits the Condenser at a temperature of about 38 °C and is delivered to storage, the oleum on the tubeside of the Condenser being returned to the oleum pump tank.

5.2.2 Storage Tanks for SO_3. Liquid SO_3 can be stored in carbon steel tanks, but extreme precautions are necessary in system design and operation in order to ensure a safe plant. Liquid SO_3 must be maintained at a temperature of 35–43 °C at all times, in order to prevent freezing. This is normally achieved by locating the storage tank in a temperature controlled room. These storage rooms are usually connected to an external scrubbing system to prevent the possibility of SO_3 emissions to the atmosphere. As the SO_3 is stored close to its boiling point, it is difficult to pump, and transfer is usually by gravity or displacement by gas padding. All these factors have been instrumental in promoting on-site generation for direct use in downstream processes.

5.3 Uses of Sulfur Trioxide

The main uses for liquid sulfur trioxide are in sulfonation and sulfation reactions. Its principal applications are in the production of detergents. It is also used as a raw material for the production of 65% oleum and chlorosulfuric acid. A number of the other important industrial applications for which SO_3 is used are

- Air/SO_3:
    Sulfonation high molecular weight raffinates;
    Sulfation fatty alcohols;
    Sulfation ethoxylated alcohols;
- Solvent/SO_3:
    Sulfonation lube oil;
    Sulfonation toluene;

- Direct addition:
    Liquid – Sulfonation nitrobenzene;
            Sulfonation asphalt;
            Sulfonation phthalic anhydride;
    Vapour – Sulfonation phenol;
            Surface treatment of polyethylene.

CHAPTER 5

Sodium Carbonate

R. D. A. WOODE

1 INTRODUCTION

Sodium carbonate has been used by mankind for many thousands of years, although large-scale industrial production has a history of a few hundred years only. Alkali deposits occur naturally in various parts of the world, and these have been exploited in the 'old world' to make products such as glass. Alkali can also be obtained by leaching the ashes formed by burning plant life, which is probably why potassium carbonate is also known by the name potash. By analogy, the common name for sodium carbonate is soda ash.

Sodium carbonate has a wide variety of uses in an industrialized society and its consumption per capita has even been used as a measure of the degree of industrialization of a society. In general the demand for soda ash has been increasing at an average rate of 2% per year over many decades, with higher and lower short-term rates of growth reflecting changes in the general world economic situation. Individual countries have tended to have growth rates which reflected the growth rates of their own economies. Current world demand is about 30 million tons year^{-1} at a capacity utilization of some 80–85%. Approximately two-thirds of the sodium carbonate is produced by the 'Synthetic' (or Ammonia–Soda) route, with the remainder being produced by the 'Natural Ash' route.

The uses of sodium carbonate are widespread and, as will be seen in the next section, some are increasing in importance whilst others diminish. However, sodium carbonate is constantly in competition with another sodium alkali, sodium hydroxide, in the market place. In many of the end uses the two products are interchangeable, though possibly at some inconvenience to the user due to their different chemical and handling properties. However, these objections can be overcome economically when the price differential between the products is sufficient. Sodium hydroxide is manufactured primarily by the electrolysis of brine, $i.e.$

$$\text{Electricity}$$
$$2NaCl + 2H_2O \rightarrow 2NaOH + H_2\uparrow + Cl_2\uparrow$$

It is produced, therefore, as a co-product of chlorine production. Because chlorine is expensive to dispose of, the rate of sodium hydroxide production by this route is normally set by chlorine demand. This means that at times of high chlorine demand there can be an excess of sodium hydroxide to dispose of, and spot prices are low, whereas when chlorine demand is low the reverse applies. Thus, there is a fluctuating tendency for sodium hydroxide to replace sodium carbonate in some markets for a period followed by another period when sodium carbonate replaces sodium hydroxide.

Up to the 1960s, the world sodium hydroxide demand was generally less than the supply available from brine electrolysis. The shortfall was made up by using the Lime–Soda process for caustic soda production. The overall process can be represented by the following equation:

$$Na_2CO_3 + CaO + H_2O \rightarrow 2NaOH + CaCO_3 \downarrow$$

During the 1960s and 1970s many of the Lime–Soda–Caustic plants were shut down because of the increasing availability of electrolytic caustic soda. However, in recent years, partly due to the environmental pressures on chlorinated products, sodium hydroxide production has not always kept up with demand and some new Lime–Soda–Caustic facilities have been commissioned in the USA.

Traditionally sodium carbonate was sold in $2\frac{1}{2}$ cwt (127 kg) jute sacks. Because of the scale of present-day operations, most of the product leaves the factory in bulk in specially designed road or rail tankers or in river/ocean going vessels capable of carrying many thousands of tonnes. A small, decreasing proportion is still sold in bags of paper or poly(ethene)/poly(propene) – usually packed in lots of 25 or 50 kg.

2 USES

Soda ash is manufactured in two grades. 'Light Ash' and 'Heavy Ash', the difference being the much greater volume occupied by a given mass of Light Ash. Heavy Ash, when transported in bulk, has a pouring density equal to or slightly above that of water, whereas the pouring density of Light Ash is only half this value. Heavy Ash is clearly the preferred form for economic transportation over long distances.

The main customers for Heavy Ash in the UK are glass manufacturers who take over 90% of the total. This grade is chosen for glass because its particle size is similar to that of sand, ensuring a homogeneous mixture. Other users of Heavy Ash are steel works, foundries, and enamellers and it has a limited application in laundry detergents. About two-thirds of sodium carbonate manufactured in the UK is in the form of Heavy Ash.

For general chemicals production Light Ash is widely used because of its traditionally lower price, its higher rate of dissolution and reaction, and its freedom from calcium ions, which are unacceptable in certain processes, for example soap manufacture (see Figures 1 and 2).

2.1 Glass Manufacture

The majority of commercial glasses are made from three prime components: sand,

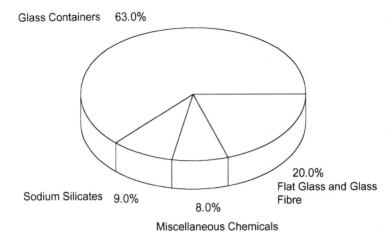

Glass Containers 63.0%

20.0%
Flat Glass and Glass
Fibre

Sodium Silicates 9.0%

8.0%
Miscellaneous Chemicals

Figure 1 *Uses of Heavy Ash*

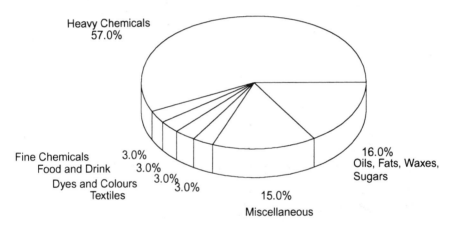

Heavy Chemicals
57.0%

Fine Chemicals 3.0%
Food and Drink 3.0%
Dyes and Colours 3.0%
Textiles 3.0%

16.0%
Oils, Fats, Waxes,
Sugars

15.0%
Miscellaneous

Figure 2 *The Major Uses of Light Ash*

limestone, and soda ash. The soda ash tends to be the single most expensive component, so its cost and quality are of prime importance to the glass manufacturer. In particular, impurities must be kept to a low level to avoid damage to refractories and undesirable coloration; also particle size must be adequate to avoid unmixing in the batch and minimize dust losses during handling.

Sodium carbonate is used to make sheet and plate glass for use in the building industry, car industry, *etc.*, for containers, for fibre-glass – both insulating and reinforcing grades – and in a multitude of special glasses. Recently there has been a fundamental change in the usage of sodium carbonate for glass manufacture due to increasing recycling of cullet, to improved technology allowing the production of thinner containers, to changes in the container market, and due to minor changes in manufactured glass composition. Nevertheless, glass manufacture still remains the largest single usage of sodium carbonate.

2.2 Sodium Silicates

There are two major manufacturing routes for sodium silicate, the hydrothermal route from caustic soda solution and sand for the more alkaline grades, and the more traditional and more widespread glass-fusion route from sodium carbonate and sand for manufacturing the full range of grades.

2.3 Phosphates/Polyphosphates

Caustic soda and soda ash are even more interchangeable in phosphates manufacture, which, though currently under increasingly severe environmental pressures, still represents a significant use of sodium alkali.

2.4 Detergents

Most detergents contain sodium carbonate, or sodium carbonate/bicarbonate products in addition to other chemicals such as silicates and phosphates which involve the indirect use of sodium carbonate. Indeed the current trends of reducing phosphate levels in detergents are associated with an increase in the use of sodium carbonate either alone or together with zeolites, which are often manufactured using sodium silicate.

2.5 Chromium Chemicals

One major user of sodium carbonate is the chromium industry, particularly in the production of chromates and dichromates. Again the quality of the product is critical in minimizing potential formation of Cr^{III} compounds.

2.6 Pulp and Paper

Much sodium alkali is used in this industry, primarily for pulp digestion. Traditionally this has been an area where chlorine, or chlorinated products, have been used for bleaching purposes. With increasing environmental pressures, this is one area of usage which is changing the caustic/chlorine balance in favour of sodium carbonate sales.

2.7 Water Treatment and Brine Purification

Hot or cold lime–soda processes have been used for water treatment for many decades. Calcium and magnesium hardness is removed by precipitation of the insoluble calcium carbonate and magnesium hydroxide, *e.g.*

$$Na_2CO_3 + CaSO_4 \rightarrow CaCO_3 \downarrow + Na_2SO_4$$
$$Ca(OH)_2 + Na_2CO_3 + MgSO_4 \rightarrow CaCO_3 \downarrow + Mg(OH)_2 \downarrow + Na_2SO_4$$

The method has the advantage of co-precipitating other impurities at the same time (silica and heavy metals in particular). A parallel process is used for removing the same types of impurities from crude brines, which can then be used for salt, chlorine, and alkali manufacture. These two types of usage account for a significant proportion of sodium carbonate sales.

2.8 Flue Gas Desulfurization

With the increasing requirement to remove SO_2 and NO_x from flue gases, the market for sodium alkalis is rapidly expanding in this area. Sodium carbonate, sodium bicarbonate, and sodium sesquicarbonate are all used for this duty in various applications. Their advantage is that they can be used for dry treatment of the flue gas, and give rise to soluble, virtually harmless, and easily disposed of end products. The major alternative, limestone, is frequently used in wet systems in order to achieve efficient utilization and gives rise to an insoluble waste product.

2.9 Miscellaneous Uses

Sodium carbonate is a cheap, soluble, basic chemical and so is used in a very wide variety of industrial processes. Some types are indicated above, but a complete list would be difficult to compile because it would be so extensive.

2.10 Caustic Soda

In the introduction the Lime–Soda–Caustic process was identified as the major manufacturing process for providing hydroxide to make up the short fall between world demand and production by the electrolytic route. In fact the Lime–Soda–Caustic process used to be the main process for providing caustic soda, and has only diminished in importance due to the very rapid increase in demand for chlorine in the early part of the century. Thus caustic soda production used to be one of the major uses of sodium carbonate.

However, sodium carbonate in its normal finished and saleable form was not the material generally used for caustic soda manufacture. It was normally more economic to use an intermediate product, commonly a sodium carbonate solution, for the reaction with lime. A sodium hydroxide solution of about 12–14% was obtained as the primary product; insoluble products were separated from the solution; the solution was evaporated to strengthen the product and crystallize out impurities, and finally sold as a variety of products from 45% NaOH solutions to anhydrous solid. Thus, in practice, the sodium carbonate manufacturer used also to be one of the main users of his own product and a major manufacturer of sodium hydroxide.

3 THE MANUFACTURE OF 'SYNTHETIC' SODA ASH

3.1 The Leblanc Process

During the eighteenth and nineteenth centuries the major process for alkali production was the Leblanc process, in which sodium chloride was fused in a furnace with concentrated sulfuric acid to give sodium sulfate and hydrogen chloride gas:

$$2NaCl + H_2SO_4 \rightarrow Na_2SO_4 + 2HCl\uparrow$$

The sodium sulfate was subsequently reacted with calcium carbonate, in the form of limestone, and with carbon, as coal, to give sodium carbonate:

$$Na_2SO_4 + CaCO_3 + 2C \rightarrow Na_2CO_3 + CaS + 2CO_2\uparrow$$

The hot fused product was discharged from the furnace and allowed to cool. The sodium carbonate at this stage was contaminated with calcium sulfide and coal ash. It was separated from the contaminants by leaching with water and was recovered as a solution. The resulting sodium carbonate solution was decolorized and concentrated.

The process had several serious disadvantages which were:

(i)   large consumption of energy at the fusion stage;
(ii) intensive labour requirements because it was a multi-stage, batch process; and
(iii) environmental problems.

The environmental problems included the release of untreated hydrogen chloride to the atmosphere and the unsightly slag heaps of the worthless calcium sulfide and coal ash produced in the second stage of the reaction. Atmospheric action on the calcium sulfide produced, amongst other decomposition materials, hydrogen sulfide and calcium sulfate:

$$CaS + 2H_2O \rightarrow H_2S\uparrow + Ca(OH)_2\downarrow$$
$$CaS + 2O_2 \rightarrow CaSO_4\downarrow$$

The pollution created by the Leblanc process led directly, in the mid-nineteenth century, to the introduction of the Alkali Act, which laid down regulations for emissions from factories and created the Alkali Inspectorate to enforce them.

From the 1880s, the Leblanc method of manufacture was gradually replaced by the more efficient and cleaner ammonia–soda process, the first continuous process to be introduced by the chemical industry. It was pioneered by the Solvay family in Belgium in 1865, who seven years later licensed John Brunner and Ludwig Mond to operate the process in the UK – at Winnington in Cheshire.

By 1920 all the factories in the UK using the Leblanc process to make soda ash had been closed.

3.2   The Modern Ammonia Soda Process

3.2.1   Chemistry. The modern process is designed to produce sodium carbonate from cheap raw materials – salt and limestone. The overall reaction may be represented as follows:

$$2NaCl + CaCO_3 \rightarrow CaCl_2 + Na_2CO_3$$

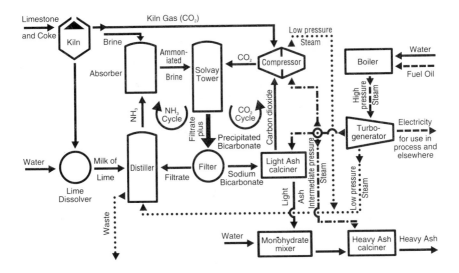

Figure 3 *Simplified Flow Chart of the Ammonia–Soda Process*

However, a reverse reaction – competing and thermodynamically more favourable – takes place between calcium chloride and sodium carbonate to give the original raw materials. This reverse reaction is favoured to such an extent that if the products are mixed together, sodium chloride and calcium carbonate are immediately formed.

A one-stage process in which salt and limestone are reacted directly together is, therefore, impossible and a more complicated sequence of reactions must occur. The reaction sequence is as follows:

$$NaCl + NH_3 + H_2O \rightleftharpoons NH_4^+ + Cl^- + Na^+ + OH^-$$
$$\text{(ammoniacal brine)}$$

$$NaCl + NH_3 + H_2O + CO_2 \rightleftharpoons NH_4Cl + NaHCO_3\downarrow$$

$$\overset{\text{Heat}}{2NaHCO_3 \rightarrow Na_2CO_3 + CO_2\uparrow + H_2O\uparrow}$$

$$\overset{\text{Heat}}{CaCO_3 \rightarrow CaO + CO_2\uparrow}$$

$$CaO + H_2O \rightarrow Ca(OH)_2$$

$$Ca(OH)_2 + 2NH_4Cl \rightleftharpoons CaCl_2 + 2NH_3\uparrow + 2H_2O$$

Although the process involves all these stages, they may be resolved to the overall reaction given earlier.

3.2.2 The Overall Operating Process. It is helpful to consider the overall process in terms of two cycles – an ammonia cycle and a carbon dioxide cycle (see Figure 3).

Limestone in kilns is decomposed by heating to give lime and carbon dioxide. The lime (calcium oxide) is slaked with water to give a hydrated lime slurry. The resulting calcium hydroxide (milk of lime) is then reacted with ammonium chloride to produce ammonia, which is distilled over for absorption in brine. The ammoniacal brine passes into Solvay towers, where it is reacted with carbon dioxide from the compressors. Sodium hydrogen carbonate is precipitated and filtered, and the filtrate recycled to the distillers. This completes the ammonia cycle.

The filtered sodium hydrogen carbonate is heated in the Light Ash calciners and decomposes to sodium carbonate, water, and carbon dioxide. This carbon dioxide is recycled and fed, together with CO_2 from the lime kilns, into the Solvay towers. This completes the carbon dioxide cycle.

The overall reaction to produce sodium carbonate is slightly endothermic.

$$2NaCl_{(aq)} + CaCO_{3(s)} \rightarrow Na_2CO_{3(s)} + CaCl_{2(aq)} \qquad \Delta H = +12\,kJ\,mol^{-1}$$

3.2.3   Lime Burning Stage. A mixture of stone and coke in the correct proportions is fed into the top of the kiln. The heat which decomposes the limestone is supplied by burning the coke in air. The temperature in the burning zone is about $1100\,°C$ and may reach $1350\,°C$; decomposition rates are insignificant below $900\,°C$. Hot lime falls down through the kiln and is cooled to ambient temperatures by the incoming air (see Figure 4). The overall fuel efficiency of the process (*i.e.* the conversion of coke into useful chemical energy) averages 85% – an extremely good figure for such a high temperature process.

The decomposition of limestone is highly endothermic, but ample chemical energy can be stored inside the compounds (lime and carbon dioxide) and released later in the process. In fact the amount of energy available in these later stages is in excess of requirements and has to be removed as heat.

	ΔH
	$(kJ\,mol^{-1})$
$CaCO_3 \rightarrow CaO + CO_2\uparrow$	$+178$
$C + O_2 \rightarrow CO_2\uparrow$	-393
$C + \frac{1}{2}O_2 \rightarrow CO\uparrow$	-110

The first equation shows that decomposition of one mole of limestone into lime and carbon dioxide requires $+178\,kJ$. From the second it is evident that more than twice this amount can be released by burning coke in air. Because of the 85% efficiency of the process, one mole of coke can be burnt to produce sufficient heat to decompose about two moles of limestone. The reaction therefore can be summarized as follows:

$$2CaCO_3 + C + O_2 + 4N_2 \rightarrow 2CaO + 3CO_2\uparrow + 4N_2\uparrow$$

It is important to note that both of the reaction products are vital to subsequent stages of manufacture. Conditions in the kiln must therefore favour:

(i)   the production of highly reactive lime;
(ii) the highest possible concentration of carbon dioxide in the gas mixture which passes to the Solvay towers; and
(iii) the suppression of the reaction producing carbon monoxide.

In order to achieve these conditions vertical shaft mixed-feed kilns are preferred.

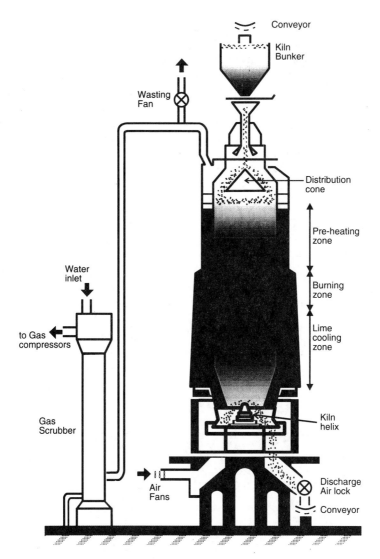

Figure 4 *A Lime Kiln*

Although this requires the use of a relatively expensive low-volatiles fuel, such as coke, the gas quality and composition that is obtained generally makes this option the most economic.

The equations show that approximately 50% more carbon dioxide is produced than is stoichiometrically necessary. However, most of this surplus is needed to compensate for losses in the gas absorption stages and for the manufacture of co-products.

3.2.4   Lime Slaking Stage.

$$CaO + H_2O \rightarrow Ca(OH)_2 \qquad \Delta H = -69\,\text{kJ}\,\text{mol}^{-1}$$

In contrast to the combustion of limestone, the slaking or dissolution of lime in water is an exothermic reaction. Initially a solution of calcium hydroxide is produced by dissolving lime, from the kilns, in water. But this solution is too dilute to be used economically so a large excess of lime is added. The solution then becomes super-saturated and crystals of calcium hydroxide are precipitated to give a concentrated suspension of calcium hydroxide in water. The reaction is so fast, and the supersaturation so great, that the crystals formed are very small. Typically the diameter of the ultimate particles is about $0.05\,\mu$. However, surface forces on particles of this size are so high that they generally form aggregates of $5\text{--}20\,\mu$ in diameter, so become visible as a 'milk-of-lime'.

3.2.5 Filtration, Distillation, and Ammonia Absorption. In the Solvay towers sodium bicarbonate is precipitated from the ammoniacal brine solution by reaction with carbon dioxide. The resulting suspension is continuously filtered and the solid washed with water to remove residual filtrate.

The filtrate contains mainly ammonium chloride – coproduced with sodium bicarbonate – along with substantial quantities of unreacted salt, ammonia, and carbon dioxide. Once the carbon dioxide has been removed by counter-current ammoniacal steam, the valuable ammonia can be recovered in the distillation plant by reaction with calcium hydroxide (see Figure 5).

The mixture of gases passing from the distillers to the absorbers contains primarily ammonia (55%), carbon dioxide (33%), and water vapour (12%). The water vapour readily condenses in the brine, giving up its latent heat of condensation:

$$H_2O_{(g)} = H_2O_{(l)}\ \Delta H = -41\,\text{kJ mol}^{-1}$$

The ammonia also readily dissolves in brine, again evolving large quantities of heat:

$$NH_{3(g)} + H_2O \rightarrow NH_{3(aq)}\ \Delta H = -34\,\text{kJ mol}^{-1}$$

Carbon dioxide is not particularly soluble in brine but, being acidic, it reacts with the ammonia in solution. Once again the reaction is exothermic:

$$CO_2 + 2NH_{3(aq)} + H_2O \rightarrow (NH_4)_2CO_{3(aq)}\ \Delta H = -79\,\text{kJ mol}^{-1}$$

When the partial vapour pressure exerted by a gas in solution equals the partial vapour pressure of the incoming gas no further absorption can occur. This situation would be reached rapidly if the solution were not cooled. Hence cooling is an essential part in the control of the absorption process.

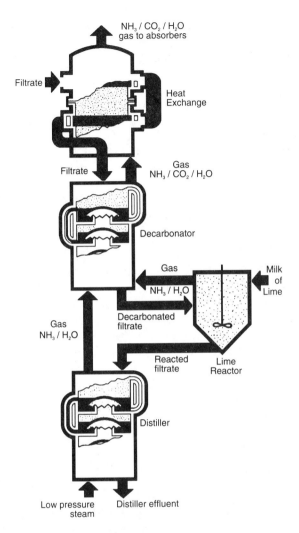

Figure 5 *An Ammonia–Soda Distillation Plant*

3.2.6 Brine Preparation. Brine for the ammonia soda process is commonly obtained by solution mining. Water is pumped into salt strata often many hundreds of metres thick, *e.g.* 180 m thick deposits are widespread in Cheshire. The salt dissolves and the saturated brine solution is removed to the surface. Another major method of providing brine is from salt produced by solar evaporation. The evaporation is done in stages to minimize contamination by calcium sulfate and magnesium chloride. The harvested solid salt is washed with a small amount of water and then dissolved in more water.

At this stage the brine contains certain impurities – notably calcium and magnesium sulfates – which would react with the ammonia and carbon dioxide. The insoluble salts precipitated would eventually block the process equipment.

These were once significant problems, but modern purification methods and anti-scaling agents have virtually eliminated them. Magnesium ions are precipitated from the brine by the addition of sodium hydroxide to give magnesium hydroxide:

$$MgSO_4 + 2NaOH \rightarrow Mg(OH)_2\downarrow + Na_2SO_4$$

The sodium hydroxide can be supplied either directly, or *in-situ* by reaction of lime and sodium carbonate as described in Section 2.7.

Calcium ions are removed by reaction with sodium carbonate, precipitating calcium carbonate:

$$CaSO_4 + Na_2CO_3 \rightarrow CaCO_3\downarrow + Na_2SO_4$$

Most ammonia soda plants consume 2–3% of their output in brine purification.

3.2.7 The Solvay Tower. Sodium bicarbonate is produced in the Solvay tower, where carbon dioxide, entering near the base at a pressure of approximately three atmospheres, reacts with incoming ammoniated brine from the absorbers. The reaction evolves considerable quantities of heat. As the solubility of sodium bicarbonate increases sharply with increasing temperature there is no precipitation of the product until a large amount of CO_2 has been absorbed. To obtain a high yield of precipitated sodium bicarbonate the solution must be cooled; cooling is also necessary to permit further absorption of carbon dioxide.

The temperature profile of the Solvay tower is carefully controlled to give sodium bicarbonate crystals of the required size. If the crystal size is too small filtration of the product is more difficult and other problems arise at the calcining stage (see Figure 6).

After a few days operation the tower's heat transfer surfaces, in contact with the process solution, become so scaled with $NaHCO_3$ that output decreases. To overcome this the input of carbon dioxide is drastically reduced until it is just sufficient to agitate the solution. At the same time, the ammoniated brine from the absorption section is fed into the towers at a much higher rate. The scale ($NaHCO_3$) then dissolves in the process solution and the mixture is fed into other Solvay towers for sodium bicarbonate precipitation. At any given time a few of the towers in an ammonia soda works will be undergoing de-scaling, while the remainder are used to precipitate sodium bicarbonate, thus maintaining continuous operation of the process.

3.2.8 Calcination to Produce Light Ash. The filtered and washed sodium bicarbonate (see Figure 7) is fed into a rotating calciner 30 m long and 3 m in diameter. It is heated in modern units, by contact with steam-heated tubes, to give the Light Ash grade of sodium carbonate (Figure 8):

$$\overset{\text{Heat}}{NaHCO_3 \rightarrow \underset{\text{Light Ash}}{Na_2CO_3} + CO_2\uparrow + H_2O\uparrow} \qquad \Delta H = +129\,kJ\,mol^{-1}$$

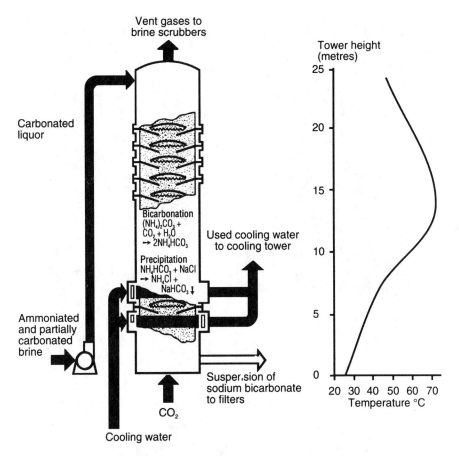

Figure 6 *A Solvay Tower with its Associated Temperature Profile*

As the reaction is endothermic heat is required to decompose the bicarbonate. Part of this heat is represented by the chemical energy in the carbon dioxide released and, since it is recycled, this energy is retained in the process. Much of the heat in the gas is also recovered by heat exchangers, and can be usefully used, for example to heat the feed water for the boilers which raise the steam.

Because the filter cake is wet, direct heating of the material would cause scale to deposit on the heat transfer surfaces. To avoid this, the wet filter cake is first mixed with some hot product recycled from the discharge end of the calciner. This recycle is achieved either by the use of external conveyors or using a small concentric reverse screw in the calciner itself.

It is also practical to calcine the filter cake in fossil-fuel heated rotary calciners. This can be done either by indirect heating from a combustion chamber external to the calciner, or by total combustion of gasifiable fuels within the calciner. Such calciners have tended to be replaced by steam-heated ones in the ammonia–soda process, but are still used in 'Natural Ash' production where there are special reasons for requiring higher temperatures.

Figure 7   *Crude Bicarbonate Crystals*

3.2.9   Heavy Ash Production. For the most part Heavy Ash is produced by the so-called 'monohydrate' route. To produce heavy ash, with particle size and bulk density to match those of the glassmaker's sand, Light Ash is first reacted with water to give sodium carbonate monohydrate crystals (Figure 9). These are then dehydrated by heating in a calciner either similar in design to the Light Ash calciner, or in a steam-heated fluid-bed calciner. The latter is practicable for Heavy Ash because the product particles are much larger and the product much less dusty (Figure 10).

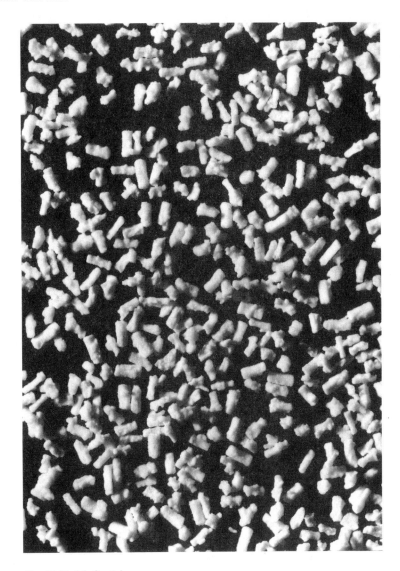

Figure 8 *Light Ash Crystals*

$$\text{Na}_2\text{CO}_3.\text{H}_2\text{O} \xrightarrow{\text{Heat}} \text{Na}_2\text{CO}_3 + \text{H}_2\text{O}\uparrow \qquad \Delta H = +55\,\text{kJ}\,\text{mol}^{-1}$$
$$\text{Heavy Ash}$$

Both of the calcination stages can be carried out at relatively low temperatures, *e.g.* 150–180 °C.

Heavy Ash is also manufactured by a process which does not involve the crystallization of sodium carbonate monohydrate. Light Ash is compacted directly using mechanical pressure; the compacted material is broken to suitable size,

Figure 9   *Sodium Carbonate Monohydrate Crystals*

graded, and dedusted for sale. Although superficially this method may appear to be more economic and less energy intensive, this is not necessarily so. Considerable capital is needed for the compaction and grading equipment, and the energy used is electricity, which is relatively expensive. The product quality is not identical with that produced by the monohydrate route.

3.2.10   Steam and Power Generation. The ammonia–soda process is a substantial user of energy in the form of heat. Much of the heat is required at relatively low temperatures, 100–200 °C, and so can be supplied as low- and intermediate-pressure steam, after first using the high pressure steam to generate electricity in

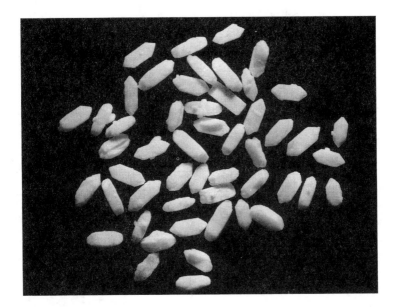

Figure 10 *Heavy Ash Crystals*

turbo-alternators. Heat can also be recovered from various process streams, so fresh boiler feed water can be pre-heated this way prior to feeding to the boilers. In a typical ammonia–soda utility it is possible to generate an excess of electricity above the internal needs of the factory whilst supplying the required process heating. This electricity can be profitably exported to external customers, as it is effectively generated at a fuel efficiency of about 80%, as is the rest of the energy for the process.

Because of the situation described above, most ammonia–soda production facilities are designed to include, or to be associated with, steam and power generation plant (see Figure 3).

4 AMMONIA SODA CO-PRODUCTS

The relationship between soda ash production and the production of the co-products outlined below is summarized in Figure 11.

4.1 Refined Sodium Bicarbonate – NaHCO$_3$

It is unusual to find a chemical made on a large scale which complies with stringent purity requirements such as that of the British Pharmacopoeia. Refined sodium bicarbonate (NaHCO$_3$) is one of the purest of all industrial chemicals, containing less than 30 g of salt and insoluble material per tonne. Its main uses are in the food trade, for instance in self-raising flour and baking powder, in antacid preparations both for humans and in animal feeds, and, because of its high purity, in the

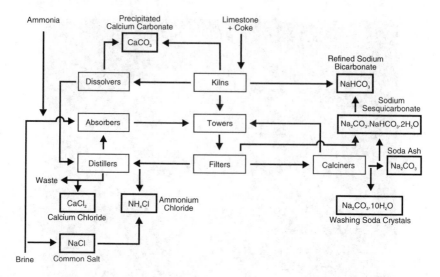

Figure 11 *Ammonia–Soda Co-products*

manufacture of fine chemicals. It is now also increasingly being used for flue gas treatment, though high purity is not required for this use.

Sodium bicarbonate is made by taking the crude product from the filtration stage and partially decomposing it to a sodium carbonate solution using steam. It is then re-precipitated as sodium bicarbonate by reaction with carbon dioxide from the lime kilns. The resulting product is filtered, dried, and graded for use.

4.2   Sodium Sesquicarbonate – $Na_2CO_3.NaHCO_3.2H_2O$

Sodium sesquicarbonate is a weak alkali (used in water softening and in detergent manufacture) and two grades are produced; one takes the form of fine silky needles and is manufactured by the precipitation of the partially decomposed bicarbonate solution mentioned above; the other grade (used in bath cubes, *etc.*) is more finely divided and is prepared by reacting Light Ash with sodium hydrogen carbonate.

4.3   Washing Soda – $Na_2CO_3.10H_2O$

Washing soda (sodium carbonate decahydrate) is produced by dissolving sodium carbonate in hot water, cooling the solution, and separating off the $Na_2CO_3.10H_2O$ crystals. These are then crushed and graded. The product, used in the home for hundreds of years, is now declining (albeit very slowly) with the introduction of more modern cleansers.

Sodium carbonate decahydrate is unstable above 35 °C and so cannot be used in tropical and sub-tropical regions conveniently. Instead the monohydrate is crystallized as a replacement product.

4.4 Ammonium Chloride – NH$_4$Cl

The filtrate from the bicarbonate filters is treated with ammonia (NH$_3$) and solid salt. The resulting solution is cooled to give crystallized ammonium chloride (NH$_4$Cl). The crystals are filtered and dried and part of the output is mixed with a non-caking additive. The main uses of ammonium chloride are in dry batteries and non-ferrous metal processing. Less pure grades are also used as a fertilizer, especially in rice-growing regions.

4.5 Calcium Chloride – CaCl$_2$

When the distiller effluent is concentrated by evaporation, the less soluble salts, such as sodium chloride, crystallize out leaving a concentrated solution of calcium chloride. Further evaporation at high temperatures produces a solution containing > 70% w/w of CaCl$_2$. This solution can be cooled to obtain solid flakes or prills of the product. Calcium chloride is primarily used for refrigeration, concrete curing, soil consolidation, in the alginate industry, and in oil drilling suspensions.

4.6 Precipitated Calcium Carbonate – CaCO$_3$

Very fine grades of calcium carbonate can be made by reacting the lime slurry from the dissolvers with carbon dioxide from the kilns. The resulting precipitated product has a particle size of approx. 0.07×10^{-6} m. It is coated with stearic acid or other specially-tailored organics, in solution, before drying. The main uses of precipitated calcium carbonate are as fillers for PVC, as reinforcing fillers for rubber, and in plastisols production.

5 VARIATIONS AND DEVELOPMENTS IN SYNTHETIC ASH PRODUCTION

5.1 The Combined Process

One major variation of the ammonia–soda process is the combined production of sodium carbonate and ammonium chloride in stoichiometrically equivalent quantities. This is a practical proposition for part of the world's production, especially in areas where there is a large demand for ammonium chloride for use as a fertilizer in rice growing.

To produce the ammonium chloride, ammonia is added to the filtrate from the Solvay towers, the filtrate is cooled to about 10 °C, and solid salt is added together with further cooling. Ammonium chloride crystallizes out and is separated for sale. The mother liquor, now enriched in salt and depleted in ammonium chloride, is returned to the absorption section of the plant. The filtrate from the Solvay towers is not treated with lime, so no lime kilns or dissolvers are employed. Most of the distillation section is also eliminated; all that is required is the recovery of ammonia from wash liquors and liquors used for gas scrubbing. The overall process reaction is:

$$2NaCl + CO_2 + 2NH_3 + H_2O \rightarrow Na_2CO_3 + 2NH_4Cl$$

From the reaction above it is seen that a supply of both ammonia and carbon dioxide is needed in addition to the salt. This is usually achieved by siting the facility near to a synthetic ammonia plant, which can be the source of both materials. This is one constraint in respect of the process. A second, of course, is establishing a market for the ammonium chloride. A third is that the salt is required as a solid rather than in the form of brine, which is usually the cheaper option. This final point is not necessarily a problem if there is no local source of brine available anyway; in which case solar salt is usually produced or purchased.

The major advantages of the combined process are the reduced number of process steps, reduced capital requirements, and substantially lower energy needs.

5.2   General Developments

Over the years many improvements have been made in the process equipment and economics. Because of the corrosion problems associated with hot carbonated ammoniacal brines, cast iron was the traditional material of construction. With developments in plastics, in aluminium, and in titanium technology, many cumbersome units have been replaced by much more compact and process intensive devices. Designs have been changed to minimize deposition on heat transfer surfaces, and the use of sophisticated additives in scale prevention and crystal size control is extensive. Methods of converting sodium bicarbonate directly into Heavy Ash without the intermediate Light Ash stage have also been developed, although they are not applied widely.

In terms of efficiency, control techniques have improved to an extent where raw materials efficiencies are very near the theoretical limits; computer control is the norm in modern plants. Heat recovery between process streams is also very extensive. Fundamental energy studies have shown that the modern plant configurations are such that energy usage can be within 2% of the theoretical optimum for the given chemistry of the particular process.

5.3   Other 'Synthetic' Processes

To date there has been no commercial exploitation of other synthetic ash processes similar in scale to the replacement of the Leblanc process by the ammonia–soda process. Much work has been done in an attempt to develop alternatives, and there are many processes which can be used to form sodium carbonate from salt, but none has proved commercially attractive.

Amongst the processes which have come nearest to commercial practicability are those where an organic amine has been used to replace the ammonia. In some cases the use of such an amine can lead to a better salt utilization. In other cases the energy to regenerate and recover the amine is much lower than for ammonia, especially when the amine is of low solubility. However, although the amines are chemically suitable, their cost is high and loss rates during operation have generally made their use uneconomic. Loss rates can be reduced by using amine/solvent systems, but only at the expense of an additional solvent recovery stage.

One way of minimizing the amine loss is to fix it to a long molecular chain, such as in an ion exchange resin or membrane. Investigations into such processes have been pursued, but none has yet reached the stage of commercial application.

6 'NATURAL ASH' PROCESSES

In fact there really is no such thing as 'Natural Ash', if by this term we mean geological deposits of solid sodium carbonate, in spite of the very common usage of the term. The so-called sources of 'Natural Ash' soda are generally either Trona deposits (sodium sesquicarbonate dihydrate, $Na_2CO_3.NaHCO_3.2H_2O$), or deposits of nahcolite (sodium hydrogen carbonate, $NaHCO_3$), or brines containing sodium carbonate and bicarbonate. These materials require to be separated from the impurities they are associated with in their deposits, and all have to be subjected to at least one calcining stage, to convert the material into saleable sodium carbonate. Thus the materials require significant processing before they yield Soda Ash.

'Natural' ash deposits have been exploited by man since prehistoric times. In certain areas of the world there are lakes containing alkaline brine, from which both salt and sodium sesquicarbonate dihydrate, $Na_2CO_3.NaHCO_3.2H_2O$, can be obtained by solar evaporation. Some such lakes are in Egypt, and were exploited by the ancient Egyptians many centuries ago.

6.1 'Dry' Lake Processes

However, large scale commercial operation first started at Lake Magadi, in Kenya, in 1916. It was based upon surface deposits of natural trona in a dried lake bed. This factory, considerably modernized, is still producing soda ash by the same basic process of dredging the trona deposit and calcining to sodium carbonate.

The lake is fed by streams which are the run-off from volcanic rocks and contain both sodium chloride and sodium bicarbonate in small, but significant, quantities. During the evaporation processes in the lake much of the carbon dioxide is lost with the water vapour. The result is a lake liquor rich in salt and sodium carbonate, and a significant deposition of trona, *i.e.*

$$3NaHCO_3 + H_2O \rightarrow Na_2CO_3.NaHCO_3.2H_2O \downarrow + CO_2 \uparrow$$

This reaction represents the decomposition of the soluble bicarbonate to give the solid trona product. The remaining lake liquor also loses carbon dioxide and water vapour as illustrated in the equation below.

$$2NaHCO_3 \rightarrow Na_2CO_3 + H_2O \uparrow + CO_2 \uparrow$$

Because of the solubility relationships and the large quantity of alkali present, the lake liquor becomes rich in sodium chloride and sodium carbonate, whereas sodium bicarbonate concentrations fall to less than 1% of the solution.

Where the trona deposit is relatively pure it is recovered by dredging from the lake; it is washed to remove the major soluble impurities, such as sodium chloride, and it is calcined in a fossil-fuel calciner. Temperatures significantly in excess of 200 °C are usually employed in order to burn off organic contamination associated with the solid trona.

$$2Na_2CO_3.NaHCO_3.2H_2O \rightarrow 3Na_2CO_3 + CO_2\uparrow + 5H_2O\uparrow$$

6.2 'Wet' Lake Processes

Sodium alkali is found in a number of lakes around the world in the form of sodium carbonate/bicarbonate solutions together with other soluble minerals. The exact method of recovering the alkali is very dependent upon the nature of the other minerals present and upon the relative concentrations of the soluble phases. The brines in Searles Lake, California, are both quite different in composition from the various structures and contain major amounts of sodium, potassium, chloride, sulfate, and borate as well as the carbonate present; they also contain significant quantities of lithium, bromide, and phosphate as well as many minor impurities. Thus, the recovery of sodium carbonate from this mixture is a very complex process, requiring extensive knowledge of the relevant solubility data, and is usually associated with the co-production and sale of other chemicals derived from the lake. Potentially there are a great many viable process options, the preferred option depending critically upon the nature of the feed 'brine'.

For Searles Lake type liquors the so-called Burkeite process has been used, especially where the recovery of other minerals for sale was also desired. This involves boiling and evaporating lake liquor to co-precipitate sodium chloride and burkeite (a sodium carbonate/sodium sulfate double salt). The salt gives rise to relatively large crystals, whereas the burkeite crystals are quite fine, and the two solids are separated using this property. The crudely separated solid is treated with recycled liquors and solid salts, in order to redissolve the sodium chloride present and precipitate more burkeite. The burkeite is then dissolved in water, cooled to about 22 °C and seeded with solid sodium chloride and glauber salt $(Na_2SO_4.10H_2O)$ in order to precipitate much of the sulfate in this form.

The solid sodium sulfate is removed, the mother liquor is enriched with sodium carbonate by dissolving some more burkeite, and the liquor is cooled to 10 °C to precipitate sodium carbonate decahydrate. The decahydrate crystals are separated, melted, and the slurry evaporated in multiple effect units to crystallize sodium carbonate monohydrate. Finally the monohydrate is calcined to give anhydrous Heavy Ash, the required product.

The process outlined above is very much an oversimplification of the situation. In addition there are streams to separate potassium and lithium salts, and to isolate chloride, phosphate, sulfate, and borate products. There are also many recycles within the process to optimize product recoveries and process economics.

In a lake where the chemistry is less complicated (as is also the case with some of the Searles Lake brines), and particularly where the major constituents are sodium chloride and sodium carbonate, the bicarbonation route is preferred. It is generally a simpler process, thus more commonly used for sodium carbonate manufacture than the burkeite process. The lake liquors, or recovered liquors from boreholes, are usually concentrated by solar evaporation before the bicarbonation stage. The bicarbonation is carried out in cooled towers, similar in principle to the ammonia–soda process, using carbon dioxide gas from the calciners and from lime kilns or from flue gas (preferably enriched in CO_2) as a make-up supply. Sodium bicarbonate is relatively insoluble, but not all of the alkali can be recovered this way. However, if the operation is combined with further solar evaporation and sodium chloride recovery, then much of the residual alkali can be recycled to the bicarbonation plant.

The solid sodium bicarbonate is filtered and washed and then calcined to give Light Soda Ash. Either this calcination stage, or a post calcination treatment, is carried out at temperatures of about 500 °C in order to burn off the organic contaminants in the product.

Alternatively the material can be hot bleached with a small quantity of oxidizing agent, such as sodium nitrate. If Heavy Ash is required then the options already described for ammonia–soda are available for densification.

6.3 Deep Mined Processes

From about 1950 onwards very large scale production units were established in the USA, based upon deep trona deposits in Wyoming. This area is now the source of the major proportion of the world's Natural Ash production, and is the largest local concentration of sodium carbonate production anywhere.

The trona layer is some 500 m below the surface so cannot be mined by cheap open-cast methods, nor can it easily be won by straight forward solution mining processes because of its incongruent solubility. The ore is, therefore, mined by relatively expensive deep-mining techniques and brought to the surface for further processing.

The crude trona is first calcined at the relatively high temperature of 500 °C to convert it into crude soda and to destroy some of the organic impurities in the ore:

$$\text{Heat}$$
$$2Na_2CO_3.NaHCO_3.2H_2O \rightarrow 3Na_2CO_3 + CO_2\uparrow + 5H_2O\uparrow$$

The crude product from the calciner is then leached with water so that the insoluble mineral contaminants in the original ore can be separated. A concentrated sodium carbonate solution is obtained and the remaining insoluble impurities are removed by settling/filtration. Some soluble organics are also removed using activated carbon. The treated solution is then evaporated using multi-effect evaporators, or evaporators with recompression, to cause the crystallization of sodium carbonate monohydrate. The monohydrate crystals are centrifuged, washed, and calcined in a second operation, at about 150 °C, to produce the desired product, Heavy Ash. This is the major chemical process used for the production of sodium carbonate from mined trona and is known as the 'Monohydrate' process.

Sodium carbonate has also been produced by the so-called 'Sesquicarbonate' process. In this process the mined trona ore is first leached with hot water to dissolve as much alkali as possible; some carbon dioxide is given off during this operation, thus 'calcining' some of the bicarbonate present. The insoluble material in the leach liquor is flocculated and settled out, some soluble organics are removed from the liquor using activated carbon, and sodium sesquicarbonate dihydrate is crystallized from the liquor by cooling, usually evaporative cooling. The solid product is separated and washed, whilst the mother liquor is recycled to the leaching process. The solid product is calcined to give a product more like the Light Ash of the ammonia–soda producers, rather than Heavy Ash. This is due to primarily to the needle-like shape of the sodium sesquicarbonate crystals and to the high proportion of mass lost during the calcination process.

Production of sodium carbonate from nahcolite ores ($NaHCO_3$) is much less widely developed than that from trona and there is no major facility yet in commercial production. In principle the process could be very similar, especially to that of the 'Monohydrate' route. Calcination of the crude nahcolite ore would yield a product quite similar to calcined trona.

Recovery of alkali by solution mining, rather than deep-shaft mining, is potentially attractive but is difficult with an incongruent material such as trona. Pumping water underground to dissolve the ore would result in a sodium carbonate-rich liquor and precipitation of sodium bicarbonate; the latter would soon cause blockages in the process and prevent practical operation. Recovery by injecting sodium hydroxide solutions is practical, but involves the cost of providing a lime-soda causticization process to provide the reagent. Alternatively, leaching with a sodium carbonate solution is possible, but the quantity of trona dissolved per pass is low. Likewise solution mining of nahcolite yields weak liquors, due to the low solubility of the sodium bicarbonate. In general, these options are not yet widely pursued for sodium carbonate manufacture. However, the final process, solution mining of nahcolite, is used to produce sodium bicarbonate. Good quality liquors can be obtained from such operations. The bicarbonate is crystallized by cooling and carbonation, and the depleted mother liquor is then recycled to the solution mining operation.

All of the 'Natural' ash processes involve the recovery of crude alkali contaminated with organic materials. Much of these are removed by high temperature calcining or by treatment of solutions with activated carbon. However, commonly other organics still remain in small quantities, and often these materials are crystal habit modifiers or frothing agents. It is therefore standard practice to control crystal form and scale growth rates in various parts of the process, and to control foaming, by the addition of appropriate proprietary additives.

7 ENVIRONMENTAL ASPECTS

The environmental problems encountered by the sodium carbonate industry are those associated with any large industrial complex, *e.g.* the emission of unwanted gases, liquids, or solids, noise, and visual impact.

In general, the gaseous emissions arise from burning fossil fuels in the power plants. Most fuels contain sulfur compounds which produce sulfur dioxide (SO_2)

when they are burned and most fuel-burning operations also give rise to oxides of nitrogen. These problems of emissions in flue gases are common to domestic and industrial users of fuels, with the exception of those who burn natural gas which is virtually sulfur-free. To date no economic method of removing these contaminating oxides has been discovered, but environmental pressures are such that producers are having to spend increasing amounts to contain the problem. Although the coke used in the lime kilns produces sulfur dioxide, the emerging gas mixture is scrubbed prior to use. Hence no sulfur dioxide need be discharged from this source.

Liquid effluent can be categorized as process effluent and water used for cooling. The latter constitutes by far the greater volume – in most sodium carbonate processes. Chemically this effluent differs little from the original water except for its higher temperature. The main effect of its discharge would be to warm the river (or nearby sea) by a few degrees. Nowadays, however, much of the inland cooling is done using closed-circuit recycled-water which is itself cooled in towers; thus the thermal effect upon the local rivers has been diminished significantly, the heat load being transferred to the atmosphere.

In the case of the traditional ammonia–soda process the bulk of the process effluent (approximately $10 \, m^3 \, t^{-1}$ of soda ash) emerges from the distillers. It is highly saline and contains a mixture of sodium and calcium chlorides, as well as suspended solids originating from impurities in the limestone. If these solids were not removed, they would cause rapid silting of local rivers, hindering navigation and necessitating constant dredging.

Originally, effluent from the distillers was stored untreated in large artificial lagoons. But since the mid-fifties in some plants the solids have been separated out, mixed with brine, and returned to the bore holes from which the brine was obtained. A settling process is used to achieve the separation, with various additives to promote agglomeration of the solids. The remaining liquid effluent still contains a small quantity of soluble calcium hydroxide, which may be removed either by precipitation with CO_2 or by neutralization with HCl. The hot, neutralized liquid is run into lagoons to cool, before discharge into the river.

In the ammonia–soda processes the main source of the insoluble solids for disposal are due to impurities in the limestone and salt (brine) and fuel used. These can vary typically from 0.1 to 0.4 tons t^{-1} product. In Natural Ash processes the main source of insoluble solids is the rock associated with mineral deposit, and the fuel ash. Also, with complex lake processes, unsaleable co-products will present disposal problems. Levels are similar to those encountered in the ammonia–soda processes.

For the most part the solids from Natural Ash operations are accumulated above ground, and confined in lagoons, so that the alkaline liquors leached from the waste material by weather action can be contained and disposed of (partly by recycling to the process). More recently the possibility of returning the solids to disused sections of the trona mine has been explored.

Problems of noise arise mainly from the handling of solids at various stages of the processes. These stages are enclosed and sound-proofed where possible. Plant operators must wear protective ear coverings in particularly noisy areas. Siting of the plants relative to the local communities is important in minimizing the impact, both visual and in respect of noise. The visual impact of the factory is improved by planting trees and establishing lawns and gardens were possible.

8 SITING

Because the production of sodium carbonate is a large-tonnage low-cost operation, the plant is best sited close to raw materials, customers, and an efficient transport network. In the UK Brunner Mond and Company's main plants are within a few miles of salt deposits, and approximately 25 miles from limestone quarries. They are also located near the glass, detergent, and chemical industries, important users of sodium carbonate. They are situated on a navigable river, connected to a canal system, and have good road and rail links. The river provides cooling water for the process as well.

This situation is quite typical of the older ammonia–soda plants in Europe, and indeed the ones that used to operate in the USA. Limestone and salt deposits are very widespread world-wide, so it is possible for synthetic sodium carbonate producers to situate their factories near to raw materials and major customers. The cost of transport can be a significant proportion of total product cost if they do not.

Natural Ash, although potentially cheaper than the current synthetic routes, suffers from a more restricted distribution of economically viable deposits. For the most part the most favourable deposits have been found in areas remote from other industrial activity. Thus, the development of such deposits can be quite costly where townships and transport connections have to be built, and product transport costs to customers are relatively high. For this reason the two types of processes are still competitive world-wide, geographical separation compensating substantially for differences in processing costs.

CHAPTER 6

Ammonia, Nitric Acid, Ammonium Nitrate, and Urea

S. P. S. ANDREW

1 INTRODUCTION

The fixation of atmospheric nitrogen initially in the form of ammonia is one of the foundations of the modern chemical industry. It is also the centre of the fertilizer industry. Of the 100 million tons of nitrogen fixed each year world-wide some 80 million are used as fertilizer. The remaining 20 million is split in use between explosives, fibres, resins, *etc*. Although the processes and catalysts employed in nitrogen fixation and fertilizer production had their origins over 75 years ago, ammonia synthesis resulting from the work of Haber and Bosch at the beginning of the twentieth century, development has been continuous since then and even in the last fifteen years there have been very significant advances in the catalysts and equipment employed and also in our understanding of the role of nitrogen fertilizer in soil chemistry and plant growth.

2 NITROGEN FERTILIZERS IN AGRICULTURE

In order to appreciate the role of nitrogen fertilizers it is first necessary to have a roughly quantitative picture of the main constituents of agricultural soils and the manner in which oxidation–reduction chemistry affects the location and availability of nitrogen compounds in the soil. A simple view of the nitrogen chemistry is that the soil consists of relatively unreactive inorganic solids (sand and silt), clay which reacts with ammonia to form an ammonium-clay; nitrogen-containing organic residues from dead plant roots and plant debris ploughed in; nitrogen-containing residues from dead microbial life; live roots and microbes; water and dissolved nitrogen compounds. In normal agricultural soils, of the total fixed nitrogen by far the greatest fraction is in the residues from dead plants. A much lower fraction is as ammonium-clay, an even smaller fraction is in the live and dead microbial life, and the least is in the soil solution despite the application of water-soluble nitrogen fertilizers. The greater part of the dead plant residues has a very long half-life in temperate climates (around 50 years), with the result that the total nitrogen in this store is determined by farming practice over many decades. In particular, perma-

149

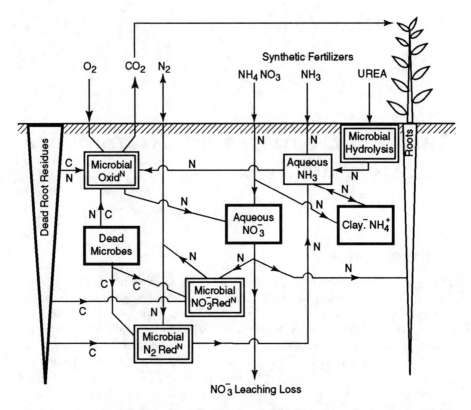

Figure 1 *Simplified Nitrogen Cycle for Nitrate, Ammonia, and Urea Fertilizers. (Nitrogen flows marked 'N', reductant flows marked 'C')*

nent grass leads to an increase in plant residues whilst annual crops such as cereals result in a fall. The general system of nitrogen stores linked by the very active oxidation–reduction reactions caused by microbial action is shown in Figure 1.

Atmospheric oxygen is the chief natural source of oxidant, whilst reduced carbon (as carbohydrates) translocated into the soil from the plant leaves where they are formed is the source of reductant. In the regions of the soil where oxygen has ready access, oxidation is the norm, for instance near the surface in light soils; whereas in regions where oxygen is starved, for instance in water-logged dense and deep soil, reduction is most likely.

The living plant roots primarily take up nitrogen as the nitrate ion and they can tolerate a very wide range of soil solution concentrations. Dissolved ammonia can also be taken up but the plant can not tolerate high concentrations. Because of the rapidity of the microbial catalysed oxidation, reduction, and hydrolysis reactions, the original molecules of an added water soluble fertilizer such as ammonia, urea, or ammonium nitrate almost all disappear from the soil solution in two or three weeks once the fertilizer has become dispersed in the soil. These molecules have not in this short interval been taken up by the plant roots, but rather they have become scrambled by reaction with the clay and by the agencies of the microbial life.

There is thus little difference in location of added nitrogen in the soil after a few weeks, whether it was originally supplied as gaseous ammonia, as urea, or as ammonium nitrate.

The mode of action of added nitrogen is to increase the concentration of nitrate ion in the soil solution. The growing plant does not then have to explore as much soil with its roots to find a nitrogen supply. As a result, a lesser fraction of its mass is devoted to roots and a greater fraction is devoted to leaves. For a given plant mass the plant thus intercepts more light and grows faster in consequence.

The loss of nitrate by leaching into the ground water reservoirs and into rivers is in part due to the build-up of soil solution nitrate in the warmth of early autumn, caused by the vigorous oxidation of dead root residues and dead microbial matter, and in part due to a supply of fertilizer nitrogen in excess of the crop uptake. Because of the partitioning of the added nitrogen between several soil stores of which only nitrate directly affects the growth rate of the crop, in order to secure maximum growth rate a substantial excess of total nitrogen must be added. Once the crop is harvested in late summer and the soil is ploughed, there is no substantial new growth of crop to take up these sources of nitrate which are then washed away by the rains of winter. This effect chiefly occurs with land on which seasonal crops are grown. With a permanent crop, such as grass, plant growth and microbial activity remain in step throughout the year and excessive fertilizer application is not financially rewarding.

In addition to loss by leaching nitrogen is also lost, in regions of the soil where gaseous oxygen can not gain access, by nitrate being used by microbial life as an oxidizing agent. Gaseous nitrous oxide and nitrogen are produced.

When nitrate is low in the soil solution, a population of bacteria grows which is capable of fixing atmospheric nitrogen and, if the crop is uncontrolled, a plant population containing some species having nitrogen-fixing ability will seed themselves. However, all growth will be at a low rate due to the low nitrate level.

Fertilizers are usually spread by the farmer in the form of 2 mm diameter granules. A contractor is employed to distribute gaseous ammonia which is injected below the surface of the soil through hollow tubes. Contractors are also used to distribute fertilizers in concentrated solution, such as urea plus ammonium nitrate. When urea granules are spread it is important that the urea is soon washed into the soil by rain otherwise quite significant losses to the atmosphere of ammonia produced by hydrolysis occur. Although a check can be made by the farmer using soil analysis of the available nitrogen level in spring and this quantity used as a guide to the amount of fertilizer to be applied, the subsequent growth of the crop and its uptake of nitrogen is as uncertain as the prediction of the weather over the next four months.

3 AMMONIA PRODUCTION

3.1 Feedstocks

The production of ammonia requires nitrogen from the air and a source of reductant to convert into hydrogen. In the past thirty years, the increasing

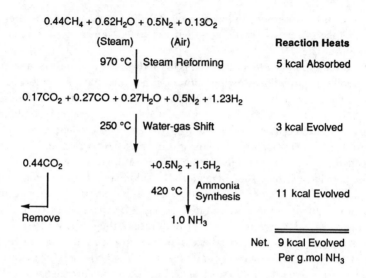

Figure 2   *Methane as Feedstock for the Production of Ammonia via Steam Reforming. Ideal Stoichiometry and Heat Requirements*

availability of either natural gas (methane) from gas wells or produced by letting down the pressure on crude oil leaving oil wells has become the reductant of choice for the production of ammonia world-wide. This is primarily a result of the low cost of producing natural gas coupled with its lack of ready use in some parts of the world for other purposes. Its price in such regions is consequently low. Ammonia production is thus favoured in these regions, which export urea often made in combined ammonia plus urea plants.

Any fossil or renewable carbonaceous material in bulk supply can be used as a feedstock, but their ease of use decreases as the fraction of reductant in the material is diminished and the level of unusables is increased. Thus the value of coal as a feedstock is not simply the positive value of the carbon and hydrogen it contains, but this value must be reduced because of the negative value of the ash which must be removed and the sulfur and chloride trapped. For this reason coal, which was the almost universal feedstock forty years ago, is now only used in regions where natural gas is not available and the importation of fertilizer is undesirable. Renewable feedstocks, such as plant debris, are generally very expensive to transport and store, which more than counterbalances their zero value on the field.

3.2   Ammonia Production Using Methane

3.2.1   The Ideal Process. Ideally the production of ammonia from methane and air involves three equilibrium limited reactions and a single selective removal process, as shown in Figure 2. The first reaction, steam reforming, involves the equilibration of methane, an amount of air chosen so as to introduce the

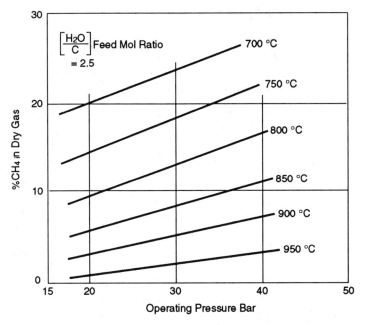

Figure 3   *Equilibrium in the Steam Reforming Reaction for Methane*

stoichiometric amount of nitrogen, steam, carbon monoxide, carbon dioxide, and hydrogen. This equilibration must be effected over a catalyst which is always metallic nickel on a ceramic support. The required equilibrium can be considered as the result of the steam reforming equilibrium (Figure 3), which controls the extent of conversion of the methane and the water gas shift equilibrium (Figure 4), which controls the ratio of carbon monoxide to carbon dioxide in the equilibrated gas mixture. The positions of both these equilibria are dependent on the ratio of steam to methane in the feed gas. This ratio is chosen to be as low as is practically possible, as steam generation requires heat.

   In modern plants advances in the catalyst formulation enable operation with a steam divided by methane molecular ratio as low as 2.5. Below this steam ratio the rate of steam gasification of any solid carbon which might form on the catalyst is inadequate compared with the rate at which methane would dehydrogenate to form this carbon. In consequence the catalyst would become coated with carbon and deactivated. Whereas the shift reaction equilibrium, Figure 4, is pressure independent, the methane steam reaction, Figure 3, is pressure dependent, forcing high temperature to be employed to secure good conversions of methane at elevated gas pressures. At these high temperatures the shift equilibrium unfortunately favours only a poor conversion of carbon monoxide into carbon dioxide. The gas leaving steam reforming is thus a mixture as shown in Figure 2. In order for the carbon monoxide to be shifted to carbon dioxide by reaction with steam and with the production of further hydrogen, the gas mixture must be reduced in temperature to around 250 °C and then passed over a catalyst capable

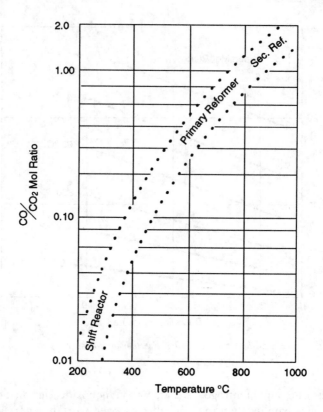

Figure 4 *Effect of Temperature on the Carbon Monoxide to Carbon Dioxide Ratio as Determined by the Water-Gas Shift Reaction*

of selectively promoting the shift reaction but not the steam reforming reaction (which at this temperature would run in reverse and form methane, a step known as methanation). It is also necessary to so formulate this catalyst that it does not promote methanol synthesis. A supported copper is used.

Following the shift reaction, ideally the gas would consist of a mixture of carbon dioxide, water vapour (which can be condensed and so removed), and the remainder being nitrogen and hydrogen in the proportions required to make ammonia. The carbon dioxide would next be removed by a solvent wash and the remaining gas passed to ammonia synthesis. The equilibrium of the synthesis reaction is pressure and temperature dependent, as seen from Figure 5. It is far from complete under all conditions practicable with current catalysts, iron or ruthenium, which need a temperature of over 350 °C. It is therefore necessary to place the reactor in a synthesis gas recirculation system known as a 'synthesis loop'. The ammonia made on each pass through the reactor is removed by condensation brought about by chilling before the synthesis gas together with a make-up of fresh gas is reheated and passed yet again over the catalyst.

From Figure 2 it can be seen that the ideal stoichiometry requires the use of 0.44 mol methane mol^{-1} of ammonia produced. The overall process is exothermic.

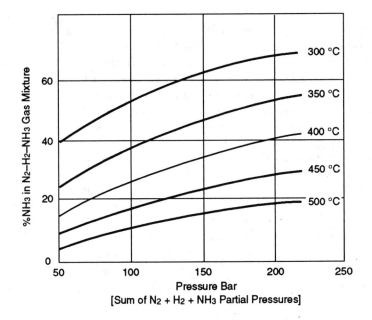

Figure 5 *Equilibrium in the Ammonia Synthesis Reaction*

3.2.2 The Commercial Process. (i) *Heated Primary Reforming.* An inspection of the thermochemistry of the ideal process in Figure 2 shows that there is no possibility of running such a process as an independent unit, as the two heat evolving reactions both have to operate at temperatures many hundreds of degrees lower than the heat-absorbing steam reforming reaction. There is therefore no possibility of heat feedback. Until 1988 the only way of running this process was to split the steam reforming operation into two stages. The first stage or primary reformer was only fed with methane and steam, and the reaction was in consequence highly endothermic. To supply the necessary heat, both sensible and reaction, the catalyst was packed in a large number of parallel tubes through which the process gases passed and these tubes were suspended in a furnace heated by burning further methane. This solution to the thermochemical problem still remains the norm though a completely different solution has recently been commercialized and will be described later.

After leaving the primary reformer the hot gases, still containing a substantial quantity of methane, receive the necessary addition of air to supply the nitrogen demanded by the stoichiometry of ammonia production. The combustion of the oxygen of this air in the gases leaving the primary reformer increases their temperature by several hundred degrees, so passage over further steam reforming catalyst in a secondary reformer allows equilibration at a markedly higher fraction of methane reformed. A simplified quantitative flowsheet for this process is shown in Figure 6. Note that compared with the ideal stoichiometry of Figure 2 the methane requirement is raised to 0.60 mol^{-1} of ammonia, primarily a result of 0.12 mol of methane being burnt in the primary reformer furnace. A further inefficiency in the real process arises from the inability to convert 0.01 mol methane because of equilibrium limitation in the secondary reformer, and also a failure to shift all the carbon monoxide due to the equilibrium limitation set by the shift equilibrium of Figure 4. This last inefficiency is made more detrimental as any carbon monoxide must be removed from the synthesis gas before it contacts the ammonia synthesis catalyst which is poisoned by oxygen-containing compounds. Residual carbon monoxide is removed after carbon dioxide has been washed out by passing the synthesis gases over a further bed of nickel catalyst, where it is converted into methane and water which is readily scrubbed out. Final water removal is often simply washed out by introducing the make-up gas into the synthesis loop before the gases circulating in the loop enter ammonia condensation. The hydrogen used in methanating the 0.01 mol carbon monoxide requires yet a further 0.01 mol of methane to be reformed in the front part of the process. Finally, the methane entering the synthesis loop together with inerts such as argon entering in the air feed to the secondary reformer must be purged from the loop before their concentration becomes excessive and carry with them some hydrogen. This purge gas is used as fuel alongside the fresh methane in the furnace, as shown in Figure 6.

The overall process thus uses 36% more methane than the ideal. However it is possible, by raising high pressure steam using the waste heat, to drive air and synthesis gas compressors and other machinery as well as supplying the process steam for reforming thus eliminating any further major requirements for power or heat.

Before leaving the real flowsheet of Figure 6, it should be noted that as all the metal catalysts are poisoned by hydrogen sulfide it is necessary to install a sulfur removal process on the feed to the reformer. This step, termed hydrodesulfurization, serves to hydrogenate the very small amounts of sulfur-containing compounds in the methane, using a small amount of recycle synthesis gas. The resultant hydrogen sulfide is then removed by reacting stoichiometrically with solid high surface-area zinc oxide. The zinc sulfide thus formed is replaced by fresh zinc oxide when the bed is used up, an infrequent occurrence.

It is also important that any small amount of chlorine-containing compounds be removed before the methane enters the reformer. Chlorine as well as acting as a catalyst poison also results in a rapid loss in catalyst surface area due to growth in metal crystal size. Chlorine compounds are hydrocracked in the hydrodesulfuriz-ation stage along with the sulfur compounds using a permanent bed of cobalt molybdenum sulfide catalyst which precedes the replaceable zinc oxide. Following the zinc oxide, a bed of replaceable alkalized alumina is installed in order to absorb the hydrogen chloride produced by the hydrocracking of the organic chlorine compounds.

The air feed to the secondary reformer in current ammonia plants has neither sulfur nor chlorine removed from it, an omission which markedly shortens the life of the copper shift catalyst unless it is also preceded by a replaceable bed of alkalized alumina. Chlorine in the air can come from very small drops of sea water resulting from bubbles bursting and the drops being carried many miles inland on the wind. Or the chlorine may merely have come from the nearby water cooling tower where periodic chlorination is used to suppress algal growth.

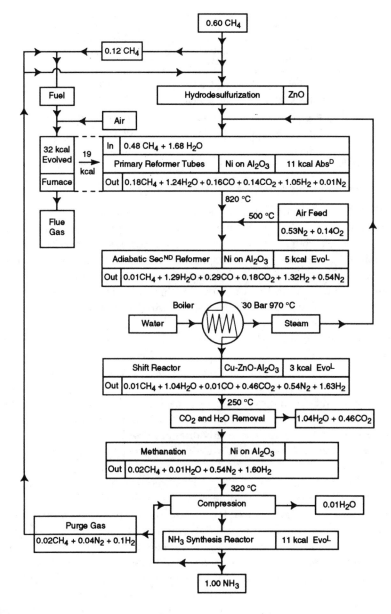

Figure 6 *Flowsheet for Ammonia Production using Furnace Heated Primary Reforming*

(ii) *Carbon Dioxide Removal.* The greatest change that has occurred recently in the process of ammonia production has been a result of developing a new method of carbon dioxide removal which at the same time can remove unwanted nitrogen. In the past, carbon dioxide removal was invariably by gas washing using either a physical solvent for the carbon dioxide or a chemically-reacting wash. Examples are shown in the left-hand two columns of Figure 7.

CO$_2$ Removal		CO$_2$ and Excess N$_2$ Removal
Physical Solution Wash	Chemical Solution Wash	
Water	Aq. K$_2$CO$_3$-KHCO$_3$	Pressure Swing Adsorption
Chilled MeOH	Aq. Amines	
Propylene-Carbonate	$\begin{cases} \text{M.E.A} \\ \text{D.E.A.} \end{cases}$	Beds $\begin{cases} \text{Silica Gel} \\ \text{Zeolite} \end{cases}$ Of
'Selexol'		

Figure 7   *Methods of Removal of Major Components from Synthesis Gas*

Of the physical solvents, water, used in all the early plants, has the least solubility for carbon dioxide. Chilled methanol has 2.8 times the solubility of water, propylene carbonate 3.7 times, and 'Selexol', a dimethylether of polyethylene glycol, has 3.3 times. In all cases the spent wash liquid is regenerated by flashing some of the carbon dioxide at pressure let-down, and the remainder is stripped off with air or steam. The regenerated liquid is then recycled to the carbon dioxide absorber. A similar system is employed when using the chemical washes listed in the second column of Figure 7.

The new method referred to above is the result of using an adsorption system on place of a wash. Pressure Swing Adsorption (PSA) employs a number (about ten) of packed beds of zeolite adsorbent which alternate between adsorbing at the gas making pressure and desorption at much lower pressures. The whole operation is adiabatic and quite complex cycles are employed allowing a number of gas streams of the stripped gas. Thus streams high in carbon dioxide and streams high in nitrogen can be produced. The disadvantage of PSA is mechanical wear as there are very many frequently operating valves. (PSA procedures are also reviewed in Chapter 9).

(iii) *Autothermic Steam Reforming.* The most recent major development in the steam reforming flowsheet is autothermic reforming (the ICI LCA process) in which air in excess of that required for introducing the nitrogen for incorporation in the product ammonia is fed into the secondary reformer. Whereas 0.67 mol of air is added per mol of ammonia in the stoichiometric flowsheet of Figure 6, leading to a reaction heat evolution of 5 kcal in the secondary reformer (which is less than the 11 kcal of reaction heat absorbed in the primary), by adding 1.19 mol air, as in the flowsheet of Figure 8, the secondary reaction heat evolution now becomes 13 kcal and the primary reaction heat absorption only 9 kcal. In this flowsheet it is thus possible to heat the tubes of the primary reformer using a counter-current flow of gases exit the secondary. Clearly, however, this extra heat is produced at the cost of, in effect, transferring the methane burnt in the furnace heating the primary reformer in the conventional flowsheet of Figure 6 into the feed of the primary in the flowsheet of Figure 8, where it is then burnt as it enters the secondary. The methane used per unit amount of ammonia made thus does not differ significantly between the two processes. What is changed is the replacement of the very bulky furnace of the old process by a very compact high pressure gas/gas

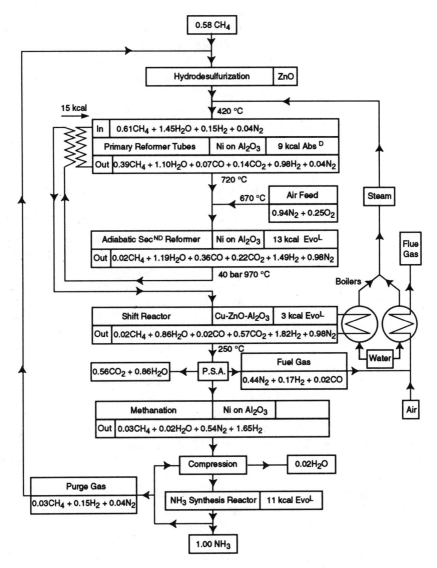

Figure 8 *Flowsheet for Ammonia Production using Autothermic Steam Reforming*

heat exchanger which is completely fabricated in the shop. Rendering this change possible, the carbon dioxide removal wash of the old flowsheet is replaced by a PSA system removing both the carbon dioxide and the excess nitrogen.

In a combined ammonia and urea production plant in which all the ammonia is reacted with carbon dioxide to make urea, a mol ratio of carbon dioxide divided by ammonia of 0.5 is required for the urea production. An advantage of the flowsheet of Figure 8 compared with that of Figure 6 is that the former produces sufficient carbon dioxide from the PSA, whereas the conventional flowsheet does not produce enough from the wash.

(iv) *Reforming Pressure.* So far in this chapter the chief factor in determining the pressure of the gas-making front section of the ammonia processes has not been mentioned. Indeed the reader might think that the process designer should use a gas-making pressure equal to the ammonia synthesis pressure. This very desirable objective would eliminate the costly and power-consuming synthesis gas compressor which is so significant in ammonia plants.

The process designer finds, however, that metallurgical limits in the primary reformer restrict gas-making pressures. This limitation has quite different causes in the radiant furnace of the conventional flowsheet and in the new secondary exit gas heated primary reformer. In the old flowsheet of Figure 6 the high differential pressure across the wall of the tubes in the furnace, coupled with their high temperature, results in a slow cracking open of the tube metal, a phenomenon known as creep. The stress resulting from the high differential pressure can be relieved by the use of thicker wall tubes. Unfortunately, during operation a marked temperature gradient then exists across the tube wall due to the heat flux and every time the load on the furnace is changed this temperature gradient changes, imposing high tensile stresses due to the differential thermal expansion between the inside of the tube wall and the outside, resulting in further creep. With currently available tube materials, these phenomena set a practical limit to the relation between gas-making pressure and the fraction of methane feed to the primary reformer exit the secondary. For the conventional flowsheet of Figure 6, the temperature and methane conversion exit the secondary are tied to those exit the primary through the fact that the air feed to the secondary is fixed by the ammonia stoichiometry. The relation between operating pressure and the fraction of methane feed entering the primary unreformed exit the secondary is shown in Figure 9.

Whilst the limitations due to creep are well known in the conventional flowsheet as a result of thirty years practical experience in a large number of plants, the new flowsheet of Figure 8 brings with it new opportunities for increasing the gas-making pressure as well as new metallurgical problems. The writer estimates that it will require another eight to ten years for the pressure limits of the new process to be accurately defined. Creep is no longer the problem, as the primary reformer tubes have to stand only a low differential pressure and are relatively thin. No pressure limitation therefore arises from mechanical stress induced effects. However, there is a problem due to what could be called 'chemically-induced' stress in the primary reformer tube walls. This stress results from the thermodynamic potential for depositing graphite inside the metal in contact with the exit gases from the secondary as they cool in the primary. This phenomenon is a result of the reaction between two molecules of carbon monoxide to form one of solid carbon and one of carbon dioxide. Thermodynamically the formation of carbon is favoured by high pressures, high ratios of carbon monoxide divided by carbon dioxide, and lower temperatures. Though the formation of carbon is not possible at the high temperature of the gases as they leave the secondary reformer, as these gases cool past 800 °C in the primary, the deposition of carbon becomes possible. This reaction can not take place in the absence of a metal catalyst at a suitably elevated temperature. It thus does not take place on the gas side of the boiler tubes

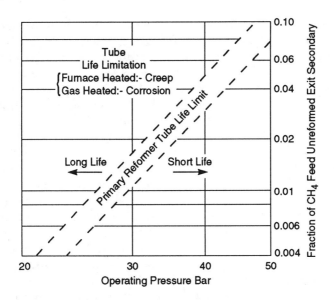

Figure 9 *Effect of Primary Reformer Tube Life Limitation on the Relation Between Operating Pressure and Unreacted Methane Leaving the Secondary Reformer*

in the conventional flowsheet of Figure 6 where, although the tube metal would be catalytically active at a higher temperature, it is kept too cold by the boiling water. This circumstance does not arise in the flowsheet of Figure 8 as the gases being heated on the opposite side rise in temperature to over 700 °C resulting in tube wall temperatures of over 800 °C. Under these conditions it is essential that the metal of the tube in contact with the secondary exit gases does not catalyse the carbon formation reaction, otherwise the graphite produced between the metal crystals that make up the tube wall forces the crystals apart and the tube fails due to a type of corrosion called metal dusting.

Clearly the task of obtaining a non-catalytic metal having suitable strength becomes more difficult as the thermodynamic potential of carbon formation increases. Higher gas-making pressures require higher secondary reformer exit temperatures to secure a good conversion of the methane, and these high temperatures cause the shift equilibrium in Figure 4 to move to higher ratios of carbon monoxide to carbon dioxide. The carbon-forming potential is thus increased both as a consequence of increased pressure and changed gas composition.

No quantitative data on these effects on currently-used tubes are available, and no reformers of this design have yet operated for as many years as those with the conventional furnace (the first full scale plant only having started up in 1988). The writer guesses (foreseeing the future in such a scientific puzzle is always a guess as accelerated life tests are not 100% certain) that the operating pressure *versus* unreformed methane fraction limitation is not significantly different from that for the conventional furnace heated primaries (see Figure 9).

In general it appears that in current practice the highest practical reforming pressure is around 40 bar with a 95% conversion of the feed methane.

(v) *The Ammonia Synthesis Reaction.* Since 1960 there has been on average a steady fall in the synthesis pressure which was once around 250–350 bar but is now around 80 bar. This has primarily been brought about by improved techniques for fabricating and inspecting welded pressure vessels, permitting much larger ammonia synthesis reactors to be constructed. Dropping the pressure from 200 bar to 80 bar necessitates a four-fold increase in the catalyst volume as seen from Figure 10. In this figure typical exit ammonia values and outputs per unit volume of catalyst are shown as a function of synthesis pressure. Also indicated are typical average operating temperatures.

Until very recently the iron catalyst has been the only commercial catalyst on offer. This catalyst is a direct descendent of the catalyst produced on an industrial scale by Fritz Haber three-quarters of a century ago. Scientists are still arguing as to the roles of the various components. The writer's view is that metallic iron is the sole species having catalytic action in these formulations and comprises some 95% by weight of the active reduced catalyst. Amorphous alumina forms some 4% of the weight and serves like a dusting powder to keep the iron crystals, which are about 40 nm in size, apart so they do not fuse together in the reactor with consequent loss in surface area and hence catalytic activity. Under operating conditions a small amount of a potassium-containing species formed from potassium added to the formulation in the ratio potash/alumina = 1/4 serves to pin the crystal faces of the iron into a structure favourable to promoting fast ammonia synthesis.

Only in the past ten years has a possible rival to the iron catalyst been developed. This is ruthenium metal supported on a high area graphitic carbon. Ruthenium appears to be around 45 times as active as iron per unit of exposed metal area. However, it is vastly more expensive, with the result that it has to be distributed on a support in very small crystals, some 1–1.5 nm, in order that a high fraction of the metal atoms are exposed and hence are catalytically active. This high state of subdivision is only possible when a low concentration of metal is dispersed on the support. The end result is that, because of the low concentration of ruthenium 1–2%, the metal area per unit volume of catalyst is roughly only a third of that of the iron catalyst. The ruthenium catalyst formulation thus has about 15 times the activity of the iron. On this basis its performance as a function of pressure should be as shown by the broken lines in Figure 10.

Operation at synthesis pressures as low as 40 bar is thus technically feasible, eliminating the gap between reforming pressure and synthesis pressure. Whether this simplification of the flowsheet counterbalances the great increase in cost of the catalyst and its as-yet novelty and restricted availability remains for the future to decide. A cautious purchaser of a process plant may not value increased catalyst activity very highly if it means that he is tied to a single supplier of this catalyst by the process flowsheet or reactor sizing.

(vi) *Ammonia Production and Heterogeneous Catalysis.* Ammonia production was the birthplace of heterogeneous catalysis. It is instructive to compare the various metals used in the production process on the basis of the amount of metal surface required to make ammonia at a fixed rate (say 1 g mol day^{-1}) and to compare them also on the ability to pack this surface in the form of supported crystals into

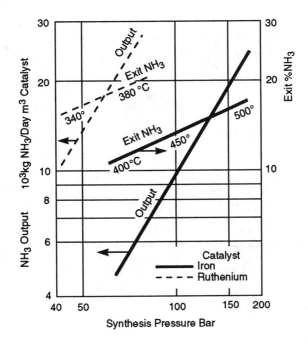

Figure 10 *Typical Parameters of the Ammonia Synthesis Reactor Using Either an Iron or a Ruthenium Catalyst*

unit volume of catalyst (square metres of surface per cubic centimetre of catalyst packed volume). Figure 11 gives these comparisons, plotting the various catalysts as a function of the average operating temperatures of the reactors in which they are employed. A great range of temperature occurs, from 250 °C in the shift reactor to 1000 °C in the secondary reformer. Corresponding to this temperature range are production rates from as low as 0.03 g mol ammonia m^{-2} day^{-1} to as high as 20 g mol m^{-2} day^{-1}. Yet just as high temperatures secure high surface kinetic velocities, they also result in high diffusional mobility of the refractory oxides which maintain the catalytic metals in a dispersed state. It is only possible to maintain coarse dispersions at high temperatures for many weeks. Thus the metal surface per unit catalyst packed volume falls from around 30 m^2 cm^{-3} to 0.03 m^2 cm^{-3} over the same temperature range. The consequence is that, irrespective of reactor operating temperature, very roughly the same volume of catalyst is used in every reactor. This volume is about 1 cm^3 (g mol)$^{-1}$ ammonia production day^{-1}. In well-run plants all these catalysts operate for many years before they need replacing due to loss in activity. This loss is usually a result of a loss of metal surface area caused by a coarsening of the catalyst structure which takes place over a long period of time at the operating temperature in the presence of the catalytic reaction.

The above broad regularity in structure of the operating catalysts is despite the three quite different methods of preparation used for these catalysts.

The steam reforming catalysts are prepared by impregnating a pre-formed and

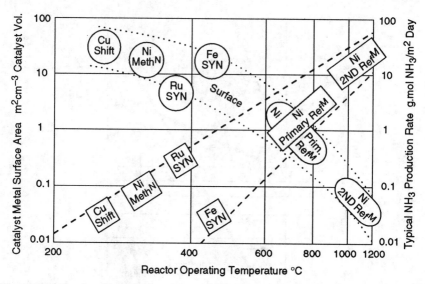

Figure 11 *Catalyst Metal Surface Areas and Their Relation to Catalyst Volumes in a Typical Ammonia Production Plant*

pre-fired ceramic support with a concentrated nickel nitrate solution. After drying and calcining to drive off the nitrous gas and convert the nickel into oxide, the nickel is then reduced to the metallic state in the reformer.

The shift catalyst is prepared by co-precipitating the copper and zinc as basic carbonates and alumina as the hydroxide by mixing a copper–zinc–aluminium nitrate solution with a sodium carbonate solution, washing out the sodium nitrate from the filtered-off precipitate, and drying and calcining it to give the mixed oxides. This powder is then tabletted. The final reduction of the copper oxide to metal again takes place in the shift reactor.

The iron ammonia-synthesis catalyst is prepared by melting together iron oxide (magnetite), alumina, and potash at a temperature of 1600 °C. The melt is then quickly frozen by pouring it onto a thick cold metal tray, and the solid oxide sheet is then broken up to form lumps. The iron oxide in these lumps is reduced to metallic iron in the ammonia-synthesis reactor. During this reduction the alumina and the potash migrate to form the structure described above.

As all the main catalytic reactions except reforming are exothermic, it is normal to use some of the temperature rise during reaction to supply a temperature difference driving potential for an exit gas to inlet gas heat exchanger, if this is required to preheat the inlet gas to reaction temperature as with the methanator and the ammonia synthesis reactor. When the heat liberated by reaction is considerable, as in the shift reactor and the ammonia synthesis reactor, it is usual to maintain a falling temperature pattern in the catalyst bed by removing reaction-generated heat by heat-exchange in order to obtain a favourable exit equilibrium. Thus the catalyst of the shift reaction can be placed in the heating tubes of a boiler, and steam generating tubes can be buried in the catalyst bed of

$$NH_3 + 1\tfrac{1}{4}O_2 \longrightarrow NO + 1\tfrac{1}{2}H_2O$$

900 °C Catalytic Oxidation

Reaction Heats

70 kcal Evolved

$$1\tfrac{1}{2}NO + \tfrac{3}{4}O_2 \longrightarrow 1\tfrac{1}{2}NO_2$$

35 °C Slow Homogeneous

20 kcal Evolved

$$1\tfrac{1}{2}NO_2 + \tfrac{1}{2}H_2O \longrightarrow HNO_3 + \tfrac{1}{2}NO$$

35 °C Absorption Reaction

8 kcal Evolved

Net. 98 kcal Evolved
per g.mol HNO_3

Figure 12 *Nitric Acid Production from Ammonia. Ideal Stoichiometry and Heat Release*

the ammonia synthesis reactor. The steam raised by these boilers supplies the process-steam requirements of the steam reformer.

Process designers spend considerable time on optimizing the thermal and mechanical power aspects of the flowsheet of ammonia plants so as to avoid any substantial inputs of further fuel or electric power during normal running of the plant. An extra input is, however, necessary in some flowsheets during plant start-up, an operation that must not be forgotten by an over enthusiastic flowsheeter who is anxious to save the last kcal.

4 NITRIC ACID PRODUCTION

The industrial production of nitric acid is by the catalytic oxidation of ammonia. This is a thermally very inefficient process as all the hydrogen in the ammonia molecule is burnt to no thermochemically-useful purpose. Indeed, dilute nitric acid could in theory be made from air and water at normal temperature without the expenditure of any energy were a suitable catalyst available.

The route from ammonia involves three reaction (see Figure 12). The first, the selective oxidation of ammonia to give nitric oxide, can be catalysed by a number of metal oxides such as cobalt and iron, and it can also be catalysed by platinum metal. The key to obtaining a good selectivity (*circa* 94–98%) at complete conversion appears to be to operate the catalyst at such a high temperature that the reaction rate is gas film diffusion limited. Under these conditions, provided excess air is used, the ammonia partial pressure at the surface of the catalyst is very low. The reactions on the catalyst which lead to unwanted nitrous oxide and to nitrogen appear to be of higher order in ammonia than the reaction producing nitric oxide. Lowering the activity of the catalyst by any means, for instance by substantially reducing its temperature or by poisoning it, results in an ammonia partial pressure increase at the catalyst surface and loss of selectivity towards nitric oxide production. Because of the much greater sensitivity of oxide catalysts to poisoning, these cheaper materials are normally rejected in favour of platinum–rhodium gauzes which operate at around 900 °C.

The ammonia mol fraction in the ammonia–air stream entering the burner is

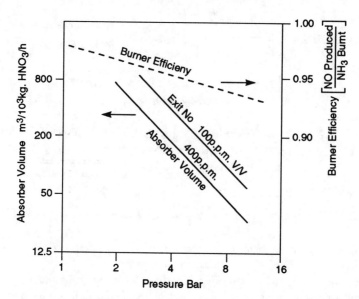

Figure 13   *Effect of Pressure on Conversion Efficiency of Ammonia into Nitric Oxide and on the Size of the Nitrous Gas Absorption Towers*

some 11%, to maintain it below the flammability limit. The rise in temperature of the gas mixture is around 70 °C per 1% ammonia burnt. As it is undesirable for the gauze temperature to exceed 900 °C (otherwise it disintegrates too rapidly with loss of some platinum) the gas mixture fed to the burner is only preheated to a little more than 130 °C. Because of the reaction order effect mentioned above the selectivity of ammonia burners falls with increase in operating pressure. The extent of this effect is shown in Figure 13.

The second of the three reactions in Figure 12 is a classical third-order homogeneous gas phase reaction which proceeds rather slowly to completion in the presence of excess oxygen. Its velocity is *increased* by operating at *lower* temperatures. No suitable heterogeneous catalyst has been found.

The third of the reactions involves the absorption of nitrogen dioxide, or more correctly, of its dimer dinitrogen tetroxide, into water where it reacts to form an equimolar quantity of nitric acid and nitrous acid. The nitrous acid then desorbs and decomposes into nitric oxide, nitrogen dioxide, and water vapour. The overall stoichiometry then appears as shown in Figure 12.

Because both the gas phase oxidation of nitric oxide and the absorption reaction proceed at a rate roughly proportional to the square of the partial pressures of the nitrous gas, the size of the oxidation plus absorption section of the nitric acid plant is almost inversely proportional to the square of the operating pressure (see Figure 13). A further incentive to the process designer for employing high-pressure absorption (and, if economic, as low a temperature as convenient) is the desire to produce a high-strength acid. The acid strength is controlled by the equilibrium of the third reaction in Figure 12. This equilibrium is shown in Figure 14, from which

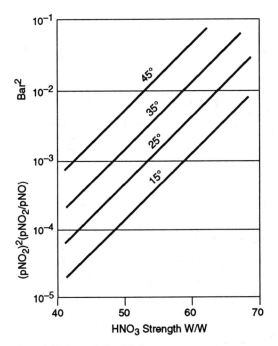

Figure 14 *The Nitric Acid Strength Equilibrium*

it can be seen that a high degree of oxidation of the nitric oxide as well as high pressure and low temperature favour high acid strength.

Many modern plants, in order to obtain both the advantage of high burner efficiency and also of lower capital cost absorption, use a dual-pressure system. The burner operates at 3 to 5 bar whilst the absorption section runs at 10 to 12 bar. The effluent gas from the absorber is expanded to 1 bar after reheat in order to supply the mechanical power required for inlet air compression and for burnt gas compression (rather after the design of a gas turbine). Figure 15 shows such a flowsheet. Note also in this flowsheet a tower marked 'acid bleacher'. The strong acid leaving the absorption system contains a high concentration of dissolved gases, brown nitrogen dioxide and blue nitrogen sesquioxide giving the acid a green colour. This gas mixture is stripped from the nitric acid using air and recycled into the process. The two cooler condensers should also be noted. These serve to condense the water vapour produced in the burner. For reasons associated with the water mass-balance these cooler condensers are designed so that they produce the weakest acid possible, and this acid is then passed to the appropriate position in the absorption tower to absorb further nitrous gas.

The inverse square dependence of absorption rate on gas partial pressure makes it very expensive to remove the nitrous gas down to the level suitable for discharge into the atmosphere simply by sufficiently enlarging the absorber. Frequently a catalytic reduction stage is used to remove most of the residual nitrous gas. A reductant such as methane is added to the waste gases which are then heated to around 300 to 400 °C and passed over a catalyst. Platinum will reduce nitrogen

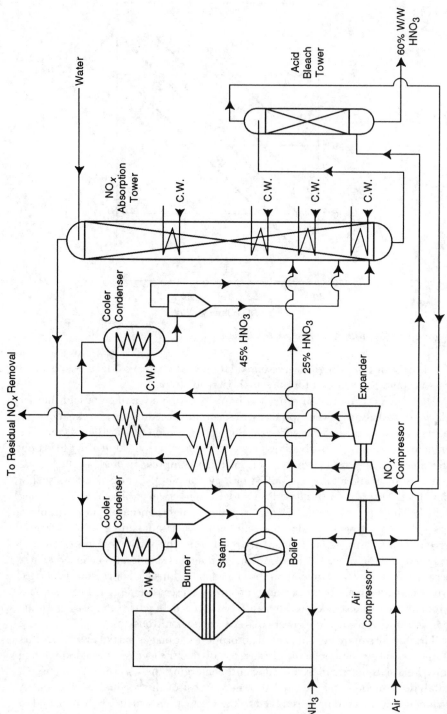

Figure 15 *Simplified Flowsheet of a Dual-pressure Nitric Acid Plant*

dioxide in preference to oxygen thereby decolouring the gas, but nitric oxide is only reduced after oxygen resulting in a markedly higher reductant demand. The obvious method of catalytically decomposing the dilute nitrous gases to nitrogen and oxygen, though thermodynamically possible, awaits the discovery of an effective catalyst that works at low temperatures in a damp gas.

5 AMMONIUM NITRATE PRODUCTION

The process for the production of ammonium nitrate is simply the neutralization of aqueous nitric acid with gaseous ammonia followed by the evaporation of water from the strong ammonium nitrate solution leaving the neutralizer. The molten ammonium nitrate at a temperature of about 170 °C is then showered in roughly 2 mm drops down a tall tower up which cooling air flows. The product is removed from the base of the tower as 2 mm spheres suitable for spreading as a fertilizer as seen in Figure 16. Though simple in essence, very considerable care has to be taken in the detailed design and operation of this process as it has to be remembered that the largest non-nuclear explosion was of an ammonium nitrate plus ammonium sulfate mixture produced for fertilizer use! The detonation of 4000 tons in store at Oppau (Germany), killing over a thousand people, and leaving a crater 250 ft in diameter and 50 ft deep, is a powerful reminder that the ammonium nitrate molecule contains both oxidizing agent and reductant ready and in close proximity to react. Very much work has been undertaken since the 1921 explosion to determine the conditions necessary to trigger ammonium nitrate decomposition and detonation. Modern plants are designed so that the composition of the ammonium nitrate is such at all points to eliminate the possibility of decomposition; and also to restrict the quantity of material to the minimum in the more hazardous regions in case the other precautions fail. The main chemical requirements for reducing the sensitivity of ammonium nitrate to decomposition are as follows.

(i)   Molten ammonium nitrate must not fall below a pH of 4.5. This is secured by the injection of ammonia gas at a number of places in the production process (see Figure 16).

(ii) The neutralizer must be well vented so the pressure can not build up.

(iii) Various decomposition catalysts, such as chloride, metals like cobalt, and readily combustible organics like oil, must be excluded.

When the ammonium is to be used as a fertilizer and purity is not essential, an addition of 0.5 to 1.0% of magnesium ions inside the ammonium nitrate crystals both represses phase-changes in the crystals, leading to break up of the fertilizer particles, and also acts as a chemical sink for water which may enter during storage of the fertilizer on the farm before application. Magnesium ions hydrate, unlike ammonium ions which do not.

The precautions necessary in producing ammonium nitrate have to be born in mind during the storage of the finished product in the depot and on the farm. Oil must be excluded and other fertilizers which could be acid or contain chloride kept separately.

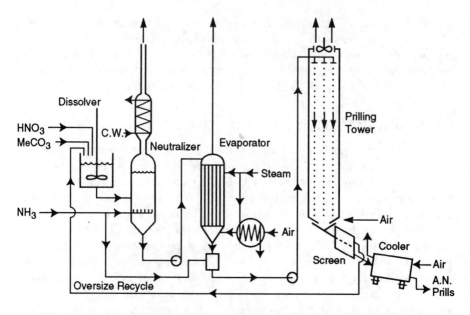

Figure 16 *Simplified Flowsheet of a Fertilizer Ammonium Nitrate Plant*

Before leaving ammonium nitrate it should be pointed out that, so far as changing the soil pH, it is an acid containing, in effect, two moles of nitric acid for each mole of ammonium nitrate. The first of these moles is the nitrate ion in the fertilizer and the second is formed by the reaction of the air with the ammonia catalysed by microbial life. Some cation leaching is thus to be expected together with an acidification of the soil if there is no balancing alkali addition in the long run. Calcium nitrate (CN) and calcium ammonium nitrate (CAN) fertilizers which are also large-scale industrial products, though lower in nitrogen content, may therefore have value for application on soils which already have a low pH.

6   UREA PRODUCTION

The explosive potential of ammonium nitrate, in politically unstable times, is a strong argument for the use of urea as a straight nitrogen fertilizer. Urea is an even more concentrated source of nitrogen than ammonium nitrate and thus has a lower cost of distribution of nitrogen. Its disadvantages arise from the possibility of substantial loss of gaseous ammonia into the atmosphere from urea hydrolysis if it is not washed into the soil by rain promptly (10 to 15%), and a further possible disadvantage is a reduction in the maximum crop yield at the highest urea application rates due to a small amount of poisoning by high concentrations of ammonia. This last effect can amount to some 5%.

Urea is produced by the uncatalysed liquid phase reaction between liquid ammonia and carbon dioxide at around 190 °C. A high pressure is required to maintain the reagents in the liquid phase and upflow reactors are invariably used, otherwise any unreactive insoluble gas such as nitrogen would, in time, fill up the

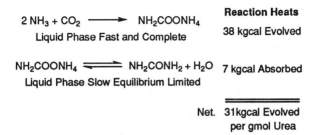

$$2\,NH_3 + CO_2 \longrightarrow NH_2COONH_4$$

Liquid Phase Fast and Complete

Reaction Heats

38 kgcal Evolved

$$NH_2COONH_4 \rightleftharpoons NH_2CONH_2 + H_2O$$

Liquid Phase Slow Equilibrium Limited

7 kgcal Absorbed

Net. 31 kgcal Evolved
per gmol Urea

Figure 17 *Urea Synthesis. Ideal Stoichiometry and Heat Release*

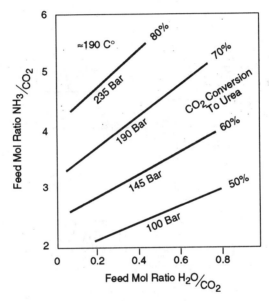

Figure 18 *Equilibrium of Urea Synthesis in the Liquid Phase. Typical Operating Pressures Indicated*

reactor had it been downflow. The reaction proceeds in two stages (see Figure 17). The first is very fast, complete, and strongly exothermic. It results in the formation of ammonium carbamate. The second reaction is quite slow, having a first-order rate constant of about $0.0015\,s^{-1}$; it is also markedly equilibrium-limited (see Figure 18). Long residence times of between one-half and one whole hour are normal in the urea synthesis reactor, and means must be employed to recycle unconverted ammonia and carbon dioxide.

A simplified flowsheet for a urea plant in which the recycle is as a concentrated aqueous solution is given in Figure 19. To effect recycle of the ammonia and carbon dioxide with the minimum proportion of water, which is detrimental to a favourable equilibrium (see Figure 18), the product from the reactor is reduced in pressure in two stages and the unreacted ammonia and carbon dioxide boiled off at each stage and condensed prior to recycle. The urea is finally separated by vacuum crystallization followed by centrifugation. It is important to perform the

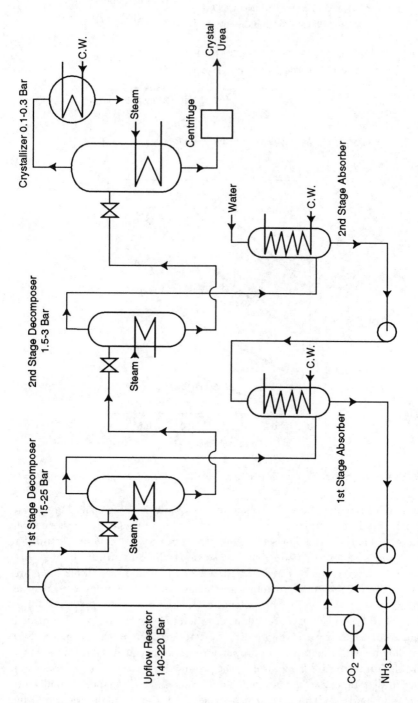

Figure 19 *Simplified Flowsheet of a Solution-recycle Urea Plant*

crystallization at as low a temperature as is economically feasible. Otherwise, in the absence of ammonia, the urea becomes excessively impure due to build up of biuret, formed by the combination of two molecules of urea with the loss of one molecule of ammonia. Biuret is a poison of plant growth and hence must be excluded from fertilizer urea. When the urea is to be sold as a fertilizer it is usual to add a further two stages onto the flowsheet of Figure 19. The crystal urea is melted and showered down a prilling tower to form 2 mm diameter spheres in the same manner as ammonium nitrate. Unlike ammonium nitrate there is no internal desiccant to guard against caking of the fertilizer whilst in store. Urea has therefore to be coated with a thin dusting of clay treated with a hydrophobic agent.

7 SLOW-RELEASE NITROGEN FERTILIZERS

There are a number of commercial straight nitrogen fertilizers which do not suffer the disadvantage of the fast hydrolysis of urea leading to loss and possible crop poisoning at very high application rates. The simplest are merely urea prills with a poorly soluble coating which delays the ingress of water. Sulfur and wax are typical coatings. More complex and hence more expensive are compounds of urea. A urea–formaldehyde compound (Ureaform), crotonaldehyde diurea (CDU), and isobutylidene diurea (IBDU) have all been produced for some years on a commercial scale. Their supposed advantages compared with ammonium nitrate or urea are difficult to measure on the farm and, in consequence, they have a very restricted sale. After all, the main slow nitrogen release system is the soil itself.

CHAPTER 7

Hydrogen Peroxide and Inorganic Peroxy Compounds

B. BERTSCH-FRANK, A. DORFER, G. GOOR, AND H. U. SÜSS

1 HYDROGEN PEROXIDE

1.1 Introduction

Hydrogen peroxide was discovered in 1818 by J.L. Thenard who reacted barium peroxide with nitric acid.[1] Until the establishment of electrolytic processes in the twentieth century, commercial production of hydrogen peroxide was based on this reaction. The very diluted hydrogen peroxide produced by the barium peroxide process found only limited use because of high production costs, low hydrogen peroxide content, and unsatisfactory stability due to impurities.

The disadvantages of the barium peroxide process were overcome by the introduction of electrochemical processes. In 1878, Berthelot showed that during electrolysis of sulfuric acid solution peroxysulfuric acid is formed which slowly hydrolysed to give sulfuric acid and hydrogen peroxide.[2] The first hydrogen peroxide plant based on this new technology went on stream in 1908 at the Österreichische Chemische Werke, Weißenstein in Austria. The Weißenstein process was modified by using ammonium sulfate instead of sulfuric acid (Münchner process; Riedel–Löwenstein process). After the beginning of electrolytic manufacture, production rose steadily and, in 1950, reached approximately 30 000–35 000 tons annually (calculated as 100% hydrogen peroxide).[3]

In 1901 Manchot observed that autoxidizable compounds like hydroquinones or hydrazobenzenes react quantitatively under alkaline conditions with oxygen to form peroxides.[4] In 1932 Walton and Filson discovered that hydrazobenzene in an alcohol or benzene solution could be oxidized with either air or oxygen to yield hydrogen peroxide and azobenzene.[5] Subsequently, Pfleiderer developed a process for production of sodium peroxide by alkaline autoxidation of hydrazo-

[1] L. J. Thenard, *Ann. Chim. Phys.*, 1818, **8**, 306.
[2] H. Berthelot, *Compt. Rend.*, 1878, **86**, 71.
[3] W. M. Weigert, H. Delle, and G. Käbisch, *Chem. Ztg.*, 1975, **99**, 101.
[4] W. Manchot, *Liebigs Ann. Chem.*, 1901, **314**, 177; *Liebigs Ann. Chem.*, 1901, **316**, 331.
[5] J. H. Walton and G.W. Filson, *J. Am. Chem. Soc.*, 1932, **54**, 3228.

Figure 1   *Part of Degussa's H₂O₂ Facility at Rheinfelden*

benzene and sodium amalgam was used to reduce azobenzene.[6] The two major technical drawbacks of the azobenzene process, *i.e.* hydrogenation of azobenzene with sodium amalgam and oxidation of hydrazobenzene in alkaline solution, were overcome by Riedl in using polynuclear quinones.

Based on the investigations of Riedl and Pfleiderer, I.G. Farbenindustrie developed, between 1935 and 1945, the anthraquinone process (so-called AO-process) in a small plant using 2-ethylanthraquinone as the autoxidizable compound. At the end of World War II this development was stopped in Germany because of Allied Control Commission Regulations. In 1953, E.I. du Pont started up the first hydrogen peroxide plant using the anthraquinone process. Several other companies began producing hydrogen peroxide by the AO-process and world-wide production increased rapidly. In 1991 world capacity stood at 1 545 000 tons of H_2O_2 100%.[7] Major H_2O_2 producers are Solvay, Degussa, DuPont, FMC, Eka Nobel, Oxysynthese, Mitsubishi Gas Chemical, and Kemira. Part of Degussa's H_2O_2 facility at Rheinfelden is shown in Figure 1.

Another autoxidation process, based on oxidation of 2-propanol, was developed and used by Shell in its 15 000 tons facility in Norco between 1957 and 1980.

1.2   Manufacture of Hydrogen Peroxide

1.2.1   AO-process. In the AO-process a 2-alkyl-9,10-anthraquinone is hydrogenated in the presence of a catalyst to the corresponding hydroquinone. After separation of the catalyst the hydroquinone is oxidized with oxygen, usually air, to yield hydrogen peroxide and the starting 2-alkyl-9,10-anthraquinone. Hydrogen

[6] BASF, DRP 649234, 1934 and DRP 658767, 1935 (G. Pfleiderer).
[7] *Inf. Chim.*, December 1991, 133–144.

peroxide is extracted with water, whereas the quinone is recycled to the hydrogenator (Scheme 1).

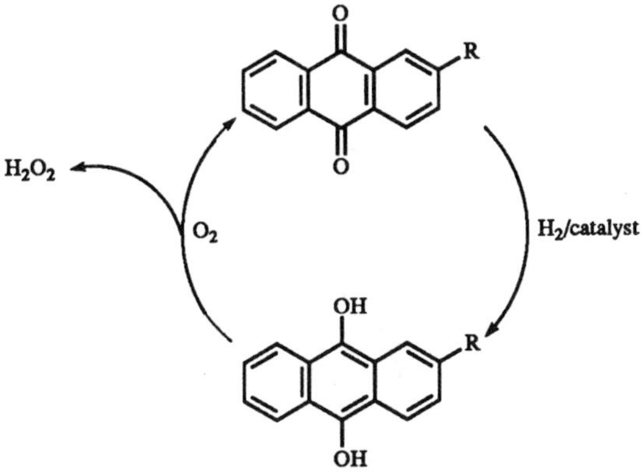

Scheme 1

The anthraquinone is dissolved in a solvent or solvent mixture for hydrogenation, oxidation, and extraction. This anthraquinone solution is referred to as the working solution. During the cyclic AO-process both quinone and hydroquinone must remain dissolved. Because quinone and hydroquinone have different solubilities, solvent mixtures are mostly used. Quinones dissolve well in non-polar aromatic solvents, *e.g.* benzene,[8] t-butylbenzene,[9] t-butyltoluene,[10] trimethyl benzene,[11] methylnaphthalene,[12] and polyalkylated benzenes.[13] Hydroquinones on the contrary dissolve well in polar solvents like esters of phosphonic[14] or phosphoric acid,[15] nonyl alcohols,[12] esters of methyl cyclohexanol,[16] N,N-dialkylamides of carboxylic acids,[17,18] tetraalkyl ureas,[19–21] N-alkyl-2-pyrrolidones,[22] and alkylated carbamates.[23]

The solvent or solvent mixture must fulfil a number of criteria such as:

- good solubility both for quinone and hydroquinone;
- chemically stable during hydrogenation and oxidation;
- slightly soluble in water and aqueous hydrogen peroxide solutions;
- low density and viscosity to facilitate phase separation in the extractor;

[8] I. G. Farbenindustrie, Patent No. GB 508081, 1939.
[9] Degussa, Patent No. DE 1945750, 1969 (G. Käbisch and R. Trübe).
[10] Edogawa, Patent No. DE 1112051, 1958 (K. Hiratsuka *et al.*).
[11] Laporte Chemicals Ltd., Patent No. DE 953790, 1953 (C.W. Lefeuvre).
[12] E. I. du Pont de Nemours and Co., Patent No. DE 888840, 1950 (C.R. Harris and J.W. Sprauer).
[13] Degussa, Patent No. DE 1261838, 1963 (G. Käbisch).
[14] Becco, Patent No. US 2537516, 1950 (L.H. Dawsey, C.K. Muehlhausser, and R.R. Umhoefer).
[15] Becco, Patent No. US 2537655, 1950 (L.H. Dawsey, C.K. Muehlhausser, and R.R. Umhoefer).
[16] Laporte Chemicals Ltd., Patent No. DE 933088, 1953 (C.W. Lefeuvre).
[17] E. I. du Pont de Nemours and Co., Patent No. BE 819676, 1974.
[18] Air Liquide, Patent No. EP 287421, 1988 (C. Pralus).
[19] Degussa, Patent No. DE 2018686, 1970 (G. Giesselmann, G. Schreyer, and W. Weigert).
[20] Degussa, Patent No. US 4349526, 1982 (G. Goor and W. Kunkel).
[21] EKA Nobel AB, Patent No. EP 284580, 1988 (E. Bengtsson and U.M. Andersson).
[22] FMC, Patent No. EP 95822, 1983 (W. Ranbom).
[23] Kemira Oy, Patent No. DE 4026631, 1990 (E. Suokas and R. Aksela).

- high boiling point and flash point;
- high distribution coefficient of hydrogen peroxide between water and the working solution in favour of the aqueous phase;
- low toxicity; and
- availability at a reasonable price.

The choice of quinone mainly depends on its solubility and that of the corresponding hydroquinone. Besides 2-ethylanthraquinone, which was initially used by I.G. Farbenindustrie,[24] 2-t-butylanthraquinone,[25] mixed 2-amylanthraquinones,[26] and 2-neopentylanthraquinone[27] are claimed in the patent literature. Improved solubility is also reported for eutectic mixtures of anthraquinones.[28] Although in most cases the substituent of monoalkylated anthraquinones is located at the 2-position, polyalkylated anthraquinones, where the alkyl groups contain a total of three to fifteen carbon atoms, were also described.[29]

The quinone used in the AO-process should be chemically stable. In the hydrogenator a number of secondary reactions occur. On hydrogenation of the aromatic ring, preferably the unsubstituted one, 2-alkyl-5,6,7,8-tetrahydro-9,10-dihydroxyanthracene is formed which on oxidation with oxygen quantitatively yields hydrogen peroxide and 2-alkyl-5,6,7,8-tetrahydroanthraquinone (commonly called 'tetra') (Scheme 2).

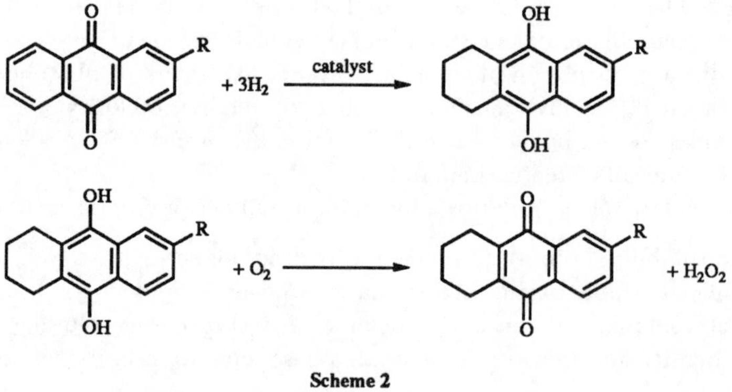

Scheme 2

Whereas the tetra is more easily hydrogenated than the corresponding anthraquinone, the resulting hydroquinone of the tetra is much more difficult to oxidize than the anthrahydroquinone. To overcome the problems arising from the poor oxidation rate of the hydroquinone of the tetra, the tetra fraction in the working solution can be kept low by either minimizing its formation or by dehydrogenation into the anthraquinone, Tetra formation is suppressed by using

[24] I. G. Farbenindustrie, Patent No. US 2158525, 1939 (H. J. Riedl and G. Pfleiderer).
[25] E. I. du Pont de Nemours and Co., Patent No. US 2689169, 1954 (W.S. Hinegardner).
[26] Edogawa, Patent No. DE 1085861, 1957 (K. Hiratsuka *et al.*).
[27] BASF, Patent No. DE 3732015, 1987 (M. Eggersdorfer, J. Henkelmann, and W. Grosch).
[28] Degussa, Patent No. DE 1063581, 1957 (L.R. Darbee and D.F. Kreuz).
[29] Degussa, Patent No. DE 1051257, 1957 (G. Käbisch and E. Richter).

selective catalysts,[30] special solvents[31] or quinones,[29] and mild hydrogenation conditions.[32] For permanent dehydrogenation of tetra, use of olefins in the presence of activated alumina has been recommended.[33] If no precautions are taken to minimize the tetra content, an equilibrium will result in which the hydroquinone fed to the oxidizer consists exclusively of the hydroquinone of the tetra. This status of the working solution is referred to as 'all-tetra-system'.

During oxidation of the tetra hydroquinone, small amounts of an epoxide are formed in a side reaction. As this epoxide does not participate in hydrogen peroxide formation, and therefore has to be considered as a loss of quinone, steps were suggested to regenerate tetra from epoxide, *e.g.* treatment of the working solution with activated alumina[34] (Scheme 3).

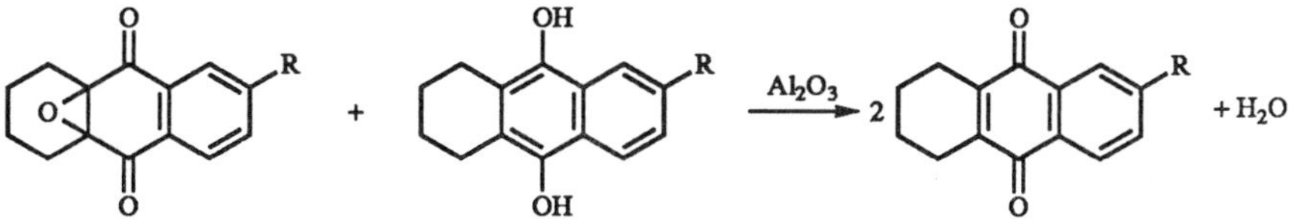

Scheme 3

At high tetra level, further ring hydrogenation generates the 1,2,3,4,5,6,7,8-octahydroanthrahydroquinone.

All operating AO-plants are based on the Riedl–Pfleiderer process. A schematic flow sheet is summarized in Figure 2.[35] Working solution from the storage tank (a) is fed into the hydrogenator (b), where it is hydrogenated in the presence of a catalyst at a slightly elevated pressure and at a temperature below 100 °C. Riedl and Pfleiderer used Raney nickel as a catalyst. Besides its pyrophoric character, Raney nickel is also deactivated by hydrogen peroxide and shows excessive tetra formation. Today, most AO-plants prefer palladium catalysts because of higher selectivity and simpler handling. Palladium can be used as palladium black[36] or supported on a carrier for slurry[37] or fixed-bed[38] operation. Several hydrogenator types have been described: tubular,[36,39] draft-tube,[40] and fixed-bed reactors.[41,42]

Hydrogenated working solution entering the oxidizer (c) should not contain any remaining catalyst: both palladium and Raney nickel catalyse hydrogen peroxide decomposition. Even a small amount of these catalysts in the oxidizer or extractor (f) would cause considerable hydrogen peroxide loss and severe process

[30] Laporte Chemicals Ltd., Patent No. GB 741444, 1953 (W. R. Holmes *et al.*).
[31] Solvay and Cie., Patent No. DE 1041928, 1956 (M. Gonze and E. Leblon).
[32] E. I. du Pont de Nemours and Co., Patent No. US 2673140, 1954 (J. W. Sprauer).
[33] Edogawa, Patent No. FR 1319025, 1962.
[34] Degussa, Patent No. DE 1273499, 1964 (G. Schreyer, O. Weiberg, and G. Giesselmann).
[35] 'Ullmann's Encyclopedia of Industrial Chemistry', 5th Edn., Verlag Chemie, Weinheim, 1988, Vol. 13, 450.
[36] Degussa, Patent No. DE 1542089, 1965 (G. Käbisch and H. Herzog).
[37] Laporte Chemicals Ltd., Patent No. GB 718306, 1953 (C.A. Morgan).
[38] Degussa, Patent No. DE 1079604, 1959 (D.H. Porter).
[39] Degussa, Patent No. US 4428923, 1984 (W. Kunkel, J. Kemnade, and D. Schneider).
[40] Laporte Chemicals Ltd., Patent No. GB 718307, 1952 (J. A. Williams and C. W. Lefeuvre).
[41] FMC, Patent No. US 3009782, 1961 (D.H. Porter).
[42] EKA AB, Patent No. EP 102934, 1983 (T. Berglin and W. Herrmann).

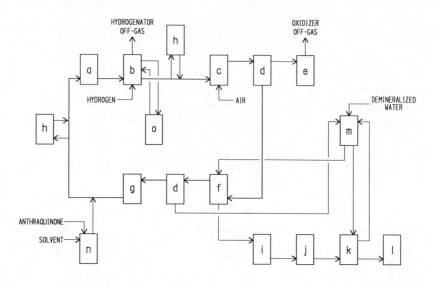

Figure 2   *Flow Sheet of AO-Process: a, working solution storage tank; b, hydrogenator including catalyst filtration; c, oxidizer; d, separator; e, activated carbon adsorber; f, extractor; g, drier; h, regeneration and purification of working solution; i, crude product purification; j, crude product storage tank; k, concentrator; l, hydrogen peroxide storage tank; m, demineralized water feed tank; n, working solution make-up tank; o, catalyst regeneration*

upsets. In the oxidizer, the catalyst-free working solution is oxidized with oxygen or air. The reaction is not catalysed (autoxidation) and the overall rate depends on temperature, oxygen partial pressure, and phase-boundary surface area. Several oxidizers have been proposed in the literature: liquid and gas may flow either cocurrently or countercurrently in a single- or multi-column system with or without packings.[43-45] To recover solvent from the excess air, the oxidizer off-gas can be purified with active carbon (e) which periodically is regenerated, *e.g.* with steam.

Hydrogen peroxide is extracted from the oxidized working solution with water (f). Besides sieve-tray extraction columns, used in the I.G. Farbenindustrie process, a number of other extractors, *e.g.* packed columns, pulsed packed columns, spray columns, and mechanical extractors, have been suggested. Free water leaving the extractor is separated (d) and the water content in the working solution is adjusted in the drier (g) to prevent agglomeration of the catalyst.

During the cyclic reduction and oxidation of the working solution, degradation products are formed from the quinone and the solvent. A large variety of processes have been proposed to remove these degradation products and to regenerate active quinone (h), *e.g.* treatment with sodium aluminum silicates,[46] activated

[43] *Br. Chem. Eng.*, 1959, **2**, 88.
[44] Degussa, Patent No. DE 2003268, 1970 (M. Liebert, H. Delle, and G. Käbisch).
[45] EKA AB, Patent No. DE 2419534, 1974 (B.G. Franzen).
[46] PPG, Patent No. US 3055838, 1962 (J.W. Moore).

alumina or magnesium oxides,[47] calcium, magnesium or zinc silicates,[48] and aqueous solution of sodium or potassium hydroxide.[49] Treatment of hydrogenated working solution with alkaline substances converts epoxide into the corresponding tetra compound.[34]

The crude aqueous hydrogen peroxide from the extractor with a H_2O_2 concentration of 15–30 wt. % has to be purified (i) to remove solvent and quinone which may have been co-extracted. This can be done by washing with hydrocarbons[50,51] or treatment with activated carbon[52] or adsorber resins.[53,54] The purified crude product is then concentrated (k) to the commercial grades of 50–70 wt. %,[55-58] stabilized,[59,60] and stored (l).

Continuous AO-operation is further sustained by additional activities. For example, to maintain activity of the hydrogenation catalyst, part of it is withdrawn from the hydrogenator and regenerated (o). To compensate for quinone and solvent losses, the working solution is periodically made up with quinone and solvents (n).

1.2.2 Other Processes. Electrolytic production of hydrogen peroxide is largely of historical value. Those plants still using this technology contribute only to a small extent to the world's H_2O_2 capacity. All processes have the anodic oxidative coupling of sulfate or bisulfate ions to peroxodisulfate ions in common.

$$2H_2SO_4 \rightarrow H_2S_2O_8 + H_2$$
$$H_2S_2O_8 + H_2O \rightarrow H_2SO_5 + H_2SO_4$$
$$H_2SO_5 + H_2O \rightarrow H_2SO_4 + H_2O_2$$

Subsequent hydrolysis of the peroxodisulfate via peroxomonosulfate yields hydrogen peroxide.[61] A new electrolytic process for production of hydrogen peroxide in alkaline solution by cathodic reduction of oxygen has been developed by H-D Tech. Inc.[62]

Based on the discovery of Harris that primary and secondary alcohols are oxidized with oxygen to give hydrogen peroxide and an aldehyde or ketone

[47] E. I. du Pont de Nemours and Co., Patent No. US 2739875, 1956 (J. W. Sprauer and T. V. Williams).
[48] Laporte Chemicals Ltd., Patent No. GB 928784, 1959 (B. K. Howe).
[49] Laporte Chemicals Ltd., Patent No. GB 991413, 1962 (B. K. Howe).
[50] E. I. du Pont de Nemours and Co., Patent No. DE 1036225, 1956 (J. W. Sprauer).
[51] Laporte Chemicals Ltd., Patent No. GB 841323, 1957 (J. G. Lait and H. B. Hopkins).
[52] E. I. du Pont de Nemours and Co., Patent No. US 3016290, 1962 (A. A. Elston and W. S. Hinegardner).
[53] Oxysynthese, Patent No. FR 1539843, 1967 (P. Thirion).
[54] Mitsubishi Gas Chemicals, Patent No. US 4792403, 1988 (S. Togo, Y. Sugihara, and T. Ikehe).
[55] Laporte Chemicals Ltd., Patent No. US 3073755, 1963 (R. H. Banfield and A. Hildon).
[56] Columbia-Southern Chemical Corp., Patent No. US 2960448, 1960 (W. S. Straub and C. J. Sindlinger).
[57] Degussa, Patent No. US 3060105, 1962 (J. Müller).
[58] FMC, Patent No. US 3152052, 1964 (T.M. Jenney, D.H. Porter, and R.H. Rice).
[59] Laporte Chemicals Ltd., Patent No. GB 893069, 1959 (P. V. Rutledge and D. S. Bell).
[60] Columbia-Southern Chemical Corp., Patent No. US 2904517, 1959 (P. H. Baker).
[61] W. C. Schumb, C. N. Satterfield, and R. L. Wentworth, 'Hydrogen peroxide', Reinhold Publ. Co., New York, 1955.
[62] A. Clifford, D. Dong, E. Giziewicz, and D. Rogers, *Proc. Electrochem. Soc.*, 90–10, 259–74.

respectively, the 2-propanol process (also referred to as Shell process) was developed.[63]

$$RCH_2OH + O_2 \rightarrow RCHO + H_2O_2$$

$$\begin{array}{c} R \\ \diagdown \\ CHOH + O_2 \rightarrow \\ \diagup \\ R \end{array} \begin{array}{c} R \\ \diagdown \\ C=O + H_2O_2 \\ \diagup \\ R \end{array}$$

Oxidation of the alcohol can be performed in the gas phase or in the liquid phase. Liquid phase reaction of 2-propanol with oxygen does not require a special catalyst but small amounts of hydrogen peroxide recycled with 2-propanol reduce the induction period. The technology of the Shell process is extensively described in reference 64.

1.3 Properties of Hydrogen Peroxide

1.3.1 Physical Properties. Hydrogen peroxide is a clear, colourless liquid which is completely miscible with water. Concentrated aqueous solutions are considerably more soluble in a variety of polar solvents, *e.g.* carboxylic esters, than water. Important physical properties of pure hydrogen peroxide and water are summarized in Table 1 and those of aqueous solutions in Table 2. Numerous other physical data appear in the literature.[65]

1.3.2 Chemical Properties. As a weak acid, with a dissociation constant of 1.78×10^{-12} at $20\,°C$ (pK_a 11.75), hydrogen peroxide reacts with metal oxides and hydroxides to form inorganic peroxides and hydroperoxides.

Hydrogen peroxide can behave both as an oxidizing and as a reducing agent. Most of its applications are based on its strong oxidizing character and the compounds oxidized range from iodide ions to organic colour bodies of unknown structure in cellulosic fibres (see Section 1.4.1). For hydrogen peroxide acting as an oxidizing agent the half-cell reaction in acidic solution is:

$$2H_2O \rightarrow H_2O_2 + 2H^+ + 2e^- \quad E_0 = -1.76 \text{ V}$$

whereas in alkaline solution it becomes:

$$3OH^- \rightarrow HO_2^- + H_2O + 2e^- \quad E_0 = -0.87 \text{ V}$$

The electrode potential depends on the hydrogen peroxide concentration and pH. Hydrogen peroxide reduces stronger oxidizing agents such as chlorine, ceric sulfate, and potassium permanganate. The latter reaction is used for the volumetric determination of hydrogen peroxide.

The decomposition of hydrogen peroxide:

$$2H_2O_2 \rightarrow 2H_2O + O_2$$

[63] E. I. du Pont de Nemours and Co., Patent No. US 2479111, 1949 (C. R. Harris).
[64] 'Ullmann's Encyclopedia of Industrial Chemistry', 4th Edn., Verlag Chemie, Weinheim, Vol. 17, 694.
[65] 'Gmelins Handbuch der Anorganischen Chemie', Verlag Chemie, Weinheim, 1966.

Table 1 *Physical Properties of Hydrogen Peroxide and Water*

	H_2O_2	H_2O
Melting point, °C	0.43	0
Boiling point, °C	150.2	100
Heat of melting, $J g^{-1}$	368	334
Heat of vaporization, $J g^{-1}$		
25 °C	1519	2443
boiling point	1387	2258
Specific heat, $J g^{-1} K^{-1}$		
liquid (25 °C)	2.629	4.182
gas (25 °C)	1.352	1.865
Relative density, $g cm^{-3}$		
0 °C	1.4700	0.9998
20 °C	1.4500	0.9980
Viscosity, mPa s		
0 °C	1.819	1.792
20 °C	1.249	1.002
Critical temperature, °C	457	374.2
Critical pressure, MPa	20.99	21.44
Refractive index, n_D^{20}	1.4084	1.3330

is important not only during manufacture but also during storage, shipment, and handling because the generation of oxygen and heat may cause safety problems. Several parameters like temperature, pH, and impurities influence the rate of decomposition. Thus, with increasing temperature or pH, decomposition increases. Furthermore, transition metal compounds, especially those of iron, copper, manganese, nickel, and chromium, can catalyse either homogeneously (dissolved metal ions) or heterogeneously (suspended metal oxides and hydroxides) H_2O_2 decomposition. Addition of a proper stabilizer or a combination of stabilizers helps to minimize decomposition. Sodium pyrophosphate and sodium stannate are very effective stabilizers under acidic conditions and are added separately or together.[59,60] Organic substances are preferred for dilute solutions as they are slowly oxidized by concentrated hydrogen peroxide.[66] In alkaline solutions magnesium silicate is the major stabilizing agent.

Hydrogen peroxide adds on to a number of compounds to form crystalline peroxyhydrates which are analogous to crystalline hydrates formed with water. Well known examples of peroxyhydrates are those from sodium carbonate (see Section 2.1.3) and urea.

1.4 Uses of Hydrogen Peroxide

Prior to 1939 the principal uses of hydrogen peroxide were in bleaching and as an antiseptic, with a total world capacity of 12 000 tons per year.[67] The development of the AO-process permitted the large scale production.

In 1990 about 1 million tons of hydrogen peroxide were consumed. Table 3

[66] Shell Oil Company, Patent No. US 3053634, 1962 (D. B. Luton and R.E. Meeker).
[67] C. A. Crampton, G. Faber, R.Jones, J. P. Leaver, and S. Schelle, in 'The Modern Inorganic Chemicals Industry', Special Publication No. 31, ed. R. Thompson, The Royal Society of Chemistry, London, 1977, 246.

Table 2   *Physical Properties of Aqueous Solutions of Hydrogen Peroxide*

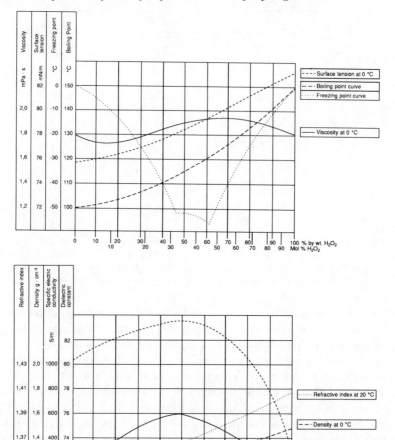

gives a breakdown of the different application areas. By far the most important volume is used for bleaching processes. Perborates and percarbonate are similarly applied for bleaching purposes in laundry agents. Their production normally represents a captive use of hydrogen peroxide.

The chemical industry has various sectors for the consumption of hydrogen peroxide.

1.4.1   Bleaching. Bleaching is generally understood as the chemical destruction of chromophores in organic or inorganic compounds. Usually the aim is to increase the brightness by weakening or removing the colour of the material being bleached.[68] Industrially, decolorization is often combined with cleaning. Chemical

---

[68] 'Ullmann's Encyclopedia of Industrial Chemistry', 4th Edn., Verlag Chemie, Weinheim, Vol. 4, 191–199.

Table 3   *Hydrogen Peroxide Consumption in North America, Western Europe, and Japan, 1990*[69]

Application	% of Total Consumption
Chemicals and Laundry Products	41
Pulp and Paper	38
Textiles	9
Environmental	6
Others	6

components that affect material quality, handling properties, or marketability of a product are removed. This is especially important for natural products, which almost always contain impurities. For example, cotton is contaminated with residual seed capsule parts, wax, fatty constituents, and inorganic compounds. Bleaching cotton simultaneously removes these impurities and increases the brightness.

Hydrogen peroxide, under alkaline conditions, is a very efficient bleaching agent. It is nowadays applied to bleach fibres for textiles like cotton. Its application in the bleaching of products for the pulp and paper industry has increased sharply in recent years, due to the growing environmental difficulties in the use of the earlier widely-applied chlorine containing bleaching agents, like chlorine, hypochlorite, or chlorine dioxide.

Of minor economical importance are bleaching processes for vegetable products, *e.g.* fatty acids, resins, wax, flour, wood veneer, or animal products like hair, bristle, feathers, and sponges. However, these products are similarly bleached with hydrogen peroxide.

Commercial processes use slightly to strongly alkaline conditions for hydrogen peroxide bleaching.[70] Weakly acidic conditions are appropriate only in special cases. Usually sodium hydroxide is the alkali source. The initial reaction is the nucleophilic addition of hydroperoxide anions to carbonyl groups. Therefore, the bleaching rate is increased by addition of alkali:

$$HOOH + OH^- \rightleftharpoons HOO^- + H_2O$$

The results of the bleaching process depend on activation and stabilization. The reaction rate increases with temperature and alkali concentration. This however, promotes side reactions, such as decomposition of hydrogen peroxide or yield losses because of solubilization of the material being bleached.

It is advantageous to use buffered conditions for the bleaching process. Compounds like sodium silicate, sodium carbonate (soda ash), and magnesium salts are applied to stabilize the bleaching process. Additionally, chelating agents can be added if transition metals are present in the bleached product. The effect of the stabilizers is to chelate heavy metal ions or to absorb these ions on colloidal silica or magnesium hydroxide. Transition metal ions act as catalysts that tend to

[69] 'Chemical Economics Handbook – SRI International, Hydrogen Peroxide', 1992.
[70] 'The Bleaching of Pulp', ed. R. P. Singh, Tappi Press, Atlanta, 1979.

increase both the consumption of hydrogen peroxide and the damage to the bleached material. Hydroxyl radicals, generated by the decomposition of peroxide, are most unselective oxidation agents, producing many undesired side reactions. Standard compounds for the chelation of transition metals are: diethylenetriaminepentaacetic acid (DTPA) or ethylenediaminetetraacetic acid (EDTA). The stabilization of the bleaching process is increased by removing catalysts with an acidic wash or a pretreatment with chelating agents.

The demand for hydrogen peroxide has rocketed in the pulp and paper industry. There are three application areas: mechanical pulp bleaching, chemical pulp bleaching, and waste paper recycling. Groundwood or thermomechanical pulp, produced by mechanical defibrization of wood, still contains nearly all of the lignin originally present. High brightness gains can only be achieved with a hydrogen peroxide bleaching step. Bleaching is conducted at a high consistency (concentration) level in downflow towers at temperature levels between 40 and 70 °C and with retention times of one to four hours with stabilized, buffered bleaching liquors.

Chemical pulping decreases the lignin content of wood using, *e.g.* a sulfonation process (sulfite pulping) or strong alkaline conditions (kraft pulping with $NaOH/Na_2S$).[71] The residual amount of lignin is oxidized and dissolved in a multi-stage process. In several countries the classical bleaching process for chemical pulp, a multi-stage treatment using chlorine and chlorine containing bleaching agents, has already been replaced with bleaching processes based solely on oxygen and hydrogen peroxide.[72] The background for this development is the difficulties in the biodegradation of the chlorinated compounds resulting from chlorine bleaching.

Due to the large amounts of waste paper available, the paper recycling market also shows a very high growth rate. In de-inking processes, which remove the printing ink from the waste paper, hydrogen peroxide is applied to prevent yellowing processes during the repulping and ink removing steps, and further to achieve higher final brightnesses in additional bleaching stages.

1.4.2 Environmental Uses. Environmental application of hydrogen peroxide covers a very wide area. Effluents of the cyanide leaching process in gold mines are detoxified, as well as hydrogen sulfide or sulfite in pit water. Hydrogen peroxide is added to activated sludge biodegradation plants as oxygen supply in situations of overloading. It also oxidizes effluents of film processing and is used to precipitate phenol resins from effluents applying Fenton's reagent ($FeSO_4 + H_2O_2$). Smaller amounts of H_2S or SO_2 in waste gases (*e.g.* the lime kiln of kraft pulp mills) are similarly oxidized with hydrogen peroxide. Flue gas washing with a hydrogen peroxide solution can be advantageous over alternative methods for sulfur dioxide removal under specific conditions.

[71] S. A. Rydholm, 'Pulping Processes', Interscience Publications, New York, 1965. D. Fengel, G. Wegener, 'Wood', DeGruyter, Berlin, 1984.
[72] H. U. Süss, 'Chlorine-free Bleaching – Current Status and Trends in Central and Southern Europe', Non Chlorine Bleaching Conference, Hilton Head Island, SC, USA, 1992.

1.4.3 Chemical Uses. The use of hydrogen peroxide for manufacture of organic chemicals, either directly or after conversion into a percarboxylic acid, is steadily growing. Epoxidation of oils, primarily soybean oil and linseed oil with performic or peracetic acid, yields plasticizers and stabilizers for poly(vinyl chloride). Epoxidation with performic acid is mostly carried out *in situ*; hydrogen peroxide is added in small portions to a mixture of the oil and formic acid while the solution is well stirred. In doing this, build-up of substantial quantities of detonable performic acid is prevented. To minimize epoxide losses due to ring opening the use of less than one mole of formic acid for each double bond is recommended.[73] By-product formation is further reduced by adding inert solvents.[73] Continuous epoxidation of fatty acids or esters in counter-current flow with hydrogen peroxide and formic acid is reported.[74] To epoxidize oils and fats with *in situ* peracetic acid the presence of a strong acid is required. Again, proper choice of the reaction conditions, *e.g.* addition of heptane or toluene,[75] helps to prevent cleavage of the epoxide ring. In the last two decades, several continuous processes for epoxidation of low molecular weight olefins, *e.g.* propylene, with solutions of percarboxylic acids in inert solvents, which give excellent yields of epoxide with minimal by-product formation, have been developed.[76-85] Water-soluble olefins like allyl alcohol and maleic acid are readily epoxidized with hydrogen peroxide in the presence of $NaHWO_4$.[86,87]

Catechol and hydroquinone are produced via the catalysed hydroxylation of phenol with hydrogen peroxide.[88-90] Catechol is used as intermediate for production of agrochemicals, pharmaceuticals, and flavours and fragrances whereas hydroquinone is mainly consumed in the photographic industry.

Oxidation of cyclic ketones with percarboxylic acids gives lactones in good yields. ε-Caprolactone is commercially produced by the Baeyer–Villiger reaction of cyclohexanone with either peracetic acid, in an anhydrous medium such as

[73] Rohm, and Haas, Patent No. US 2485160, 1949 (W.D. Niederhauser *et al.*).
[74] Henkel, Patent No. DE 3320219, 1983 (G. Dieckelmann *et al.*).
[75] Ashland Oil and Refining Co., Patent No. US 3360531, 1971 (W.H. French).
[76] Degussa and Bayer, Patent No. DE 2519297, 1975 (G. Prescher *et al.*).
[77] Degussa and Bayer, Patent No. DE 2519298, 1975 (G. Prescher *et al.*).
[78] Atochem, Patent No. EP 61393, 1982 (J.C. Lecoq, M. Pralus, and J.P. Schirmann).
[79] Propylox, Patent No. DE 2747761, 1977 (A. Hildon and P. Greenhalgh).
[80] FMC, Patent No. US 3954815, 1976 (W.C. Fisher *et al.*).
[81] Interox Chemicals, Patent No. DE 2602776, 1976 (A. Hildon and P. Greenhalgh).
[82] Bayer, Patent No. EP 56932, 1982 (G. Rauleder and H. Waldmann).
[83] Degussa, Patent No. DE 3528005, 1985 (A. Grund *et al.*).
[84] Solvay, Patent No. EP 19322, 1980 (R. Walraevens and L. Lerot).
[85] Degussa, Patent No. DE 3528007, 1985 (R. Siegmeier *et al.*).
[86] Degussa, Patent No. DE 2408948, 1974 (W. Heim *et al.*).
[87] Degussa, Patent No. DE 2508228, 1975 (G. Prescher and G. Schreyer).
[88] J. Varagnat, *Ind. Eng. Chem. Prod. Res. Dev.*, 1976, **15**, 212.
[89] P. Maggioni and F. Minisci, *Chim. Ind. (Milan)*, 1977, **59**, 239.
[90] U. Romano *et. al.*, 'Studies in Surface Science and Catalysis', Vol. 55, Elsevier, Amsterdam, 1990, p. 33.

ethyl acetate,[91] or distilled aqueous peracetic acid.[92] Low molecular weight polyesters, made from ε-caprolactone, are of interest for production of high performance polyurethane elastomers and coatings.[93]

Hydrogen peroxide is also widely used for oxidation of N-containing compounds. For example, lauryldimethylamine reacts with hydrogen peroxide to afford lauryldimethylamine oxide which is primarily used in light-duty liquid dishwashing detergents. Catalysed oxidation of ammonia with hydrogen peroxide in the presence of a ketone and subsequent hydrolysis of the ketazine intermediate yields hydrazine.[94,95] This new route eliminates formation of large quantities of salt as by-product by manufacture of hydrazine according to the Raschig process. Reacting hydrogen peroxide and excess ammonia with cyclohexanone in the presence of catalysts gives cyclohexanone oxime, intermediate for ε-caprolactame manufacture.[96,97] The uncatalysed reaction of excess cyclohexanone with hydrogen peroxide and ammonia gives 1,1'-peroxydicyclohexylamine.[98] Base catalysed rearrangement of 1,1'-peroxydicyclohexylamine in the presence of lithium compounds affords ε-caprolactam.[99] By vapour-phase pyrolysis of 1,1'-peroxydicyclohexylamine, 11-cyanoundecanoic acid is obtained,[100] which can be either hydrogenated to 12-aminododecanoic acid, the nylon-12 monomer, or hydrolysed to 1,12-dodecanedioic acid. Hydrogen peroxide can also be used to prepare cyanogen,[101] cyanogen chloride,[102] and formamidine sulfinic acid.[103]

An important use of hydrogen peroxide is in the manufacture of organic peroxides,[104] including t-butyl hydroperoxide, dibenzoyl peroxide, methyl ethyl ketone peroxide, and t-butyl peracetate. These compounds are used as polymerization catalysts and cross-linking agents.

2 INORGANIC PEROXY COMPOUNDS

2.1 Sodium Perborate and Sodium Percarbonate

2.1.1 Introduction. The two most important inorganic derivatives of hydrogen peroxide are sodium perborate and sodium percarbonate. The existence of perborates was first reported by the French researcher Etard in 1880, at a meeting of the Academy of Sciences.[105] This study was continued, 17 years later, by the

[91] Union Carbide Corp., Patent No. DE 1086686, 1955 (C. B. Phillips and P. S. Starcher).
[92] Degussa, Patent No. DE 1258858, 1965.
[93] R. E. Redfern in 'Specialty Chemicals', May 1982.
[94] Produits Chimiques Ugine Kuhlmann, Patent No. US 3869541, 1975 (F. Weiss, J. P. Schirmann, and H. Matais).
[95] Produits Chimiques Ugine Kuhlmann, Patent No. US 3972878, 1976 (J. P. Schirmann, J. Combroux, and S.Y. Delavarenne).
[96] Toa Gosei, Patent No. DE 1245371, 1960; Patent No. DE 1265745, 1967; Patent No. DE 1274124, 1967.
[97] Montedipe, Patent No. US 4745221, 1988 (P. Roffia *et al.*).
[98] E. G. Hawkins, *Angew. Chem.*, 1973, **85**, 850.
[99] BP Chemicals Ltd., Patent No. DE 1814288, 1969; Patent No. DE 1803872, 1971; Patent No. DE 1670246, 1971.
[100] BP Chemicals Ltd., Patent No. GB 1198422, 1966 (E. Hawkins).
[101] R. Fahnenstich, W. Heimberger, F. Theissen and W. Weigert, *Chem. Ztg.*, 1972, **96**, 388.
[102] Degussa, Patent No. US 4046862, 1977 (W. Heimberger and G. Schreyer).
[103] SKW Trostberg, Patent No. DE 2736943, 1977 (K. Büchler, L. Mittelmeier, and W. Pollwein).
[104] D. Swern, 'Organic Peroxides', Wiley-Interscience, New York, 1971.

Russian chemists Tanatar and Melikoff.[106] In 1899, Tanatar synthesized sodium percarbonate for the first time.

Sodium perborate became industrially significant in Germany in 1907, when it was employed in a laundry powder, making obsolete the very laborious process of sun bleaching that had been common up to that time.

Sodium perborate has remained the most widely used bleaching agent in laundry powders. It is sold mainly in the form of the tetrahydrate, although the monohydrate is gaining in importance. Anhydrous sodium perborate (Oxoborate) is not a bleaching agent, but its ability to liberate gaseous oxygen immediately on contact with water makes it suitable for use as a tablet disintegrating agent. The principal application of sodium percarbonate is in bleach boosters.

Production capacity for sodium perborate in 1992 is estimated at 900 000 t (as the tetrahydrate), the market around 700 000 t. Capacity for monohydrate is thought to be 275 000 t but is being expanded. Because over 80% of the world's perborate output is consumed in Europe, most of the manufacturers are located there as well. The most important producers are Interox, Degussa, L'Air Liquide, Atochem, Montefluos, DuPont, Foret, and Mitsubishi Gas. World-wide sodium percarbonate capacity in 1992 is estimated at 80 000 t, with a consumption of some 25 000 t. This capacity is centred in Japan. The most important manufacturers are Interox, Degussa, Nippon Peroxide, Mitsubishi Gas, and Tokai Denka.

2.1.2 Manufacture of Sodium Perborate. Manufacture of sodium perborate tetrahydrate. Sodium perborate tetrahydrate was formerly produced by an electrolytic process.[107] Since hydrogen peroxide became available on an industrial scale, the product has been obtained primarily by a continuous crystallization process.

The first stage of the process is the digestion of sodium borate or boron minerals with caustic soda to yield sodium metaborate, which in the next stage is reacted with hydrogen peroxide to give sodium perborate:

$$Na_2B_4O_7 + 2NaOH + 7H_2O \rightarrow 4NaB(OH)_4$$
$$NaB(OH)_4 + H_2O_2 + H_2O \rightarrow NaBO_3 * 4H_2O$$

The raw materials are the minerals ulexite (a sodium calcium borate), colemanite, and pandermite (or priceite, a calcium borate). Because the precipitation of calcium carbonate generates a substantial amount of waste, higher-purity raw materials are preferred today; examples are borax pentahydrate and higher-gangue minerals such as kernite and tincal.

The boron-containing raw material is placed in a feed vessel with recycled mother liquor and digested with caustic soda at temperatures of 60–90 °C. Any impurities present are removed by either decantation or filtration.[108] The hot concentrated solution contains 600–900 g l^{-1} NaBO$_2$. For crystallization, it is diluted with more mother liquor or water and then cooled.[109]

[105] Etard, Academy of Sciences, Paris, 1880.
[106] S. Tanatar, *Z. Phys. Chem.* 1898, **26**, 132–134. P. Melikoff and L. Pissarjewsky, *Ber. Dtsch. Chem. Ges.* 1898, **31**, 678–680.
[107] Chem. Fabrik Grünau, Patent No. DE 297223, 1912 (K. Arndt).
[108] Interox, Patent No. EP 399593, 1990 (K. Hauchecorne *et al.*).
[109] Laporte Chemicals Ltd., Patent No. GB 761371, 1953 (W. S. Wood and C. S. Godfrey).

The desired product properties, such as grain shape and bulk density, are achieved through the selection of crystallizer, reactor conditions and additives.[110–112] The stability of sodium perborate can be improved, for example, by adding magnesium salts during crystallization.[113]

The sodium perborate tetrahydrate crystallizes out in the temperature range 15–30 °C and is separated from the mother liquor by centrifugation. Drying takes place in fluid-bed dryers at temperature between 40 and 60 °C.[114] The temperature must be held below the melting point of the tetrahydrate (64 °C). The mother liquor is recycled for borax digestion (see Figure 3).

Manufacture of sodium perborate monohydrate. Sodium perborate monohydrate is obtained by the dehydration of sodium perborate tetrahydrate:

$$NaBO_3*4H_2O \rightarrow NaBO_3*H_2O + 3H_2O$$

This operation is usually carried out by continuous contact with warm air in fluid-bed or vibrating-conveyor dryers. To prevent agglomeration of the product bed, the temperature of the feed air is raised from 80 to 160 °C zone by zone.[115,116] Before it exits the drying zone, the monohydrate is cooled with cold air.

Manufacture of anhydrous sodium perborate (Oxoborate). Oxoborate can be produced by removing all the water of crystallization from either sodium perborate monohydrate or tetrahydrate. Dehydration is performed in fluid-bed equipment at air temperatures of 120–160 °C.[117]

2.1.3 Manufacture of Sodium Percarbonate. Sodium percarbonate is formed by the reaction of soda with hydrogen peroxide:

$$Na_2CO_3 + 1.5H_2O_2 \rightarrow Na_2CO_3*1.5H_2O_2$$

Both dry and wet processes are used to produce sodium percarbonate.

Wet Process. Because sodium percarbonate is highly soluble, it is usually prepared by salting out of aqueous solutions with sodium chloride. According to a Peroxid–Chemie patent,[118] a soda/NaCl suspension is added to sodium hexa-metaphosphate and spread out. Aqueous hydrogen peroxide stabilized with magnesium sulfate and water glass is added continuously with stirring and cooling. The crystallized sodium percarbonate is separated from the mother liquor by centrifugation, then dried in a fluid-bed dryer.

Dry Process. The so-called dry processes essentially involve spraying aqueous hydrogen peroxide solution on solid soda; a solid–liquid reaction yields sodium percarbonate.

[110] W. Wöhlk, Th. Adamski, *Chem. Anlagen + Verfahren*, 1973, **2**, 45.
[111] Atochem, Patent No. EP 452164, 1991 (J. Dugua).
[112] H. V. May, Patent No. GB 2188914, 1986 (S. K. May).
[113] Solvay and Cie., Patent No. DE 1138381, 1960 (E. Leblon and H. Lambert).
[114] Kali-Chemie, Patent No. DE 2004158, 1970 (H. Honig *et al.*).
[115] Noury and van der Lande, Patent No. DE 970495, 1953 (D. W. van Gelder).
[116] Kali-Chemie, Patent No. DE 1801470, 1968 (H. Honig *et al.*).
[117] Degussa, Patent No. DE 1257122, 1961 (H. Meier-Ewert *et al.*).
[118] Peroxid-Chemie, Patent No. DE 2328803, 1973 (H. Dillenburg *et al.*).

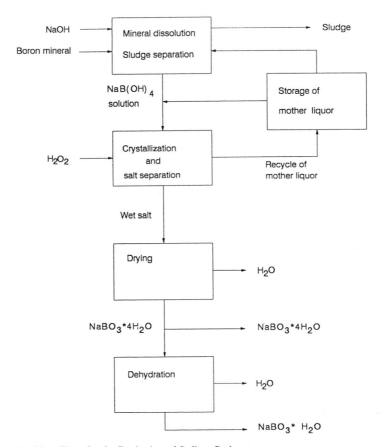

Figure 3 *Flow Sheet for the Production of Sodium Perborate*

In a recent FMC patent,[119] anhydrous soda is treated in a ribbon blender and preheated. An aqueous hydrogen peroxide solution containing hydroxyalkylidene diphosphonic acid is sprayed onto the bed with continuous agitation. In order to hold the reaction temperature around 65–70 °C, hot air is conveyed over the bed of product. Completion of the reaction is followed by cooling; a dry, free-flowing powder is obtained.

Spray Processes. In an Interox process, solutions of sodium carbonate and hydrogen peroxide are sprayed onto a bed of sodium percarbonate nuclei in a fluid-bed granulator. The resulting shell-like pellets are round, compact particles of sodium percarbonate. To prevent plugging of the nozzle and enhance the yield of active oxygen, hexametaphosphate and sodium silicate are added to the soda solution. The product bed is kept in movement by a stream of heated air. Product is continuously withdrawn from the dryer and the desired grain-size fraction is obtained by classification. The fines and ground oversize are recycled to the fluid-bed granulator as nuclei.[120]

[119] FMC, Patent No. EP 429321, 1990 (H. A. Pfeffer and C. Adams).
[120] Interox, Patent No. DE 2733935, 1977 (J. Brichard and J.-C. Colery).

Table 4 *Structural Formulae and Synonyms*

Sodium perborate tetrahydrate

$$2\mathrm{Na}^+ \left[\begin{array}{c} \mathrm{HO} \quad\quad \mathrm{O-O} \quad\quad \mathrm{OH} \\ \diagdown \diagup \quad\quad \diagdown \diagup \\ \mathrm{B} \quad\quad\quad \mathrm{B} \\ \diagup \diagdown \quad\quad \diagup \diagdown \\ \mathrm{HO} \quad\quad \mathrm{O-O} \quad\quad \mathrm{OH} \end{array} \right]^{2-} 6\mathrm{H_2O}$$

Disodium-di-μ-peroxo-bis(dihydroxoborate)hexahydrate
Sodium peroxoborate tetrahydrate

Sodium perborate monohydrate

$$2\mathrm{Na}^+ \left[\begin{array}{c} \mathrm{HO} \quad\quad \mathrm{O-O} \quad\quad \mathrm{OH} \\ \diagdown \diagup \quad\quad \diagdown \diagup \\ \mathrm{B} \quad\quad\quad \mathrm{B} \\ \diagup \diagdown \quad\quad \diagup \diagdown \\ \mathrm{HO} \quad\quad \mathrm{O-O} \quad\quad \mathrm{OH} \end{array} \right]^{2-}$$

Disodium-di-μ-peroxo-bis(dihydroxoborate)
Sodium peroxoborate monohydrate

Anhydrous sodium perborate
Oxoborate $\mathrm{NaBO_3}$
Structure not defined
Sodium borate-boron oxygen radical?

Sodium percarbonate
                                                 $\mathrm{Na_2CO_3 \cdot 1.5H_2O_2}$
Sodium carbonate peroxohydrate

2.1.4 Properties. Structural formulae and synonyms, physical and chemical properties of the various forms of sodium perborate and those of sodium percarbonate are summarized in Tables 4 and 5.

Note: Because hydrogen peroxide decomposes to oxygen and water, this nascent (atomic) oxygen was long thought to be the actual bleaching agent. The expression 'active oxygen' dates from that time and continues in use in product specifications for inorganic peroxy compounds. Anhydrous perborate does not generate hydrogen peroxide in aqueous solution; instead, it spontaneously liberates gaseous (effervescent) oxygen.

2.1.5 Uses. Perborate in aqueous solution liberates hydrogen peroxide, which is a good bleaching and oxidizing agent. In Europe, by far the largest market for heavy-duty powdered detergents, perborate has been the preferred bleaching component for decades. In the washing process, it cleans laundry of bleachable soiling due to coffee, tea, red wine, fruits, and vegetables (catechol tans, gallotannins, carotenoids, anthocyanidins, curcumin, chlorophyll, humic acids, *etc.*). At the same time, it harms neither the fibres nor the dyes in the fabric. But perborate develops its full bleaching power only around 90 °C, as in the laundry boiling. Laundry practices in Europe, however, have shifted to lower temperatures, both on grounds of energy conservation and because many of today's coloured fabrics can withstand temperature of only 30, 40, or 60 °C. If 'bleach activators' such as TAED (tetraacetylethylenediamine) are incorporated into detergent

Table 5 *Properties of Sodium Perborate and Sodium Percarbonate*

	Sodium Perborate			Sodium Percarbonate
	Tetrahydrate	*Monohydrate*	*Anhydrous*	
Molecular formula	$NaBO_3.4H_2O$	$NaBO_3.H_2O$	$NaBO_3$	$Na_2CO_3.1.5H_2O_2$
Molecular mass, $g\,mol^{-1}$	153.9	99.8	81.8	157.0
Active oxygen (theor.), %	10.39	16.03		15.3
Effervesc. oxygen (theor.), %			19.5*	
Bulk density, $kg\,l^{-1}$	0.7–0.85	0.5–0.65	0.30–0.40	0.9–1.1
Density, $g\,cm^{-3}$	1.73	2.12	1.65	2.14
Solubility in water at 20 °C, $g\,l^{-1}$	25.5	17.5	**	*ca.* 140
Time of dissolution at 20 °C, $2\,g\,l^{-1}$	8 min	10 s	**	1.5–2 min
pH-value at 20 °C, $10\,g\,l^{-1}$	10.1–10.4	10.1–10.4	10.1–10.4	10.4–10.6
Melting point, °C	64	NA	NA	NA
Colour	white	white	yellowish	white

*The commercial product contains approx. 13% EO and 2% AO; **immediate release of oxygen.

formulae, good results can be achieved even at these low temperatures. Most of the detergents now in use contain activators.

Chemically, what happens in the alkaline washing bath can be described as follows:

$$NaBO_3*4H_2O \rightarrow NaB(OH)_4 + H_2O_2 + H_2O$$
$$H_2O_2 + OH^- \rightarrow HO_2^- + H_2O$$

Most authors regard perhydroxyl anion as the species active in bleaching.[121-124] Perhydrolysis of the activator TAED yields two molecules of peracetic acid. The oxidation potential of peracetic acid is higher than that of perhydroxyl anion, so that spots are well bleached even at low temperatures.

In countries where cold washing is traditional, such as the USA, the highest usual washing temperature is *ca.* 55 °C. Such regions accordingly did not represent a market for perborate until bleach activators also became available. The activator most commonly used in detergents in the USA is n-NOBS (sodium n-nonanoyloxybenzenesulfonate). Many US households still use hypochlorite for bleaching.

The tetrahydrate was the only form of perborate used in detergents for decades. More recently, the monohydrate has also found use because of a trend, originating

121 D. F. Evans and M.W. Upton, *J. Chem. Soc., Dalton Trans.* 1985, 2525.
122 E. Koberstein and H. Kurzke, *Tenside Deterg.*, 1987, **24**, 210.
123 A. P. James and I. S. MacKirdy, *Chem. Ind.*, 1990, 641.
124 E. Kissa *et. al.*, *J. Am. Oil Chem. Soc.*, 1991, **68**, 532.

in Japan, to market small packets of highly effective detergent called 'compact detergents'. The higher active oxygen content of the monohydrate makes it possible to cut the weight of the bleaching agent by up to a third with no loss of performance.

A heavy-duty European detergent contains between 10 and 25% tetrahydrate or 5–17% monohydrate. Mixtures of the hydrates are often used as well. Levels of bleaching agent are much lower in the USA and Japan: detergents with active oxygen contain *ca.* 5% monohydrate.

In principle, sodium percarbonate can exert the same bleaching action as sodium perborate. Because of its structure (perhydrate addition compound), however, the active oxygen stability of the percarbonate in the detergent package is not satisfactory. Percarbonate is therefore used instead, as the main or sole component of bleach booster, which serves to intensify the bleaching action of the detergent when washing or soaking heavily soiled laundry. Despite the low temperature used in soaking, the long treatment time allows good spot removal.

Dishwasher detergents represent a growing market segment. Strongly alkaline products with active chlorine compounds used to dominate the market, but today more and more formulae are being changed to lower pH values (around 10) and perborate/TAED is being included as bleaching agent. Under these conditions, enzymes can also be included in the formula.

The volume of the denture cleaner market is smaller. Perborate monohydrate is preferred on account of its rapid solubility; in the presence of peroxymonosulfate, this form offers not only an oxidizing and disinfecting action but also effervescence (formation of oxygen from hydrogen peroxide by the stronger oxidizing agent peroxymonosulfate).

In the cosmetics industry, perborate still finds use in dry cold-wave preparations, foot-bath salts, bath salts (oxygen baths), and mouthwashes. In the polymer industry, it can serve as a hardener or an initiator/accelerator.

Anhydrous perborate (tradename Oxoborate) is present in alkaline denture cleaners as a tablet disintegrating agent.

While the bleach consumption in detergents in Europe may be falling slightly as users change over to compact detergents, new potential for growth exists in the developing countries. If machine washing replaces hand washing or active chlorine bleaches give way to active oxygen bleaches, new markets will appear for perborate as well.

2.2 Peroxysulfate Compounds

2.2.1 Introduction. The salts of peroxymonosulfuric acid ($2KHSO_5.KHSO_4.K_2SO_4$) and of peroxydisulfuric (persulfuric) acid [$Na_2S_2O_8$, $K_2S_2O_8$, $(NH_4)_2S_2O_8$] have achieved commercial importance.

The pure acids are unstable and difficult to synthesize; they are industrially significant only as intermediates in the production of the salts.

World production capacity for persulfates is estimated at 74 000 t (1992). The main producers are Peroxid–Chemie, Degussa, Mitsubishi Gas, FMC, and L'Air Liquide.

Total capacity for potassium peroxymonosulfate triple salt is estimated at 12 000 T (1992). The most important producers are Degussa, Peroxid–Chemie, and DuPont. This product reaches the market under the names Caroat (Degussa), Curox (Peroxid–Chemie), and Oxone (DuPont).

2.2.2 Peroxydisulfate. Manufacture. All persulfates are produced by electrochemical processes. The electrolysis cells were originally developed for the manufacture of hydrogen peroxide.[125]

The persulfate ion is obtained solely by the anodic oxidation of sulfuric acid:

$$2SO_4^{2-} - 2e^- \rightarrow S_2O_8^{2-}$$

$$2HSO_4^- - 2e^- \rightarrow S_2O_8^{2-} + 2H^+$$

Many parameters influence the reaction at the electrode. These include temperature, concentration, anode material, voltage, current density, and the presence of foreign ions. The most important secondary reactions generate oxygen and ozone. Oxygen formation is generally suppressed by the high electrochemical overvoltage of O_2 on the smooth platinum electrode. Another process that lowers the current efficiency is the formation of Caro's acid by hydrolysis of persulfuric acid followed by anodic decomposition to yield oxygen:

$$H_2SO_5 + H_2O - 2e^- \rightarrow O_2 + H_2SO_4 + 2H^+$$

Ozone formation depends chiefly on the current density; the loss of current efficiency due to ozone formation increases with the current density.

Ammonium persulfate, by virtue of its solubility properties and also on electrochemical grounds, is most easily obtained by crystallization from the cell liquor.[126]

The cell features a lead cathode, a platinum anode, and an alumina diaphragm. Sulfuric acid is fed into the cathode compartment; it is diluted and partially neutralized by diffusion of H_2O and in migration of ammonium ions. The catholyte is treated with ammonia to render it weakly alkaline, recycled mother liquor from crystallization is added to increase the volume, and the resulting liquor is fed to the anode compartment. Anodic oxidation yields a solution of *ca.* 26% $(NH_4)_2S_2O_8$, from which a coarse crystalline product is obtained after evaporation and cooling.

The alkali salts, like the ammonium salt, can be produced by a direct electrolytic process.[127–129] The preferred method, however, has been an indirect one via the ammonium salt:[130–132]

[125] 'Ullmann's Encyclopedia of Industrial Chemistry', 5th Edn., Verlag Chemie, Weinheim, Vol. 19, 189–191.
[126] Mitsubishi Gas, Patent No. JP 7309277, 1973 (T. Marushima *et al.*).
[127] Peroxid-Chemie, Patent No. DE 2346945, 1973 (E. Rossberger).
[128] J. Balej, Patent No. CS 145656, 1972.
[129] W. Thiele and K. Wildner, Patent No. DD 119197, 1976.
[130] FMC, Patent No. US 3954952, 1976 (M.J. McCarthy *et al.*).
[131] FMC, Patent No. DE 1057588, 1959 (G.G. Crewson and J.R. Ryan).
[132] 'Ullmann's Encyclopedia of Industrial Chemistry', 4th Edn., 17, 723.

Table 6 *Properties of Peroxydisulfate*

	Ammonium persulfate	Sodium persulfate
Molecular formula	$(NH_4)_2S_2O_8$	$Na_2S_2O_8$
Molecular mass, $g\,mol^{-1}$	228.2	238.1
Active oxygen (theor.), %	7.0	6.7
Bulk density, $kg\,l^{-1}$	0.95–1.05	1.25–1.30
Solubility in water at 20 °C, $g\,l^{-1}$	559	556
pH-value, $250\,g\,l^{-1}$ at 20 °C	2.3	4.3
Colour	white	white

$$(NH_4)_2S_2O_8 + 2NaOH \rightarrow Na_2S_2O_8 + 2H_2O + 2NH_3$$

$$(NH_4)_2S_2O_8 + 2\,KHSO_4 \rightarrow K_2S_2O_8 + 2\,NH_4HSO_4$$

Uses. The most significant technical application of peroxodisulfates is the initiation of the polymerization of organic monomers in an aqueous phase. The monomers are either sufficiently soluble in water and precipitate during polymerization (*e.g.* acrylonitrile), or polymerization takes place in an aqueous emulsion of the monomers. This applies to the polymerization of styrene–butadiene, vinyl chloride, vinyl acetate, and acrylic esters.

The peroxodisulfate initiates the polymerization process by decomposition into radicals. Special additives accelerate the decomposition, thus permitting high rates of polymerization even at low temperatures. The amount needed varies between 0.5% and 2% peroxodisulfate related to the monomer volume. Peroxodisulfates are used in large quantities for etching and pickling of metal surface (*e.g.* copper and brass). The etching of thin copper coatings is the central process of the production of printed circuit boards.

Of minor importance is the application of peroxodisulfates in the textile industry for the development of dyes and the make-up of bleaching baths in the production of colour prints.

2.2.3 Peroxymonosulfate. Manufacture. A Peroxid–Chemie patent[133] describes a continuous process for the production of potassium peroxymonosulfate triple salt.

In the first reaction stage, a solution of *ca.* 50 wt. % Caro's acid is obtained by reacting concentrated sulfuric acid with 85 wt. % aqueous hydrogen peroxide solution:

$$H_2SO_4 + H_2O_2 \rightarrow H_2SO_5 + H_2O$$

The Caro's acid is metered into a working solution that contains $KHSO_5$, H_2SO_4, and K_2SO_4, which is partially neutralized with KOH:

$$2H_2SO_5 + H_2SO_4 + K_2SO_4 + 3KOH \rightarrow 2KHSO_5.KHSO_4.K_2SO_4 + 3H_2O$$

The heat of reaction and the product water are removed in a vacuum evaporation

[133] Peroxid-Chemie, Patent No. DE 3427119, 1984 (W. Reh and S. Schelle).

Table 7 *Properties of Peroxymonosulfate*

Molecular formula	$2KHSO_5*K_2SO_4*KHSO_4$
Molecular mass, $g\,mol^{-1}$	614.8
Active oxygen (theor.), %	5.2
Bulk density, $kg\,l^{-1}$	0.95–1.25
Cryst. Density, $g\,cm^{-3}$	2.34
Solubility in water at 20 °C, $g\,l^{-1}$	260
pH-value, $10\,g\,l^{-1}$ at 20 °C	2.2
Melting point	NA
Colour	white

system. The crystal suspension is separated in a centrifuge, and the filtrate is recycled to the production loop. The filter cake is washed in the centrifuge. The moist salt is dried in a fluid-bed dryer and forwarded to packaging via a silo.

Uses. The peroxymonosulfate triple salt is a strong oxidizing and bleaching agent. Its principal application is in denture cleaners, where its oxidizing, bleaching, and disinfecting action makes it effective in removing deposits and discolorations. It can serve as a tablet disintegrating agent in the presence of sodium perborate monohydrate; in an alkaline environment, peroxymonosulfate oxidizes the hydrogen peroxide liberated from the perborate, generating gaseous oxygen, which also exerts a mechanical cleaning action on the dentures.

Other applications are in acidic household and sanitary cleansers.

In the metal fabricating industry, Caroat is employed as a mild etching and pickling agent for the surface treatment of copper and copper alloys. A major use is the micro-etching of copper conductors in printed circuit board production.

Wastewaters in the electroplating industry are polluted with heavy-metal cyanides. Treatment with Caroat quickly and completely detoxifies even complex cyanides.

2.3 Other Peroxo Compounds

There are only a few commercially important peroxides besides the peroxy compounds already described.[134–136]

Sodium peroxide, Na_2O_2, was formerly produced on a large scale but has now had its applications taken over by hydrogen peroxide. Today it is still used as a reagent in analytical chemistry and as an oxygen source in cartridges for breathing devices. Its industrial synthesis is based on the oxidation of sodium, with sodium oxide as intermediate.

Calcium peroxide, CaO_2, is poorly soluble in water and forms hydrogen peroxide only slowly in aqueous solution. For this reason, it is employed particularly when slow evolution of oxygen is desired. This peroxide finds use as a hardening accelerator in sealing compounds, for the disinfection of seeds, in

[134] Kirk-Othmer, Enc. Chem. Techn., 3rd Edn., 17, (1982), 2–6, 19.
[135] 'Ullmann's Encyclopedia of Industrial Chemistry', 5th Edn., Verlag Chemie, Weinheim, Vol. 19, 179–181.
[136] Römpp Chemie Lexikon, 9th Edn., (1989–1992), 346, 558, 2595, 2933.

environmental protection, and as a dough improver in the baking industry.

Zinc peroxide, ZnO_2 produced in a similar way to calcium peroxide, is likewise poorly soluble in water and comes into use chiefly in the pharmaceutical industry and in rubber manufacture.

Barium and strontium peroxides, BaO_2 and SrO_2, are utilized in pyrotechnics. While barium peroxide is produced by atmospheric oxidation of barium oxide at 500 °C, the synthesis of strontium peroxide involves the use of oxygen under high pressure.

Magnesium peroxide, MgO_2, is made from magnesium oxide or hydroxide with hydrogen peroxide. It finds use in veterinary and human medicine.

CHAPTER 8

Hydrofluoric Acid, Inorganic Fluorides, and Fluorine

R. L. POWELL AND T. A. RYAN

1 INTRODUCTION

Fluorine can be accurately described as the enabling element of modern technology, playing key roles in industries as diverse as nuclear power, electronic microchips, refrigerants, pharmaceuticals, agricultural chemicals, high performance polymers, and steel making. All depend upon the inorganic fluorine chemicals industry for their raw materials, notably calcium fluoride, hydrofluoric acid, and elemental fluorine itself.

2 INORGANIC FLUORINE CHEMICALS: A BRIEF HISTORY

An excellent history of early fluorine chemistry, written by R.E. Banks, appears in 'Fluorine: The First Hundred Years' which was published in 1986 to commemorate the centenary of the first unequivocal isolation of elemental fluorine by Moissan. The following section is based in part on information from this article.

The ability of fluorspar, naturally-occurring calcium fluoride, to lower the melting point of minerals, *i.e.* to act as a flux, was known as early as the Sixteenth century. Indeed the name 'fluorspar' is derived from the Latin word 'fluor' meaning 'to flow'. Crude hydrofluoric acid was first prepared by K.W. Scheele (1742–1786) in 1771 by repeating experiments of A.S. Marggraf (1709–1782). The latter was surprised that the glass retort in which he had heated a mixture of fluorspar and sulfuric acid 'was corroded into holes in several places', an observation to be repeated by generations of fluorine chemists who inadvertently have produced hydrofluoric acid in their experiments.

Following contributions by Meyer, Priestley, Wenzel, Wiegleb Buchholz, Thenard, and Gay-Lussac, the next major step forward was the clear identification by Humphry and John Davy and Ampère in 1813 of fluorine as an entity comparable to chlorine and oxygen, although it had not been isolated. Most importantly they resolved any confusion about the nature of hydrofluoric acid. However, the preparation of pure anhydrous hydrogen fluoride, on which the ultimate isolation of fluorine would depend, was only achieved over 40 years later by Frémy who heated potassium hydrogen fluoride slowly to about 500 °C.

Industrially, perhaps the earliest application of hydrofluoric acid was the etching of glass but, although in modest production in the late Nineteenth century, its use did not expand greatly until the introduction of the chlorofluorocarbon refrigerants in the 1930s. In contrast fluorspar has been used in considerable quantities from the 1860s for producing steel.

Despite periodic attempts, starting with Humphry Davy in 1813, to isolate elemental fluorine the first accepted generation of the gas was achieved by Moissan in 1886. For many years fluorine remained an academic curiosity, but perusal of old ICI records dating from the mid 1930s to the mid 1940s suggests that elemental fluorine was seen as the key long-term technology for the generation of fluorinated refrigerants rather than the Swarts chlorine/fluorine exchange reaction (*vide infra*). Presumably an analogy was being drawn with the rapidly growing chlorocarbon industry based on elemental chlorine. Clearly this did not happen for reasons which will be described later. However, fluorine found its first real large scale application in the production of uranium hexafluoride for the diffusive separation of fissionable ^{235}U from ^{238}U required to build the first atomic bombs. Code named 'the Manhattan Project' this work accelerated the pace of fluorine chemistry in the search for materials and fluids which could resist highly corrosive F_2 and UF_6. Perhaps not surprisingly perfluoro-organic compounds emerged as the most effective. The products and processes developed under the Manhattan Project have formed the basis of a significant proportion of the post-war speciality fluorochemical industries which have underpinned the nuclear, electronic, and aerospace industries.

The implication of chlorofluorocarbon and hydrochlorofluorocarbon refrigerants, aerosol propellants, and foam blowing agents in the reduction of stratospheric ozone is currently generating a revolution in industrial fluorine chemistry as profound as that engendered by the Manhattan Project in the 1940s. Whatever is written now will undoubtedly need revising when the full effects of the current changes can be properly appreciated, probably sometime in the first decade of the next century.

3 OCCURRENCE OF FLUORINE

Fluorine is widely distributed, accounting for about 0.065% of the Earth's crust, and among the elements is thought to be about thirteenth in order of terrestrial abundance. Although the major industrial source of fluorine is fluorspar, containing about 49% fluorine, other fluorine containing ores are known and are listed in Table 1.

Although the largest reserves of fluorine are in phosphate rock deposits, the low concentration makes it uneconomic to recover fluorine from this source while reserves of fluorspar are still available. The previous edition of this article, written in the mid 1970s, anticipated that by the year 2000 most of the world's fluorspar would have been consumed, necessitating a move to fluorophosphate. In fact the discovery of new fluorspar reserves over the past 15 years means that there will be sufficient fluorine from this source to take us well into the next century. For

Table 1 *Fluorine Containing Minerals*

Mineral	Composition	% Fluorine Content
Fluorspar (fluorite)	CaF_2	49
Fluorapatite	$Ca_5(PO_4)_3F$	3–4
Cryolite	$Na_3(AlF_6)$	45
Sellaite	MgF_2	61
Villiaumate	NaF	55
Bastnaestite	$(Ce,La)(CO_3)F$	9
Topaz	$Al_2SiO_3(OH,F)_4$	
Fluorohydrosilicates		

Table 2 *World Production of Fluorspar*

Country	Production/(kt)	
	1990	*1991*
Mexico	634	386
Western Europe	640	570
Central and Eastern Europe	460	425
Middle and Far East	2300	2170
Africa	565	515
South America	90	90
USA	64	58
Canada	17	0
Total	4770	4214

fluorocarbon refrigerant manufacture the spar needs to be low in impurities such as silica and carbonate. In particular the trace levels of arsenic should be low, otherwise this element can be carried forward as its fluoride to the refrigerant producing plants where it would poison the inorganic antimony halide or chromia catalysts used in these processes (see Section 5.5).

4 FLUORSPAR

4.1 Structure of the Fluorspar Industry

The world output of fluorspar is obviously dropping rapidly in response to a combination of the economic recession and the reduced demand for chlorofluorocarbons (CFCs) as they are progressively phased out in line with the Montreal Protocol. The level of production in 1991 is comparable with that in 1975, which in 'The Modern Inorganic Chemicals Industry' (RSC Special Publication No. 31, 1977) was quoted as 4280 kt.

The UK is a substantial fluorspar producer with an annual production of about 80 kt from deposits situated in the Derbyshire Pennines and Weardale in County Durham. The world's biggest producer is China with estimated output in 1991 continuing at 1500 kt, similar to its output in 1990.

Fluorspar is a translucent or transparent glassy material, varying in colour from white, amber, green, and blue to purple. The latter grades are known in the UK as 'Blue John' which is used for the manufacture of decorative items such as jewellery and ornaments. The most distinctive physical property of fluorspar is its ability to glow with a blue or purple light when illuminated by shorter wavelength light, especially UV. The mineral has thus given its name to the general phenomenon known as 'fluorescence'.

Fluorspar is invariably associated with other minerals, principally quartzite, but also calcite, dolomite, baryite, and galena. For most purposes the mining of the mineral is followed by 'benefication' to remove other mineral impurities. The degree of concentration depends upon the final application and of course determines the price. The cheapest grade is used in metallurgy and contains 60–80% calcium fluoride resulting from a crude gravity concentration. The next most expensive is the ceramic grade used in the glass industry containing 85–95% calcium fluoride, concentrated by flotation. Acid grade, used, as its name suggests, for HF production, is also produced by flotation and contains a minimum of 97% calcium fluoride.

After relatively high consumption in 1989, fluorspar demand in the early 1990s is being subjected to a variety of adverse factors. In particular the world-wide recession has led to reduction of fluorspar use in steel manufacture, exacerbating an existing long-term trend resulting from the improving efficiencies in this industry. The rapid phase-out of the CFCs is also reducing demand for acid grade fluorspar. The replacement products are likely to be needed in smaller tonnages and this, coupled with the considerable product recycling and reduced leakage, is also predicted to reduce fluorspar requirements. In the longer term this may be offset by increased demand from the developing countries, such as China and India, where more extensive refrigeration will be a significant contributor to improving their peoples' quality of life.

In view of this uncertainty, trends in fluorspar production and usage over the next ten years are difficult to predict with accuracy. For a more detailed discussion the reader is recommended to consult articles listed in the bibliography.

4.2 Fluorspar Applications

The metallurgical industry remains a major consumer of fluorspar, accounting for about 40% of the total mined. In the production of steel it facilitates the separation of the molten metal from the slag by reducing the latter's viscosity. It also helps the removal of impurities such as phosphorus, sulfur, and silicon as their oxides which are generated during oxygen treatment. The amount of fluorspar required varies according to the process. The open hearth process typically requires up to 3.5 kg t^{-1}, the basic oxygen process up to 6.0 kg t^{-1} and the electric arc process up to 2.4 kg t^{-1} of steel. The average is estimated to be 1.6 to 2.2 kg t^{-1}. The UK has a low consumption of 1.5 kg t^{-1}. In contrast Eastern Europe, which has older steel-making technologies, is still a major fluorspar consumer; but, as its industry is uprated, requirements will also fall in these countries. For example, Polish imports have dropped from a high of $20\,000 \text{ t year}^{-1}$ to 7000 t year^{-1} in 1990 as its technology has improved. Fluorspar demand tends to follow ferroalloy production which can consume up to $100 \, kg \, t^{-1}$ of alloy, obviously much higher than that of common steels.

The release of toxic fluorine compounds into the environment and an economic need to reduce the corrosion of furnace linings has led to a prolonged search for alternative fluxes, for example alumina, colemanite, lime, topaz, and ilmenite. However, none has so far proved to be as good as fluorspar in all applications; but they are replacing fluorspar for some uses.

Fluorspar is also used in fluxes in the refining of copper, tin, zinc, antimony, silver, gold, nickel, and magnesium. Other metallurgical uses of fluorspar include welding and brazing, about 15 000 tonnes being used in the USA alone in 1990 for this purpose. The fluorspar content in welding fluxes can vary from as low as 2% up to 40%.

Fluorspar has also been used since antiquity in the glass industry. When added to silicate melts, fluorides separate as fine crystals during cooling giving a milky appearance to the glass. In small amounts, the fluorides are also used to control the viscosity, coefficient of thermal expansion, and other properties of glasses.

5 HYDROFLUORIC ACID

5.1 The Production of Hydrogen Fluoride

Most industrially-produced fluorine chemicals, organic or inorganic, can be traced back to anhydrous hydrogen fluoride or its aqueous solution, hydrofluoric acid. Clearly it is the key to the fluorochemical industry, both organic and inorganic. Hydrogen fluoride is manufactured by the reaction of oleum with acid grade fluorspar. The reaction can be summarized simply by equation (1) but, in reality, the technology is very sophisticated.

$$CaF_2 + H_2SO_4 \rightarrow CaSO_4 + 2HF \qquad (1)$$

Figure 1 represents a typical modern continuous production HF plant based on an externally-heated rotary kiln. A widely-used technology is that developed by Buss AG (Switzerland). For a single unit producing up to $20\,000 \, t \, year^{-1}$, the steel kiln would be about 30 m long, the whole rotating at about one revolution per minute.

The typical raw material requirement per tonne of hydrofluoric acid is 2.2 tonne of fluorspar and 2.6 tonne of sulfuric acid which are fed continuously into a pre-reactor, from which they pass as a paste to the rotary kiln main reactor. Gases evolved in the kiln leave through the feed end into a disengaging section, while the calcium sulfate solids are discharged at the downstream end.

The remainder of the plant consists of facilities for pulverizing the feed spar, neutralizing and handling the calcium sulfate, scrubbing, cooling, condensing, and purification of the HF product, and elimination of compounds such as silicon(IV) fluoride from the vent gases.

The hot reactor gases, *i.e.* the crude hydrogen fluoride, enter a column (the pre-condenser) where they are cooled and scrubbed free of entrained solids and sulfuric acid mist using incoming sulfuric acid. This acid, after the addition of

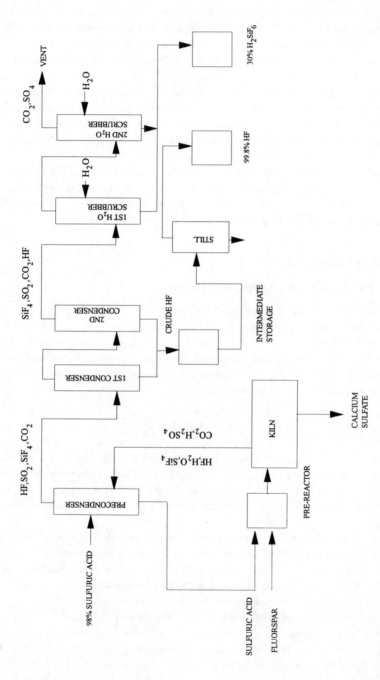

Figure 1 *Flow Diagram for Manufacture of Hydrofluoric Acid*

oleum to combine with any water, then becomes the feed to the pre-reactor. The HF vapour leaving the pre-condenser is condensed from the gas stream using refrigeration, to give a product of 99.9% purity. Gaseous effluent is now largely SiF_4, which is absorbed into water and recovered as 30% fluorosilicic acid. In the past the calcium sulfate by-product, after neutralization with lime, was mainly sent for landfill, apart from a small quantity sold for the manufacture of anhydrite cement. Significant quantities are now being used to produce a self-levelling floor coating; for example ICI Chemicals and Polymers Ltd. sell it for this purpose under the trade name 'Andricite'[TM].

5.2 Hydrogen Fluoride Producers

Table 3 summarizes the hydrogen fluoride manufacturers. The list is based only on data available in the open literature. Some tonnages may have been estimated and some producers may have been omitted. However, despite possible inaccuracies the table does provide an indication of the extensive world-wide manufacture of hydrogen fluoride.

5.3 Safety

Years of experience in the manufacture and use of hydrogen fluoride have shown that it can be handled safely, provided its hazards are recognized and the necessary precautions are taken. The anhydrous acid, together with the higher concentrations of aqueous acid, are extremely corrosive to skin, eyes, mucus membranes, and lungs. Sometimes, burns of the skin may not be immediately evident and may be manifested later by deep seated ulceration, very slow to heal, and accompanied by intense pain which may not develop until several hours after exposure. Anybody contemplating working with HF in any form, fluorine gas, or any fluoride capable of generating HF under the conditions of use, must first consult organizations expert in the handling of these compounds. When buying HF, fluorine gas, or corrosive fluorides the specific information will be provided by the supplier as required by modern legislation. Special consideration must be given to the use of appropriate protective clothing, the containment of the chemicals, equipment decontamination, and the disposal of residues. In the case of an exposure, approved first aid procedures must be followed and rapid access provided to medical experts knowledgeable in the treatment of HF burns.

5.4 The Properties of Hydrogen Fluoride

Anhydrous hydrogen fluoride is a colourless, fuming liquid with a boiling point of 19.4 °C, which is much higher than the boiling points of the other hydrogen halides (HCl −83.7 °C; HBr −67 °C; HI −36.4 °C) despite their higher molecular weights. The reason for this apparent anomaly lies in the strong hydrogen bonding in HF, comparable to that in water, which also has a high boiling point. Hydrogen fluoride has a very high affinity for water with which it forms an azeotrope boiling

Table 3 *HF Manufacturers 1990*

Manufacturer	Country	Capacity/$(kt\,year^{-1})$
AKZO	Netherlands	10
AlliedSignal	Canada	47
	USA	104
Asahi Glass	Japan	25
Elf-Atochem	USA	24
	France	40
Australian Fluorine Chemicals	Australia	5
Bayer	Germany/Brazil	70
Derivados del Fluor	Spain	30 (50% Bayer)
Central Glass	Japan	13
China	Beijing	10
	Xaingxiang	15 (inc. AlF_3)
Daikin	Japan	12
Du Pont	USA	68
Fluorex	Mexico	22
Fluorosid	Italy	18
Formosa Plastics	Taiwan	32
Gujarat Fluorochemicals	India	3
Hashimo	Japan	50
ICI Chemicals & Polymers	UK	50
Industrias Quimicas de Mexico	Mexico	10
Laporte Industries	UK	12 (aqueous)
Lap Tan	Thailand	15
Ausimont	Italy	50
Morita	Japan	28
Navin Fluorine	India	5
Nippon Light Metal	Japan	10
Pechiney	France	53
Quimica Fluor	Mexico	60
Quimobasicos	Mexico	5
Rhône Poulenc	France/UK	20 (total)
Riedel de Haan	Germany	12
Solvay Fluor	Germany	25
SGING	Greece	(captive) 10
Tanfac	India	big

at 112 °C at 1 atm. and containing 38% by weight of HF. Table 4 lists the values of selected physical properties of HF.

5.5 Uses of Hydrogen Fluoride

5.5.1 Fluorocarbon Manufacture. By far the largest use for anhydrous hydrogen fluoride is the manufacture of fluorine-containing refrigerants, foam blowing agents, solvents, halon fire-fighting agents, and, in the past, for aerosol propellants. The implication of chlorine and bromine containing fluorocarbons, especially the fully halogenated CFCs, in the destruction of stratospheric ozone has led to an international agreement, known as the Montreal Protocol, which mandates a restriction in the production of CFCs leading to their eventual complete

Table 4 *Selected Physical Properties of Anhydrous HF*

Property	Value
Formula weight	20.006
Boiling point (1 atm.), °C	19.54
Melting point, °C	−83.55
Liquid density at 25 °C, kg l^{-1}	0.9576
Vapour pressure at 25 °C, atm.	1.214
Heat of Formation, ideal gas at 25 °C, kJ mol^{-1}	−272.5
Critical temperature, °C	188
Critical pressure, bar	64.16
Critical density, kg l^{-1}	0.29
Surface tension at 0 °C, mN m^{-1}	10.1
Refractive Index, liquid at 25 °C(58 930 nm)	1.1574
Viscosity, liquid at 0 °C, Pa s	0.256
Bond length, H–F, nm	9.2

phase-out. According to the most recent revision of the Protocol (November 1992) this will occur at the end of 1995. The environmental aspects of fluorocarbons have been described extensively elsewhere, so this important topic will not be discussed further here. Instead, the fluorocarbon replacements for the CFCs and hydrofluorocarbons (HCFCs) will be described, as far as they have been identified at the time of writing.

Table 5 summarizes the major chlorine containing fluids to be replaced and their proposed replacements. HFC 134a has been accepted world-wide as the replacement for CFC 12. An HFC 32/134a/125 blend is the most likely replacement for HCFC 22.

The routes to the new refrigerants are obviously still under development; however, two, fundamentally different, routes have emerged for the production of HFC 134a and are summarized in reaction Schemes 1 and 2.

Both steps are carried out in the vapour phase using a fluorinated chromia solid-phase catalyst. However, the second step is equilibrium-limited and thus is pushed to the left-hand side by the HCl generated by the first step. Various processes based on this chemistry have been described in the patent literature, including both single-reactor and double-reactor systems. In one version of the latter 1,1,1-trifluoro-2-chloroethane is generated in the first reactor and separated from by-product hydrogen chloride, before being fed to a second reactor with more hydrogen fluoride to produce 134a. Despite being limited to conversions between 10 and 20%, which requires a large recycle, the selectivity of this chemistry is high, producing few by-products.

In the alternative 134a process (Scheme 2) the first step is a standard vapour phase fluorination over a catalyst such as fluorinated alumina.

In the second step 1,1-dichlorotetrafluoroethane is preferentially hydrogenated over a catalyst such as Pd on charcoal at 240 °C to HFC 134a, conditions where its isomer is unreactive. After separation from the reaction products the 1,2-dichlorotetrafluoroethane isomer can then be converted into an isomer mixture for re-feeding to the hydrogenator.

Table 5 *CFCs and their Proposed Replacements*

Existing Product	Applications	New Products
CFC 12 (CCl_2F_2)	Refrigerant Polystyrene foam blowing	HFC 134a (CF_3CH_2F)
CFC 11 (CCl_3F)	Polyurethane foam blowing	a) HCFC 141b, (CCl_2FCH_3)* b) HCFC 22 ($CHClF_2$)*/ 142b($CClF_2CH_3$)* c) HFC 134a d) Cyclopentane
CFC 113 ($CClF_2CCl_2F$)	Solvent	a) HCFC 225cb, ($ClCF_2CF_2CHClF$)* b) water/surfactants c) perfluorocarbon/surfactants d) perchloroethene

Refrigerant 502
Azeotrope of
HCFC 22/CFC 115
($CHClF_2/CF_3CF_2Cl$)
	Supermarket refrigeration
	Japanese domestic fridges
		a) HFC 32/HFC 125/HFC 134a
		  ($CH_2F_2/CF_3CF_2H/CF_3CF_2H/$
		  CF_3CH_2F)
		b) HFC 143a/HFC 125
		  (CH_3CH_3/CF_3CF_2H)
		c) HCFC 22*/propane/
		  PFC 218 ($CF_3CF_2CF_3$)

HCFC 22 ($CHClF_2$)	Air conditioning	a) HFC 32 b) HFC 32/HFC 134a c) HFC 32/HFC 134a/HFC 125 d) HFC 32/HFC 125 e) HFC 143a/HFC 125/HFC 134a

*Interim HCFC products which will be ultimately phased out.

$$CCl_2:CHCl + 3\,HF \longrightarrow CF_3CH_2Cl + 2\,HCl$$

$$CF_3CH_2Cl + HF \longrightarrow CF_3CH_2F + HCl$$

Scheme 1

$$CCl_2:CCl_2 + 4\,HF \longrightarrow CClF_2CClF_2 + CF_3CCl_2F + 4\,HCl$$

$$CClF_2CClF_2 + CF_3CCl_2F + H_2 \longrightarrow CF_3CH_2F + CClF_2CClF_2 + 2\,HCl$$

$$CClF_2CClF_2 \longrightarrow CF_3CCl_2F + CClF_2CClF_2$$

Scheme 2

5.5.2 Anhydrous Hydrofluoric Acid as an Electrolyte. Electrochemical fluorination (ECF), discovered in the 1940s by Simons, involves the electrolysis of an aliphatic hydrocarbon based substrate in liquid anhydrous HF in an undivided cell using nickel electrodes at typical anode potentials in the range 5 to 5.5 v and at a temperature between 0 and 20 °C. The process was pioneered industrially by 3M

(USA) and subsequently introduced by Miteni (originally Rimar, Italy) and Dai Nippon Ink (Japan).

During the process, the hydrogen atoms of the original hydrocarbon are replaced by fluorine atoms in an anodic process which does not appear to involve the production of elemental fluorine. The process is particularly used for the conversion of $C_8H_{17}SO_2F$ into $C_8F_{17}SO_2F$ which is an intermediate for surfactants, textile treatments, and other surface coatings. $C_7F_{15}COF$, also produced by ECF, is converted to $C_7F_{15}CO_2NH_4$ the key surfactant for poly(tetrafluoroethene) manufacture. Major co-products from these electrochemical fluorination processes are perfluorinated alkanes and ethers mainly boiling in the range from room temperature to $100\,°C$, which are used in a variety of applications where their high thermal and chemical stabilities are sufficiently valuable to offset their high prices relative to other, less stable organic fluids. For example, perfluorinated fluids are used for the 'thermal shock' testing of microchips, for cooling super computers, and in the formulation of synthetic oxygen carriers, so-called 'artificial blood'. The electrochemical fluorination of hydrocarbon substrates to produce specific perfluorinated fluids is also carried out, for example the conversion of $(n\text{-}C_3H_7)_3N$ into $(n\text{-}C_3F_7)_3N$.

5.5.3 Aromatic Fluorocarbons. Hydrofluoric acid is used as the solvent and reactant for the production of fluoroaromatics in the Balz–Schiemann diazotization reaction, the major method for the production of fluoro-aromatics, as shown in Scheme 3.

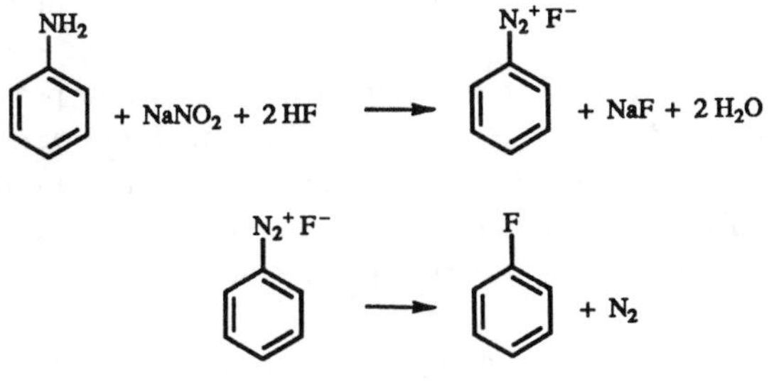

Scheme 3 *Diazotization in liquid HF*

5.5.4 Other Uses. Hydrogen fluoride is generated and used captively by aluminium producers to form 'synthetic cryolite', $Na_3(AlF_6)$ which is used as the fused salt solvent to dissolve alumina for the production of aluminium. About $3.5\,kg$ of cryolite are required for each tonne of aluminium.

Anhydrous HF is used as an acid catalyst for a variety of dimerization, disproportionation, and isomerization reactions in the petroleum industry. In particular it is used to convert C_4 feed stocks into highly branched C_8 alkanes which are essential for the blending of unleaded, high octane petroleum. Scheme 4 illustrates this type of chemistry.

$$(CH_3)_2C{:}CH_2 + HF \longrightarrow (CH_3)_3C^+ + F^-$$

$$(CH_3)_3C^+ + (CH_3)_2C{:}CH_2 \longrightarrow (CH_3)_3CCH_2C^+(CH_3)_2$$

$$(CH_3)_3CCH_2C^+(CH_3)_2 \longrightarrow (CH_3)_3CCH{:}C(CH_3)_2 + H^+$$

Scheme 4

Hydrogen fluoride is superior to other acid catalysts such as sulfuric in this application because it combines good catalytic activity with ease of product separation and low by-product formation. Only 0.068 kg of HF are required per barrel of product compared to 0.23 kg of sulfuric acid.

Miscellaneous uses of hydrogen fluoride include the removal of scale from stainless steels, oil-well acidification to increase the rate of flow as the well ages, the etching of electronic chips, and uranium hexafluoride production (Section 6). The long-standing application of hydrofluoric acid to glass etching continues, most familiarly in the production of frosted light bulbs. Hydrogen fluoride is also used in the manufacture of other inorganic fluorides (Section 6).

6 INORGANIC FLUORIDES

The vast majority of the fluorine compounds of commerce are prepared in one way or another from the primary inorganic fluoride source, calcium fluoride, the importance of which has been reflected in the earlier sections of this chapter. Owing to the dominance of hydrogen fluoride in the industry, the inorganic fluorides constitute the larger proportion of manufactured fluorine-containing materials.

Fluorine combines with more elements than any other in the periodic table, and more of the higher oxidation states can be realized with fluoride than with any other anion. In addition, because of the small size of the fluoride anion, high co-ordination numbers are frequently exhibited in fluorine-containing complexes. Thus, the number of fluorides of any particular element is potentially higher than for all other anions.

The bond strengths in fluorine compounds can range from the weakest to the strongest of any combination of elements yet known. The properties and corresponding applications of inorganic fluorides are therefore extremely diverse, ranging from highly reactive materials such as chlorine(III) fluoride and tungsten(VI) fluoride to highly stable compounds such as sulfur(VI) fluoride and graphite fluoride. Inorganic fluorides are employed in the defence industry and, by contrast, in medicine and dentistry. They are also used in traditional industries, such as ceramics and steel-making and, by contrast, in advanced, sophisticated industries, such as aerospace and laser technology.

The order of description of materials in this section follows the logic of the periodic table rather than reflecting the current commercial importance of the compounds described. Mostly, only those inorganic fluorides which are produced on any significant scale are described. The health and safety issues pertaining to inorganic fluorides are outside of the scope of this chapter and the reader is directed to other sources, such as the technical literature provided by commercial suppliers, for information concerning occupational exposure to these compounds. The intrinsic scientific and industrial interest in inorganic fluorides has been reflected in a recent book, 'Inorganic Solid Fluorides', and gives some tantalizing hints to the promising new applications of inorganic fluorides that may be expected in our increasingly technologically sophisticated world.

6.1 Group 1 Fluorides

The first member of the Group 1 series of fluorides, hydrogen fluoride, has been described in detail in Section 5.

The remaining Group 1 (alkali metal) fluorides, comprising those of lithium, sodium, potassium, rubidium, and caesium, have been most commonly prepared by the reaction of dilute aqueous hydrofluoric acid with the corresponding alkali metal hydroxide or carbonate, according to equations (2) and (3).

$$MOH + HF \rightarrow MF + H_2O \tag{2}$$
$$M_2(CO_3) + 2HF \rightarrow 2MF + CO_2 + H_2O \tag{3}$$

The fluorides of sodium, lithium, and potassium are the most important, in commercial terms, of the alkali metal series with those of caesium and rubidium being produced in only small quantities. All of the Group 1 metal fluorides are white or clear crystalline compounds in their pure state and are toxic by ingestion or by dust inhalation.

6.1.1 Lithium Fluoride. Lithium fluoride, LiF, has a low solubility in water and is hydrolysed, releasing hydrogen fluoride, only at high temperature – more akin to the Group 2 metal fluorides than the other fluorides of the alkali metals.

The global production of lithium fluoride is about 200 t year^{-1} and the material is mainly employed in the ceramics industry where it is used to reduce firing temperatures and increase resistance to thermal shock and abrasion. It is used as a flux for welding and soldering and also in cryolite baths for aluminium manufacture. Because of its inherent resistance to oxidation, lithium fluoride can be used as a high-temperature lubricant, capable of operation in excess of 500 °C, which is above the failure temperature for graphite fluoride (Section 6.12.1). Lithium fluoride is a minor, but essential, component in the KF.2HF electrolyte used for fluorine production where it reduces the tendency of the cells to depolarize and improves the wettability of the anodes. Molten lithium fluoride is employed as an electrolyte component in high temperature batteries and as a heat exchange medium in nuclear reactors. Very pure lithium fluoride crystals are used as cell windows and as refractive components in *X*-ray diffraction instruments and infrared and ultraviolet spectrometers. The material has also found utility in radiation dosimeters employed for personnel and environmental monitoring.

6.1.2 Sodium Fluoride. Sodium fluoride is produced on a scale, estimated from one source, of several million tons per year. Although traditionally manufactured from the reaction of hydrofluoric acid with soda ash (Na_2CO_3) or caustic soda (NaOH), recent production in the highly developed nations has been based upon the reaction of the cheap and abundant hexafluorosilicic acid (Section 6.12.2) with sodium hydroxide, equation (4).

$$H_2(SiF_6) + 6NaOH \rightarrow 6NaF + SiO_2 + 4H_2O \tag{4}$$

The most prominent application of sodium fluoride is for the fluoridation of municipal water and as a component of toothpaste and other dental prophylactics for the prevention of dental caries. Other uses are for the manufacture of sodium monofluorophosphate and synthetic cryolite, $Na_3(AlF_6)$; as a flux for degassing steel and the re-smelting of aluminium; as a preservative for wood and adhesives; as a pesticide; and as a disinfectant for fermentation equipment and laundry sours. It has also found utility in chemical cleaning, electroplating, and glass manufacture and in the production of vitreous enamel and coated papers. Single crystals are employed as windows for infrared and ultraviolet detection.

Sodium fluoride can be used to clean gas streams contaminated with hydrogen fluoride. This is particularly pertinent for the purification of element fluorine for specific uses where the presence of residual HF would be detrimental. At temperatures below about 100 °C, the sodium fluoride, in pellet or powder form, fixes the HF as the white, crystalline bifluoride, $Na(HF_2)$. The reaction is reversible, the bifluoride effectively releasing all of the combined HF when heated to about 300 °C. Like the monofluoride, the bifluoride is also used in laundering. Other uses of the bifluoride include leather bleaching, tin plate production, glass etching, antisepsis, and the preservation of anatomical and zoological specimens.

6.1.3 Potassium Fluoride. Potassium fluoride finds application principally in the manufacture of fluxes for a variety of metallurgical processes and as a starting material for electrolytic fluorine manufacture. In its dried and finely powdered form it is a useful fluorinating agent employed in the so-called 'Halex' reaction in which fluoride displaces labile bromide or chloride from a wide variety of organic compounds. For example, this process is employed commercially for the manufacture of fluoroaromatics (in which the aromatic ring is activated by an electron-withdrawing group) and for the manufacture of the powerful rodenticide, sodium fluoroethanoate. Potassium bifluoride, $K(HF_2)$, also known as Frémy's salt, is prepared from the further reaction of hydrofluoric acid with potassium hydroxide or potassium carbonate. It is used as a frosting agent in the glass industry and as a flux for silver solders. The related material KF.2HF, in its molten state, is the electrolyte used for the manufacture of elemental fluorine. In the laboratory it is used as an alkylating agent and as a solid source of hydrogen fluoride, which is formed when the bifluoride is sufficiently heated.

6.1.4 Rubidium and Caesium Fluorides. The manufacturing scale for rubidium and caesium fluoride is little more than about 1 t year^{-1}. Despite the fact that both of these compounds are more effective fluorinating agents than potassium fluoride they are considerably more expensive on a weight basis and contain less fluoride on a molar basis because of the high atomic weights of the metals. For this reason, the uses of rubidium or caesium fluoride is restricted to speciality applications where the other alkali metal fluorides are ineffective. Caesium fluoride has been employed as a catalyst for the oligomerization of fluoroalkenes.

6.2 Group 2 Fluorides

The Group 2 fluorides currently of any commercial significance are those of beryllium, magnesium, calcium, strontium, and barium. Of these materials, calcium fluoride is by far the most important, being the principal raw material source of fluorine world-wide. Calcium fluoride and its related minerals have been described in detail in Section 4 and will not be discussed further here.

6.2.1 Beryllium Fluoride. Beryllium fluoride, BeF_2 (a known carcinogen) is a hygroscopic solid prepared by the thermal decomposition of ammonium tetrafluoroberyllate, $(NH_4)_2(BeF_4)$, equation (5).

$$(NH_4)_2(BeF_4) \rightarrow BeF_2 + 2(NH_4)F \tag{5}$$

Beryllium fluoride is employed in the production of beryllium metal, in glass manufacture, and in the nuclear industry as a moderator and reflector. Metallic beryllium is manufactured by the reduction of beryllium fluoride with magnesium in graphite vessels, according to equation (6).

$$BeF_2 + Mg \rightarrow Be + MgF_2 \tag{6}$$

6.2.2 Magnesium Fluoride. Magnesium fluoride is manufactured by the reaction of hydrofluoric acid with the oxide or carbonate of magnesium, equations (7) and (8).

$$MgO + 2HF \rightarrow MgF_2 + H_2O \tag{7}$$
$$Mg(CO_3) + 2HF \rightarrow MgF_2 + H_2O + CO_2 \tag{8}$$

It is also a by-product in the production of metallic beryllium, referred to in Section 6.2.1 and may be employed for the lining of the crucibles used for that process. Crude magnesium fluoride is used as a flux for ceramics and metallurgy and high purity material is employed for the production of optical windows transparent in the range from the infrared to the vacuum ultraviolet. The low refractive index of this salt has been utilized in the coating of lenses to reduce back-reflection.

6.2.3 Strontium and Barium Fluoride. The binary fluorides of strontium and barium have been prepared by the reaction of hydrofluoric acid with the metal carbonate, sulfide, or hydroxide. They find limited commercial use in glass making; high temperature, dry-film lubrication; single crystals for lasers and spectroscopy; and, in the case of barium, as a component of ceramic and arc welding electrode fluxes.

6.3 Group 3 and Lanthanide Fluorides

Scandium(III) fluoride is prepared from the reaction of naturally occurring scandium(III) oxide with ammonium hydrogen fluoride. It is employed in the preparation of scandium metal by means of active metal reduction.

The fluorides of lanthanum and the lanthanides are of especial importance owing to their water insolubility. They can thus be separated from most other elements by precipitation of the fluorides by addition of hydrofluoric acid, or fluoride ions, from nitric acid solution; the elements then being separated from each other by ion-exchange following re-dissolution. Many of the lanthanide metals are made by active metal reduction of the trifluoride using calcium.

All of the anhydrous lanthanide fluorides (other than the radioactive promethium compound) are available in bulk from specialist suppliers. Lanthanum(III) fluoride is employed in fluoroglasses; as a phosphor lamp coating; in carbon arc electrodes; and in lasers. It is employed in analytical chemistry, together with a europium(II) fluoride dopant, as a single-crystal in fluorine ion-specific electrodes. The utilization of the ionic conductivity properties of LaF_3 has been further proposed for gas sensor applications. Cerium(III) fluoride, CeF_3, is used in arc carbons to increase brilliance and in the preparation of cerium metal. Likewise, gadolinium(III) fluoride, GdF_3, and terbium(III) fluoride, TbF_3, are used in the preparation of the respective metals.

6.4   Group 4 Fluorides

No significant commercial applications of titanium(III) or titanium(IV) fluoride have been disclosed, although bulk material is available. The hexafluorotitanates, *e.g.* $K_2(TiF_6)$, on the other hand, have been employed in abrasive grinding wheels and for the inclusion of titanium into aluminium alloys.

Zirconium(IV) fluoride, in its anhydrous form, is prepared by the action of anhydrous hydrogen fluoride upon zirconium(IV) chloride; by the thermal decomposition of $(NH_4)_3(ZrF_7)$; or by the action of elemental fluorine upon zirconium(IV) oxide at elevated temperatures. The material is mainly employed as a component of molten salts used in nuclear reactors, although some catalytic applications (*e.g.* alkene polymerization) have been described. Zirconium(IV) fluoride reacts with alkali metal, ammonium, or alkaline earth metal fluorides in aqueous or HF media to form the hexafluorozirconates, typified by $K_2(ZrF_6)$. This compound, like the titanium analogue, is used in the manufacture of grinding wheels. Other applications of the complex have been found in welding fluxes; catalysis; glass manufacture and metallurgical grain-refining in magnesium and aluminium. The related tin(II) complex, $Sn(ZrF_6)$, is said to be a more effective agent than tin(II) fluoride in the prevention of tooth decay, see Section 6.12.3.

6.5   Group 5 Fluorides

Niobium(V) and tantalum(V) fluorides are white, hygroscopic, and deliquescent solids which hydrolyse slowly in moist air at room temperature. They are most commonly produced either by the reaction of the corresponding pentachlorides with anhydrous hydrogen fluoride, or by the action of elemental fluorine or hydrogen fluoride on the metal.

Although the commercial production of both of these materials is probably less than $1 t year^{-1}$ they are gaining increasing usage in catalytic applications, particularly in the petroleum processing industry. Both niobium and tantalum pentafluorides are classified as superacids (TaF_5 being somewhat stronger than NbF_5 when dissolved in liquid hydrogen fluoride). These compounds are frequently capable of catalysing reactions, such as the isomerization of alkanes, the alkylation of aromatic compounds, or the cracking of heavy hydrocarbons, at temperatures much lower than is observed with conventional mineral acids or Lewis acids. Such transformations are important in the oil industry, for example, for increasing the octane number of fuels.

6.6 Group 6 Fluorides

6.6.1 Chromium Fluorides. Chromia, Cr_2O_3, is used as a fluorination catalyst for the manufacture of the chlorofluorocarbons (CFCs) and hydrofluorocarbons (HFCs). The catalyst functions as an active site for chlorine–fluorine exchange between the chlorocarbon, or hydrochlorocarbon, and hydrogen fluoride reactants (Section 5.5.1). Although the precise mechanism of the reaction is not known it is believed to involve long-lived chromium(III) oxide fluoride species which are ultimately converted into chromium(III) fluoride by the hydrogen fluoride. In this form the catalyst has very little activity and the spent material requires re-activation or re-processing before it can be further deployed.

6.6.2 Molybdenum Fluorides. Molybdenum(VI) fluoride is the most significant of the molybdenum fluorides although it is currently produced only for the small research market. It is a low melting ($17.4\,^{\circ}C$), low boiling ($35\,^{\circ}C$), highly reactive solid, prepared by the high temperature, direct fluorination of the powdered metal. MoF_6 has been used for the separation of molybdenum isotopes, and hydrogen reduction of the hexafluoride has been employed for the production of molybdenum surface films.

6.6.3 Tungsten Fluorides. Tungsten(VI) fluoride, WF_6, is a very reactive, low-boiling ($17.5\,^{\circ}C$), pale yellow liquid or colourless gas prepared on a scale of $<10 t year^{-1}$ by the direct fluorination of the powdered metal at or above $300\,^{\circ}C$. It is the heaviest gas known, with a density of $12.9 kg m^{-3}$.

Tungsten hexafluoride is a powerful fluorinating agent but its principal commercial application is as a source of pure tungsten for coating substrates. The vapour phase deposition of tungsten is achieved by the hydrogen reduction of high purity WF_6 at about $650\,^{\circ}C$ under low pressure. Tungsten coatings are used as shields against X-rays or γ-rays and composite coatings of tungsten and rhenium are employed to improve ductility and high-temperature and corrosion resistance of the substrate for speciality applications in the electronics and aerospace industries.

6.7 Group 7 Fluorides

Rhenium(VI) fluoride is the only member of the Group 7 fluorides of any real commercial significance. It is prepared by the action of elemental fluorine upon a stoichiometric excess of rhenium powder in order to inhibit the formation of rhenium(VII) fluoride. ReF_6 is a dense, low melting (18.5 °C), yellow solid used in the production of rhenium/tungsten alloys by chemical vapour deposition (Section 6.6.3). Such alloys display higher temperature superconductivity properties than materials produced by conventional techniques. World production of ReF_6 is unlikely to be more than 1 t year^{-1} although the material commands a high price owing to the cost of the metal.

6.8   Group 8 Fluorides

The action of hydrofluoric acid upon iron(III) chloride is used to produce iron(III) fluoride – the only iron fluoride species of any real commercial importance. FeF_3 is a lime-green crystalline material used in ornamental ceramics and as a fluorinating agent. It has been commonly employed as a catalyst for dealkylation, polymerization, and hydroforming reactions and as a component in flame-retarding polymers.

Cobalt(III) fluoride is a powerful and valuable fluorinating agent used on the small scale for the conversion of hydrocarbons into fluorocarbons and for the conversion of aromatic hydrocarbons into perfluorocyclic derivatives. The cobalt(II) fluoride co-produced from these reactions is readily converted back into the cobalt(III) state with elemental fluorine. It is produced readily by the reaction of fluorine with cobalt(III) chloride at 250 °C or with cobalt(II) fluoride at somewhat lower temperatures. A small, but widespread, use of cobalt(III) fluoride is in the assaying of uranium in uranium ore, equation (9).

$$U_3O_8 + 18CoF_3 \rightarrow 3UF_6 + 18CoF_2 + 4O_2 \tag{9}$$

6.9   Fluorides of Copper and Silver

Copper and silver form fluorides of some commercial importance. The existence of copper(I) fluoride is doubtful; thermodynamic studies indicate that this compound would be unstable with respect to disproportionation to copper(II) fluoride and copper.

Copper(II) fluoride, CuF_2, is prepared by the treatment of copper(II) oxide or basic copper carbonate with an excess of hydrogen fluoride. High purity material can be obtained by the direct fluorination of copper(II) hydroxy fluoride, $Cu(OH)F$, with elemental fluorine. It is an essentially white material which develops a blue coloration on exposure to moist air owing to the formation of the dihydrate. It has been employed in the manufacture of ceramics and enamels; as a flux in metallurgy (in the form of the dihydrate); as an electrode in high energy batteries; and as a catalyst and fluorinating agent.

Silver(I) fluoride is easily made by dissolving the corresponding oxide or carbonate in aqueous or anhydrous hydrofluoric acid. It is a photolytically sensitive compound and is usually obtained as a yellow or brown solid owing to surface decomposition. Unlike the other silver(I) halides it is very soluble in water. It has been used as an antiseptic; as a cathode material for batteries; and as a fluorinating agent for the conversion of bromine- or chlorine-containing organic compounds into their corresponding fluoro-derivative.

Silver(II) fluoride is a black crystalline compound prepared from the reaction of fluorine on the powdered silver metal, silver(I) chloride, or silver(I) fluoride at 200–300 °C. It is a more powerful fluorinating agent than cobalt(III) fluoride and has an extensive oxidation chemistry – frequently resulting in explosions when combined with organic compounds. Because of the greater expense of silver compared to cobalt, its uses as a fluorinating agent is much more limited.

6.10   Fluorides of Zinc, Cadmium, and Mercury

Zinc fluoride, ZnF_2, is a high-melting, ionic solid with lower water solubility. It can be produced by the action of hydrofluoric acid on the metal oxide, hydroxide, or carbonate, or by the addition of sodium fluoride to the metal acetate. The tetrahydrate so formed is then dried (> 75 °C) in a stream of hydrogen fluoride gas to inhibit hydrolysis and generate a technical grade material of about 95% purity.

Although the present production of zinc fluoride is likely to be on a small scale, it has been previously employed as a metallurgical flux. Current uses are found in phosphors, wood preservation, ceramic glazes, in electroplating, and organic fluorination where it can function as a mild fluorinating agent or as a catalyst for cyclization reactions.

Like the zinc analogue, cadmium fluoride, CdF_2, is a high-melting solid with low water solubility. It can be prepared in a similar way to that described for zinc fluoride. Despite its highly toxic properties, cadmium fluoride has found utility in electronic and optical applications. It is a high temperature, dry-film lubricant and has been employed as a starting material for crystals in lasers.

No commercial applications are known for mercury(I) fluoride and the fluoride of this element in its divalent state, HgF_2, has only limited commercial utility owing to its toxic properties. Although the material is commercially available in bulk quantity (< 0.25 t year^{-1}, prepared from the treatment of the metal(II) oxide or chloride with elemental fluorine) it can also be prepared ostensibly *in situ* by the reaction of anhydrous HF with the oxide or chloride of mercury(II). Mercury(II) fluoride decomposes at its melting point (645 °C) and is readily hydrolysed. Like the cadmium salt it crystallizes in the fluorite structure.

6.11   Boron and Aluminium Fluorides

Both boron and aluminium fluorides have an extensive commercial importance whereas the remaining fluorides of the series are not known to be commercially significant.

6.11.1   Boron Fluorides. Boron trifluoride, BF_3, is a hydrolytically stable gas produced by the reaction of borax ($Na_2B_4O_7.10H_2O$) with fluorspar in sulfuric acid, or by the reaction of boric acid with fluorosulfonic acid as shown by equation (10).

$$B(OH)_3 + 3HSO_3F \rightarrow BF_3 + 3H_2SO_4 \qquad (10)$$

Table 6   *Commercially Available Adducts of Boron Trifluoride*

BF_3-Adduct	Formula	Supplier
Acetic Acid	$BF_3.CH_3COOH$	Engelhard
Dihydrate	$BF_3.2H_2O$	Engelhard; ISC
2,4-Dimethylaniline	$BF_3.NH(CH_3)_2C_6H_5$	Engelhard
Dimethyl Ether	$BF_3.O(CH_3)_2$	BASF
Ethyl Ether	$BF_3.O(CH_2CH_3)_2$	BASF; Heyl; ISC
Methanol	$BF_3.CH_3OH$	Rhône-Poulenc
Monoethylamine	$BF_3.NH_2CH_2CH_3$	BASF; Engelhard; Heyl
Monoisopropylamine	$BF_3.NH_2CH_2(CH_3)_2$	Heyl
n-Methylcyclohexylamine	$BF_3.NH(CH_3)C_6H_{10}$	Heyl
Phenol	$BF_3.C_6H_5OH$	BASF
Piperidine	$BF_3.NHC_5H_{10}$	BASF; Engelhard; Heyl
Triethanolamine	$BF_3.NH(CH_2CH_2OH)_3$	Heyl

It is a strong Lewis acid and is employed as a catalyst in a variety of organic reactions, particularly the alkylation of benzene for the detergents industry and as an isomerization catalyst in the oil industry for the production of gasoline. It is used in the production of diborane, B_2H_6, and in the nuclear industry for the measurement of neutron intensity. BF_3 protects molten magnesium and magnesium alloys from oxidation during metal casting. Whilst soluble in a wide range of inert organic solvents, such as saturated or halogenated hydrocarbons, it tends to polymerize unsaturated compounds, such as styrene, and to co-ordinate to compounds containing available donor atoms, such as diethyl ether or pyridine to form adducts of the type $BF_3.OEt_2$ or $BF_3.py$, respectively. The common, commercially available adducts of BF_3 are listed in Table 6.

Like boron trifluoride itself, the adducts listed are frequently used as catalysts, as curing agents for epoxy resins, and they find application in the dyeing of polyesters and the preparation of phenolic resins. Global production of BF_3 and its adducts is of the order of 5000 t year^{-1}.

6.11.2   Fluoroboric Acid and Derivatives.   Fluoroboric acid, $H(BF_4)$, has never been isolated as a discrete chemical entity. It exists as an aqueous solution, most commonly available in a concentration of 48%. It is manufactured by the reaction of 70% hydrofluoric acid with boric acid, $B(OH)_3$; an excess of the boron compound being employed in the preparation to obviate the particular fume and skin-burn problems caused by HF. Alternatively, it can be manufactured by the action of boric and sulfuric acids on fluorspar. World-wide consumption of fluoroboric acid is about 10 000 t year^{-1}.

Its main use is in the preparation of fluoroborate salts and in the electroplating and electrolytic brightening of aluminium. It is also used for the cleaning or pickling of metals prior to plating or other surface treatment and it is employed to remove solder. Fluoroboric acid is used in the manufacture of synthetic cryolite for aluminium production; as an esterification catalyst; and for making stabilized diazo salts, particularly for the manufacture of fluoroaromatics by a process referred to as the Balz–Schiemann reaction. A common method used nowadays for

Table 7 *Principal Commercial Fluoroborate Salts*

Formula	Typical Manufacture	Application
$(NH_4)(BF_4)$	Reaction of ammonia with $H(BF_4)$	Flame retardant for plastics; inert atmosphere soldering flux; solid lubricant
$Cd(BF_4)_2.6H_2O$	Reaction of Cd or CdO with $H(BF)_4$	Plating solution; catalyst in the manufacture of polyesters
$Co(BF_4)_2.6H_2O$	Reaction of cobalt hydroxide with $H(BF_4)$	Polymerization catalyst; plating solution
$Cu(BF_4)_2.6H_2O$	Reaction of hydroxide with $H(BF_4)$	Curing agent for epoxy resins; plating solution
$Fe(BF_4)_2.6H_2O$	Reaction of Fe or FeO with $H(BF_4)$	Plating solution; catalyst
$Pb(BF_4)_2.H_2O$	Electro-dissolution of Pb in $H(BF_4)$	Catalyst for linear polyesters; plating solution
$Li(BF_4)$	Reaction of LiOH with $H(BF_4)$	High temperature metal flux; flame retardant for cotton and rayon; battery electrolyte
$Mg(BF_4)_2.6H_2O$	Reaction of $H(BF_4)$ with Mg metal, oxide, or carbonate	Epoxy resin catalyst; preparation of MgF_2 for the television industry
$Ni(BF_4)_2.6H_2O$	Electro-dissolution of Ni in $H(BF_4)$	Curing agent for epoxy resins; plating solution
$NO(BF_4)$	Reaction of N_2O_4 with $H(BF_4)$	Agent for the preparation of fluoroaromatics
$K(BF_4)$	Reaction of $B(OH)_3$, HF, and KOH	Sand-casting of aluminium and magnesium; grinding aid; flux for soldering and brazing
$Ag(BF_4)$	Addition of BF_3 to AgF in ethylbenzene	Catalyst; laboratory reagent
$Na(BF_4)$	Heating of NaF and $H(BF_4)$ and cooling slowly	Catalyst; flame retardant; sand-casting of aluminium and magnesium; oxidation inhibitor; flux for non-ferrous metals
$Sn(BF_4)_2.xH_2O$	Electro-dissolution of Sn in $H(BF_4)$	Catalyst for linear polyesters
$Zn(BF_4)_2.6H_2O$	Reaction of Zn or ZnO with $H(BF_4)$	Plating solution; resin curing

the manufacture of certain speciality fluoroaromatics (*e.g.* 4,4′-difluorobenzo-phenone) is based on the reaction (illustrated for fluorobenzene) in Scheme 5 (*cf.* Section 5.3).

Scheme 5

The principal fluoroborate derivatives are listed in Table 7, together with a description of their manufacture and principal uses.

6.11.3   Aluminium Fluorides. Aluminium trifluoride, AlF_3, and cryolite, $Na_3(AlF_6)$, are the two aluminium fluorides of commercial importance, accounting for some 40% of HF consumption. Aluminium trifluoride is traditionally, but decreasingly, manufactured by the reaction of alumina hydrate with hydrofluoric acid

$$Al(OH)_3 + 3HF \rightarrow AlF_3.3H_2O \rightarrow AlF_3 + 3H_2O \tag{11}$$

Where anhydrous hydrogen fluoride is available, aluminium trifluoride may be manufactured in a high temperature (400–600 °C), fluidized bed process by the reaction of the HF with partially dehydrated alumina. In many parts of the world, aluminium fluoride is also manufactured from fluorosilicic acid aqueous solution, $H_2(SiF_6)$, obtained as a by-product of the phosphate fertilizer business, equation (12).

$$2Al(OH)_3 + 2H_2O + H_2(SiF_6) \rightarrow 2AlF_3.3H_2O + SiO_2 \tag{12}$$

Cryolite, once most commonly mined (until 1990) in Greenland as the natural mineral, is now mainly produced by the reaction of alumina with HF and sodium hydroxide, equation (13).

$$Al_2O_3.xH_2O + 12HF + 6NaOH \rightarrow 2Na_3(AlF_6) + (9 + x)H_2O \tag{13}$$

Synthetic cryolite is frequently deficient in sodium and fluorine although this is not detrimental to its applications.

Both aluminium trifluoride and cryolite are principally used in the manufacture of aluminium. The aluminium trifluoride (in addition to serving as a flux and a conductivity aid) functions as a top-up material in the cryolite/alumina molten electrolyte. This then accounts for the fluoride losses in the process which operates at *ca.* 1000 °C and results in the dissociation of the cryolite to form $Na(AlF_4)$ vapour according to equation (14).

$$Na_3(AlF_6) \rightarrow 2NaF + Na(AlF_4) \tag{14}$$

Aluminium trifluoride is used as a catalyst for the replacement of chlorine by fluorine in a variety of chlorinated organic compounds by means of hydrogen fluoride. In addition, AlF_3 frequently serves as a useful catalyst for the isomerization of chlorofluorocarbons. Other, minor applications of aluminium trifluoride include its use as a flux for welding and in ceramic glazes and enamels; and for the production of aluminium silicate fibre. The capacity for aluminium trifluoride production, world-wide, is estimated to be about 5×10^5 t year^{-1}.

Cryolite, in addition to its role in aluminium manufacture but to a much lesser extent, is used as a flux in the glass and ceramics industries; as an insecticide on grape vines; as a binder for grinding wheels; and as a catalyst for the polymerization of alkenes. World production of synthetic cryolite is of the order of 4×10^5 t year^{-1}.

6.12 Fluorides of Carbon, Silicon, Germanium, Tin, and Lead

Of these, only the fluorides of germanium are not known to have any real commercial significance. Germanium(IV) fluoride is available, but only in research quantities.

6.12.1 Carbon Fluorides. The simple fluorides of carbon are located at the boundary between organic and inorganic fluorine chemistry. In this section are considered the simple, unsubstituted carbon tetrafluoride as the prototype and the commercially-important graphite fluorides; the industrially-important chloro-fluorocarbons and hydrofluorocarbons having been described in Section 5.5.1 and the perfluorocarbons in Section 5.5.2. Other fluorine-containing carbon compounds are beyond the scope of this Section. The commercially promising carbonyl fluorides, $COClF$ and COF_2, and the fluoride derivative of the new and remarkable carbon allotrope, commonly known as Buckminsterfullerene, are mentioned in Section 8.

Carbon(IV) fluoride, CF_4, more commonly referred to as carbon tetrafluoride, is a remarkably stable gaseous compound and is the thermodynamic end-point in the fluorination of organic compounds. It can be produced by the catalysed reaction of R-12 (CCl_2F_2) or R-11 (CCl_3F) with HF in the gas phase at elevated temperatures, or by the direct fluorination of almost any available organic compound. For example, the fluorination of wood charcoal initiates at ambient conditions to generate carbon tetrafluoride; the exothermic reaction which ensues resulting in high temperatures. CF_4 is used for speciality purposes as a low-temperature refrigerant and as a gaseous insulator. In its high purity form, it is used as an etchant for microchips.

Fluorine reacts under carefully controlled conditions with graphite or other high purity carbons at elevated temperatures to form the chemically and thermally stable poly(carbon monofluoride), commonly referred to as graphite fluoride $(CF_x)_n$, in which $0 < x < \sim 1.25$. These materials are used as very effective solid lubricants and release agents, and as cathode materials on high energy batteries such as the graphite fluoride/Li battery. The long shelf-life and superior discharge characteristics of these cells has led to their widespread use in many commonplace devices, such as the coin type battery found in electronic watches. Although first chemically characterized in the 1930s, it was not until the late 1960s that the application of these materials was properly realized. They are now manufactured on the industrial scale by several companies, particularly in Japan.

6.12.2 Silicon Fluorides. The silicon fluorides are probably one of the most under-utilized sources of fluorine chemicals, providing a potential source for the manufacture of hydrogen fluoride via the vast reserves of fluorapatite.

Silicon(IV) fluoride is a thermally stable, but hydrolytically unstable, gas boiling at $-90.2\,^\circ C$. It is manufactured from the reaction of silica with HF or $H_2(SiF_6)$. Very pure material can be produced from the thermal decomposition of metal hexafluorosilicates, such as $Na_2(SiF_6)$, equation (15).

$$Na_2(SiF_6) \rightarrow 2NaF + SiF_4 \qquad (15)$$

SiF_4 hydrolyses in the gas phase to give hydrogen fluoride and copious amounts of silica smoke, according to equation (16):

$$SiF_4 + 2H_2O \rightarrow SiO_2 + 4HF \qquad (16)$$

whilst in the liquid phase, hexafluorosilicic acid is the final product:

$$3SiF_4 + 2H_2O \rightarrow 2H_2(SiF_6) + SiO_2 \qquad (17)$$

Hydrolysis in a high-temperature flame results in the production of high surface area, finely divided, pyrogenic silica. SiF_4 can be used for the manufacture of disilane, Si_2H_6, and of pure silicon. It has been employed to seal water out of oil wells during drilling and the treatment of concrete or cement with SiF_4 is known to impart corrosion resistance.

Hexafluorosilicic acid does not exist in the anhydrous state. It is available in aqueous form, mostly as a by-product of the phosphate-based fertilizer industry from the action of sulfuric acid on phosphate rock (fluorapatite) containing fluorides and silica, or silicates. The hydrofluoric acid formed from the fluoride/acid interaction reacts with the silica to produce silicon tetrafluoride which is subsequently hydrolysed as previously described.

The acid (and its salts) are employed increasingly for water fluoridation; for the manufacture of fluorosilicate salts; to increase the hardness of ceramics and cement, and in electroplating. The bactericidal and fungicidal properties of the acid are apparent by its use in wood preservation; the disinfection of copper-based vessels; and the sterilization of brewing equipment. The role of $H_2(SiF_6)$ in sodium fluoride and, particularly, of aluminium trifluoride manufacture has been discussed in Section 6.1.2 and 6.11.3, respectively. Production, world-wide, is probably several hundred thousands of tons per annum, mostly for captive use.

The reaction of $H_2(SiF_6)$ aqueous solution with ammonia or basic metal oxides, hydroxides, or carbonates, results in the production of fluorosilicate salts. Sodium hexafluorosilicate is probably the most important material in industrial terms although the fluorosilicates of ammonia, potassium, zinc, magnesium, barium, copper, and lead have also been produced commercially. The common uses of the fluorosilicate salts are given in Table 8.

6.12.3   Tin Fluorides. Tin(II) fluoride is a white, crystalline salt prepared from the reaction of the corresponding oxide with aqueous hydrofluoric acid, or by the dissolution of tin metal in anhydrous HF. It was commonly employed as an additive in toothpaste until the 1980s and is still used in other dental preparations owing to its anticaries properties in which the major constituent of the tooth, hydroxyapatite, $Ca_5(PO_4)_3(OH)$, is converted into the more acid resistant fluorapatite, $Ca_5(PO_4)_3F$.

6.12.4   Lead Fluorides. Prepared by the action of hydrofluoric acid on the corresponding hydroxide or carbonate, lead(II) fluoride is used in a variety of speciality applications. It is a mild fluorinating agent used for the conversion of

Table 8 *Principal Commercial Fluorosilicate Salts*

Formula	Applications
$Na_2(SiF_6)$	Water fluoridation (particularly in the USA); laundry sour; polishing/etching additive in glass manufacture; preparation of pure SiF_4
$K_2(SiF_6)$	Manufacture of porcelain enamels; synthetic mica; insecticide compositions; opalescent glass; metallurgy
$(NH_4)_2(SiF_6)$	Laundry sours; mothproofing; disinfectant (brewing industry); glass etching; light metal casting; electroplating
$Mg(SiF_6).6H_2O$	Hardener for concrete; water-proofing; mothproofing; laundry sours; magnesium casting
$Ba(SiF_6).6H_2O$	Insecticidal compositions; ceramics; luminophor in ultraviolet lamps
$Cu(SiF_6).4H_2O$	Fungicide for the treatment of grape vines; dyeing and hardening of white marble
$Zn(SiF_6).6H_2O$	Hardener for concrete; laundry sour; mothproofing agent; preservative
$Pb(SiF_6).2H_2O$	Solution for electrofining lead

molybdenum or tungsten oxide into the oxyfluoride and as the starting material for the preparation of lead(IV) fluoride. Lead(II) fluoride is employed as a high-temperature dry film lubricant. It is used in the manufacture of special grades of glass and it has been used as a constituent in marine paints.

Lead(IV) fluoride has been employed occasionally as a laboratory reagent but is not known to have any significant commercial value.

6.13 Fluorides of Nitrogen, Phosphorus, Arsenic, Antimony, and Bismuth

The commercially important fluorides in this grouping comprise those of nitrogen (and ammonium), phosphorus, arsenic, and antimony. Bismuth(V) fluoride is not known to have any significant utility although the material is commercially available.

6.13.1 Nitrogen Fluorides. The gaseous trifluoride of nitrogen, NF_3, is a thermodynamically stable material at ordinary temperatures. At elevated temperatures it is a powerful oxidizing agent, reacting with organic compounds often explosively. It is made by the electrolysis of molten ammonium bifluoride or by the direct fluorination of ammonia. NF_3 was once employed as an oxidizer for high-energy fuels and is currently used as a fluorine source (in preference to elemental fluorine) in high-energy chemical lasers. It is used as a chip etchant and minor quantities of NF_3 are used for the production of tetrafluorohydrazine, N_2F_4, in the defence industry. Production is believed to be probably less than 20 t year^{-1}.

Ammonium fluoride, $(NH_4)F$, is a colourless, extremely water-soluble, hygroscopic, and sublimable salt readily prepared from the reaction of anhydrous ammonia with HF, or by heating aqueous solutions under reduced pressure, followed by crystallization. It decomposes readily on heating to form ammonium bifluoride, as shown in equation (18):

$$2(NH_4)F \rightarrow NH_3 + (NH_4)(HF_2) \tag{18}$$

Ammonium fluoride is a poor fluorinating agent for halogen exchange compared to most of the alkali metal fluorides. It is used as a laboratory reagent in analytical chemistry and has been employed as an antiseptic in the brewing industry; as a wood preservative; as a textile mordant; in oil-well drilling; as a mothproofing agent; and, perhaps more commonly, as a glass etchant.

Ammonium bifluoride is made, as indicated above, by dehydrating ammonium fluoride solutions followed by thermal decomposition or by combining anhydrous ammonia and HF in stoicheiometric proportions. It is used as a sour in commercial laundries and textile plants and removes iron stains by forming almost colourless complexes that are readily removed by water-rinsing the fabric. Pure binary fluorides can be prepared by reaction of ammonium bifluoride with many metal oxides or carbonates followed by decomposition of the ammonium complex. This method reduces the contamination of the fluoride salt by oxyfluorides and is especially useful in the manufacture of beryllium fluoride (Section 6.2.1).

Other uses for ammonium bifluoride include: the pickling of stainless steel and titanium to avoid hydrogen embrittlement; the activation of metals prior to nickel plating or phosphating; fungicidal treatment of wood; glass etching, renovation, and decoration; sterilization of brewery and dairy equipment; metal glazing; and the production of nitrogen(III) fluoride.

6.13.2 Phosphorus Fluorides. Most of the fluorine in the Earth's crust is present as the phosphate-containing mineral, fluorapatite. The role of this material in hexafluorosilicic acid production has been alluded to in Section 6.12.2. Most of the commercial utility of phosphorus fluorides is centred on phosphorus in the +5 oxidation state.

Phosphorus(V) fluoride, PF_5, is a colourless, readily hydrolysable gas prepared by a number of methods, in particular the exchange reaction between phosphorus(V) chloride and arsenic(III) fluoride as in equation (19):

$$3PCl_5 + 5AsF_3 \rightarrow 3PF_5 + 5AsCl_3 \tag{19}$$

It is a powerful Lewis acid and has been used as a Friedel–Crafts polymerization catalyst and as a source of phosphorus for ion-implantation in the semiconductor industry. The phosphorus oxyfluoride, POF_3, is formed by the partial hydrolysis of PF_5 whilst further hydrolysis results in the formation of a mixture of fluorophosphoric acids. The most important of the fluorophosphoric acids are monofluorophosphoric acid, H_2PO_3F, difluorophosphoric acid, HPO_2F_2, and hexafluorophosphoric acid, $H[PF]_6$. These compounds can be prepared by the reaction of phosphoric acid or phosphorus oxyfluoride with HF. They are used in aqueous form as catalysts, metal cleaners, electrolytic or chemical polishing agents, protective coatings for metal surfaces, and for the preparation of fluorphosphate salts.

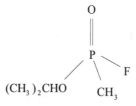

Figure 2 *Structural Formula of Sarin*

Sodium monofluorophosphate, $Na_2(PO_3F)$, commercially prepared on the scale of several hundreds of tons per annum by the fusion of sodium fluoride and sodium metaphosphate, is presently employed in toothpastes, facilitating the formation of the acid-resistant fluorapatite in teeth (Section 6.12.3). Potassium monofluorophosphate, $K_2(PO_3F)$, prepared in a similar way to the sodium salt, is used for the passivation of metal surfaces; providing a good surface for paint adhesion and corrosion protection.

The lithium and potassium salts of hexafluorophosphoric acid, prepared by the neutralization of the aqueous acid with the corresponding alkali metal hydroxide, have been employed as electrolytes in lithium anode batteries. Silver hexafluorophosphate, $Ag(PF_6)$, forms complexes with alkenes, an unusual feature which has been exploited in the separation of alkenes from paraffins.

An insidious development in the area of phosphorus–fluorine technology was the production of an isopropyl ester derivative of monofluorophosphonic acid, known as Sarin. This compound, which has the structural formula indicated in Figure 2, was made for use as a nerve gas (cholinesterase inhibition) during World War II. Several other ester derivatives of fluorophosphoric acid, however, have been employed for beneficial medical applications.

6.13.3 Arsenic Fluorides. Arsenic(III) fluoride, AsF_3, is a colourless liquid prepared on a small scale (<1 t year^{-1}) by the action of HF or fluorosulfonic acid on arsenic(III) oxide. It is employed as a mild fluorinating agent (sometimes in conjunction with $SbCl_5$) and in the preparation of arsenic(v) fluoride and hexafluoroarsenate(v) salts.

Arsenic(v) fluoride is a hydrolysable gas prepared on a similar scale to that of the trivalent fluoride by the fluorination of arsenic or arsenic(III) fluoride with elemental fluoride. It is employed in ion-implantation, as a fluorinating agent, and in the preparation of $(AsF_6)^-$ salts. Graphite/AsF_5 intercalation compounds have been noted to have electrical conductivity comparable to that of silver.

The hydrolytically-stable hexafluoroarsenate(v) salts, mostly prepared by the neutralization of $H(AsF_6)$ with the corresponding base, find numerous applications as catalysts, electrolytes, herbicides, and polymerization photo-initiators.

6.13.4 Antimony Fluorides. Antimony(v) fluoride is a colourless, fuming, viscous liquid. It is a very strong Lewis acid and a powerful fluorinating agent, capable of effecting chlorine/fluorine exchange in a number of chlorinated hydrocarbons. It is prepared in its pure form by the fluorination of antimony(III) fluoride with elemental fluorine and is employed as a catalyst in the fluorination of hydrocarbons. Antimony(v) chloride/HF mixtures are commonly employed on a large scale as catalysts for the reaction of HF with simple chlorinated hydrocarbons for the manufacture of the commercially-declining chlorofluorocarbons (CFCs) and hydrofluorochlorocarbons (HCFCs). The active catalytic species in these systems are believed to approximate to SbF_2Cl_3 and $SbFCl_4$.

The addition of SbF_5 dramatically increases the acidity of fluorosulfonic acid or hydrogen fluoride to constitute a system of superacids (Section 6.5) which can catalyse the isomerization, cracking, or alkylation of paraffins into useful components for gasoline or other petroleum derivatives at low temperatures.

Antimony(III) fluoride is a white solid prepared from the reaction of antimony(III) oxide with aqueous hydrofluoric acid. It is a mild fluorinating agent when employed alone but in the presence of chlorine or other oxidizing agents its effectiveness is considerably increased. Once used in CFC manufacturing processes (the Swarts reaction), antimony(III) fluoride is now used primarily for the preparation of the pentavalent fluoride.

6.14   Sulfur Fluorides

Of the fluorides in this Group, only those of sulfur have any current commercial value outside of the laboratory environment.

The sulfur fluorides constitute a diverse range of materials used for an equally diverse range of applications. Sulfur(VI) fluoride, SF_6, is a dense, colourless gas with a high dielectric strength and remarkably low chemical and physiological activity. It is manufactured on the thousands of tons scale ($5000-8000\,t\ year^{-1}$) from the reaction of sulfur vapour with pure fluorine and used in a variety of applications in the electrical and electronics industries, and especially as a gaseous insulator for high-voltage generators. Non-electrical uses of SF_6 include: noise and thermal insulation in three-pane glass windows; air-flow tracing; and as a low-leakage gas filling in the spare tyres of high-performance motor cars.

In contrast, sulfur(IV) fluoride, SF_4, is a highly reactive, hydrolysable and toxic gas made from the stoichiometric combination of sulfur and fluorine. The principal interest in this material stems from its ability to replace oxygen in alcohols and carbonyl compounds with fluorine. However, owing to its hazardous nature (and high price) the commercial applications of SF_4 have been very much limited to speciality purposes. Its employment in the preparation of oil and water repellents and lubricity improvers has been noted.

Sulfuryl fluoride, (SO_2F_2), boiling point $-50\,°C$, has been used as a fumigant. Fluorosulfonic acid, $SO_2F(OH)$, is a mobile, corrosive, colourless liquid prepared from the reaction of HF with sulfur(VI) oxide, using the product acid as the reaction medium. It is used as a catalyst for the alkylation of aliphatic and aromatic compounds in the petroleum industry; as a catalyst for the polymerization of alkenes; and as a fluorinating agent in the synthesis of a range of organic and inorganic fluorides. The addition of fluorosulfonic acid to antimony(v) fluoride (a mixture known as 'Magic' acid), or other strong Lewis acids, markedly increases its acidity and catalytic power. World production of fluorosulfonic acid is about $15\,000$ t year^{-1}.

6.15 Halogen Fluorides

Chlorine(III) fluoride, ClF_3, bromine(III) fluoride, BrF_3, and iodine(v) fluoride, IF_5, are the three industrially important halogen fluorides. They provide the advantage over gaseous elemental fluorine in that they can be used and stored as liquids at low pressure. They are prepared on a scale of several hundreds of tons per annum by the direct combination of the respective elements in stoichiometric proportions; the preparation of IF_5 requiring a non-reactive solvent.

ClF_3 is the most reactive of the halogen fluorides, and is mostly used by the nuclear industry for the preparation of UF_6 (Section 6.17). It is also employed for the cutting of oil-well tubes; as an oxidizer in propellants; and as a fluorinating agent for the preparation of high-oxidation state metal fluorides employed for the synthesis of perfluorocarbons. Bromine(III) fluoride is also used both in oil-well tubing cutters and as a fluorinating agent. Iodine(v) fluoride is the mildest fluorinating agent, and easiest to handle, of the halogen fluorides. Its most important application is in the addition of iodine and fluorine to tetrafluoroethene to give the perfluoroethyl iodide.

6.16 Noble Gas Fluorides

Of the noble gas fluorides, only those of xenon have any extensive chemistry; and of the xenon fluorides, only xenon(II) fluoride, XeF_2, is available in 'bulk' quantity ($ca. \leq 1$ kg) from a Russian source. Xenon(II) fluoride is a colourless, crystalline, sublimable solid prepared from the reaction of fluorine with excess of xenon gas by heating or by exposure of the mixed gases to sunlight. It has considerable utility in the laboratory syntheses of both organic and inorganic fluorides but the present apparent lack of commercial applications renders a more complete description beyond the scope of this book.

6.17 Actinide Fluorides

The fluoride compounds of the actinide elements are essentially dominated, commercially, by one compound, uranium(VI) fluoride. Since World War II the production and enrichment of UF_6 has become vital for nuclear weaponry and atomic power plants to the extent that about $40\,000$ tons, world-wide, are now consumed annually. At room temperature UF_6 is a white solid when pure. It sublimes at about $64\,^{\circ}C$ at atmospheric pressure, and the gaseous UF_6 is separated into its ^{238}U and ^{235}U isotopic components by gaseous thermal diffusion or gas centrifugation to the extent of about 5% for utilization in the nuclear power industry.

Uranium(VI) fluoride is prepared by two main processes. In the so-called dry process, natural U_3O_8 ore is reduced with hydrogen to form uranium(IV) oxide which is then fluorinated with HF to UF_4. The fluorination of UF_4 with elemental

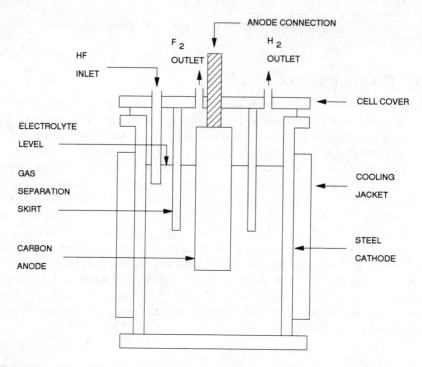

Figure 3 *Electrochemical Cell for the Production of Fluorine*

fluorine at around $220\,^\circ$C then produces UF. In the wet process, the uranium(IV) oxide precursor is prepared by the reaction of the ore with nitric acid to give $UO_2(NO_3)_2$, followed by thermal decomposition to uranium(VI) oxide and reduction to U^{VI}.

Thorium(IV) fluoride, ThF_4, is used in high temperature ceramics and for the production of the metal and magnesium/thorium alloys. The oxide fluoride, $ThOF_2$, is used on reflective surfaces as a protective coating.

The actinide metals may be prepared by the active metal (Li, Mg, Ca, or Ba) reduction of either the anhydrous tri- or tetra-fluorides at high temperature: a technique suitable for operation on a scale ranging from micrograms to tens of kilograms.

7 FLUORINE

7.1 Fluorine Production

Although the chemical generation of fluorine has been recently reported, the electrochemical process remains the only feasible method for industrial scale production. This originally started in the 1940s to meet the Manhattan Project demand for uranium(IV) fluoride. At that time three different types of cell were developed, all based on mixed KF/hydrogen fluoride electrolytes and classed, according to their operating temperatures, as low (15–50 $^\circ$C, KF.nHF), medium (70–130 $^\circ$C, KF.2HF), and high (245–310 $^\circ$C, KF.HF). The medium temperature cell emerged as the most feasible system combining relatively low corrosion rates with low HF vapour pressure.

Various designs of this cell type have been developed by different companies, but Figure 3 shows the principal components common to all cells. Although anode materials such as nickel and graphite have been evaluated in the past, all modern fluorine cells are based on carbon anodes and steel cathodes and operate around 90 °C. Effective cooling is necessary to maintain this temperature either by circulating water through an external jacket or pipes immersed in the electrolyte which can also be the cathode. The coolant enters the cell at 75 °C to avoid solidifying the electrolyte which melts at 71 °C. The copper skirt separating the anode and cathode is essential to keep the hydrogen and fluorine gases apart. To preserve the seal the electrolyte level must be maintained above the bottom of the skirt at all times by feeding liquid HF intermittently through the dip pipe. Typically the HF content is held in the range 40–45 wt %.

In a commercial cell a number of carbon anodes are suspended from the cell cover. For example, the cell developed by ICI, now operated by BNFL, has the following specifications:

Length	3 m
Width	0.8 m
Depth	0.6 m
Electrolyte charge	1000 kg
Anodes	12 assemblies each containing 2 carbon anodes
Cathodes	24 assemblies consisting of steel cooling pipes
Current	4000–6000 A
Cell voltage	8–12 V
Output	3–4 kg of $F_2\,h^{-1}$

In contrast to other designs, the ICI/BNFL cell employs porous carbon anodes rather than the hard, dense carbon used in the USA and elsewhere. The low-permeability carbon is susceptible to a phenomenon known as polarization which causes the current to fall, gradually or rapidly, to a fraction of its normal operating value while still maintaining the normal operating potential. However, porous anodes are less mechanically robust.

The cell cover has vents which allow the fluorine and hydrogen to pass into their respective manifolds which are also linked to the output of other cells. The fluorine is contaminated with 4 mol.% HF, which is reduced to 0.2 mol.% by a combination of refrigeration and absorption onto solid sodium fluoride to form sodium bifluoride. The latter can be reconverted into the former by heating to drive off the HF. The fluorine is then used immediately or stored by compressing to 400 p.s.i. Cryogenic storage as liquid has also been used in the past but is now less common. The boiling point of liquid fluorine at 1 bar is −219 °C.

7.2 Uses of Elemental Fluorine

The economics of fluorine production are dominated by high power consumption, especially in the UK where electricity prices have been traditionally high. The cost of fluorine production in the UK is estimated to be between £6000 and £10 000 t^{-1}, which compares unfavourably with the cost of HF at about £1000 t^{-1} as a source of fluorine atoms for downstream products. For this reason processes capable of using HF directly have generally been preferred for the higher tonnage, lower cost fluorine products such as the CFC and HCFC refrigerants. The same will undoubtedly be true for their hydrofluorocarbon (HFC) replacements. However, elemental fluorine has been used to produce perfluorinated fluids, for example perfluorodecalin, via cobalt trifluoride, and to complete the fluorination of perfluoro-polyethers produced from fluorinated epoxides. These products are used in specialist applications such as synthetic oxygen carriers, so-called 'artificial blood', lubricants for magnetic discs and coolants for super computers, which can accept the high cost of production in return for excellent chemical stability.

The last 15 years has seen substantial developments in taming fluorine as a chemical reagent. For example, a number of companies are offering fluorinating agents containing the N—F bond generated by reacting N—H with F$_2$. Undoubtedly this activity will grow in the coming years and perhaps the use of elemental fluorine in the production of organic compounds will overtake what is still its current major application, the production of uranium(VI) fluoride.

8   THE FUTURE FOR INORGANIC FLUORIDES

Although this chapter has been principally concerned with the present production and uses of inorganic fluorides, the authors would like to finish by speculating on possible developments which may become industrially important in the future. Undoubtedly the production of HF from fluorspar will remain the starting point for the production of fluorocarbons. However, the consumption of HF for organic fluorocarbons may well decrease significantly as the CFCs and HCFCs are replaced by the non-ozone depleting HFCs. In part this will occur because there will be no substantial aerosol market for HFCs, but major contributions will come from considerable recycling of the new refrigerants, and by reducing the leakage from large scale refrigeration and air conditioning units. Consequently less HFCs will need to be manufactured, at least in the advanced nations, to satisfy the market compared to the CFCs and HCFCs in the past. However, as developing countries improve their living standards their demands for refrigerants will increase and thus their HF production could expand.

Inorganic fluoride glasses, for example based on ZrF$_4$–BaF$_2$–LaF$_3$–AlF$_3$–NaF, may ultimately achieve their promise of high-performance optical fibres for data transmission using IR lasers. So far this work has contributed to better infrared optics, infrared imaging, precise thermal surgery, and fibre optic lasers. The fluoride glasses can be readily drawn into kilometre long optical fibres and have ultimate theoretical attenuation limits in the range 10^{-3}–10^{-2} dB km^{-1}, which is significantly better than silica fibres (0.15 dB km^{-1}). In 1989 performance was still three orders of magnitude less than the required target. Better quality glasses, a reliable technique for making long lengths of fibre and adequate protection of the moisture sensitive fibres against undersea conditions were seen as the key problems.

The growing interest in solid-state batteries has focused attention on the possibility of using fast fluoride ion conductors as solid-state electrolytes. Of especial interest for this purpose are alkaline earth fluorides containing point defects induced by the inclusion of trivalent cations such as La^{3+} or monovalent cations such as Na^+. The former produce interstitial fluoride, while the latter produce fluoride vacancies. Either type of defect can induce fluoride ion conductivity. However, many of these compounds are electronic insulators which makes them ideal as solid-state electrolytes. Until now oxides with CaF_2-type structures, which rely upon O^{2-} for their conductivity (*e.g.* ZrO_2), have received most attention as anionic conductors, although they operate at temperatures greater than 600 °C. Isostructural fluorides have been found to conduct at much lower temperatures, a consequence of the higher mobility of the fluoride ion which is a result of its smaller size and lower charge than the corresponding values for the oxide ion. Perhaps doped calcium fluoride will become the core of future high performance batteries?

Metals such as tungsten and molybdenum are highly refractory and difficult to work. However, they form volatile hexafluorides which can be reduced back to the metal with hydrogen. This offers a method of coating these metals onto surfaces where wear or chemical resistance are important.

Although not generally known to be commercially available in significantly large quantities, carbonyl difluoride, COF_2 and carbonyl chloride fluoride, $COClF$, are particularly versatile gaseous fluorinating agents for a wide variety of organic (especially aromatic) and inorganic transformations. Prepared from the reactions of phosgene ($COCl_2$) with any one of a variety of fluorinating agents, so far, these interesting materials have remained at the laboratory or pilot plant scale.

The prolonged exposure of elemental fluorine to the recently characterized carbon allotrope, 'Buckminsterfullerene', C_{60} results in the formation of heavily-fluorinated white solids with various claimed degrees of fluorination up to $C_{60}F_{60}$. These species were originally anticipated to have super-lubricity properties, but are, in fact, hydrolytically unstable, releasing hydrogen fluoride. Perhaps these remarkable new materials will have found their commercial niche when this book requires another revision in 15 or 20 years time.

Possibly the greatest breakthrough which could occur in industrial inorganic fluorine chemistry would be the development of a low-energy, low-cost process for the production of elemental fluorine itself. This would enable many present-day, speciality inorganic fluorides to be produced economically on a larger scale and thus command a wider range of applications. Indeed the large part of industrial fluorine chemistry, both organic and inorganic, would be transformed if cheap elemental fluorine became available.

Maybe our speculations will prove incorrect. However, we believe that the inorganic fluorochemical industry, which contributed to the early development of industrial metallurgy and was key to the development of the nuclear power industry and electronics, will continue to expand and diversify to meet the ever growing needs of our technological society.

9   BIBLIOGRAPHY

R.E. Banks, 'Preparation, Properties, and Industrial, Applications of Organofluorine Compounds', Ellis Horwood, Chichester, 1982.

'Fluorine: The First Hundred Years', ed. R.E. Banks, D.W.A. Sharp, and J.C. Tatlow, Elsevier Sequoia, Lausanne, 1986.

V.V. Bardin and Yu.L. Yagupolskii, in 'New Fluorinating Agents in Organic Synthesis', eds. L. German and S. Zemskov, Springer-Verlag GmbH and Co., Berlin, 1989, p. 1.

W. Büchner, R. Schliebs, G. Winter, and K.H. Büchel, 'Industrial Inorganic Chemistry', Verlag Chemie, Weinheim, 1989.

F.A. Cotton and G. Wilkinson, 'Advanced Inorganic Chemistry', 4th Edn., John Wiley and Sons Ltd., New York, 1980.

J. Griffiths, *Ind. Miner.*, July, 1990, p. 21.

J. Griffiths, *Ind. Miner.*, November, 1988, p. 77.

P. Hagenmuller, 'Inorganic Solid Fluorides', Academic Press, Orlando, 1985.

G.G. Hawley, 'The Condensed Chemical Dictionary', 10th Edn., Van Nostrand Reinhold Co. Inc., New York, 1981.

B.L. Hodge, 'Fluorspar: 1991 World Review', 1992 Metals and Minerals Annual Review, The Mining Journal Limited, London, 1992.

M. Hudlický, 'Chemistry of Organic Fluorine Compounds', 2nd (Revised) Edn., Ellis Horwood, New York, 1992.

'Kirk-Othmer: Encyclopædia of Chemical Technology', 3rd Edn., John Wiley and Sons Ltd., New York, 1980, Vol. 10.

G. Olah, G.K.S. Prakash, and J. Sommer, 'Superacids', John Wiley and Sons Ltd., New York, 1985.

'Ullmann's Encyclopædia of Industrial Chemistry', 5th Edn., Verlag Chemie, Weinheim, 1988, Vol. 11A.

CHAPTER 9

Industrial Gases

K. W. A. GUY

1 INTRODUCTION

The Industrial Gases considered in this chapter are those supplied by the world-wide Industrial Gas Industry. This encompasses a wide range of gases, derived from the atmosphere and chemical processes, and related products. The current world-wide market for industrial gas is an estimated US$16 billion, and an understanding of this market is fundamental to the understanding of the production and uses of industrial gases.

The world industrial gases industry has a number of important and permanent characteristics. It is an industry driven by technology and research, both in terms of its production technology and in the development of new gas applications for its customers. Its production and distribution methods are sophisticated and highly capital intensive. Thus a high proportion of sales are made on long-term (10–15 year) contracts, many of which have 'take-or-pay' clauses. This has led to an underlying stability, where changes in technology or application have been gradual rather than sudden in their effect. Production is now concentrated in the hands of a small number of international companies. The four principal ones, L'Air Liquide of France, BOC of the UK, PRAXAIR (Union Carbide) of the USA, and Air Products of the USA, account for nearly two-thirds of the market. Four small companies, Messer Griesheim and Linde of Germany, AGA of Sweden, and Nippon Sanso of Japan raise this total to 80%. Within this apparently stable market there is, however, intense competition, with each company seeking to advance its market share through the development of new applications or improved production technology. This has led to a wide variation in the growth and profitability of different gases, distribution systems, and regional markets. This in turn has led to long-term market growth of between one and a half times and two times capital gross domestic product (GDP).

2 DISTRIBUTION OF INDUSTRIAL GASES

There are three principal distribution methods for industrial gases, each of which has a different underlying economic basis: On-Site or Tonnage Supply, Merchant Supply, and Cylinder Supply.

2.1   On-site or Tonnage Supply

In this supply scheme, the industrial gas company builds a dedicated industrial gas plant next to its customer and supplies the gas 'across the fence' through a pipeline. This generally applies to large-volume users, and because of the dedication of the plant to that user, contracts are usually long-term, 10–15 years, on a 'take-or-pay' basis. A contract usually has two parts: (i) a fixed 'base facility charge' which provides for a return on the capital investment of the industrial gas company and (ii) a 'unit charge' which contains the costs of energy, maintenance, and labour and is subject to escalation against agreed indices. Such contracts allow the industrial gas company reasonable certainty of the return on its investment and at the same time reduces the risk to the customer and passes on lower energy costs in the form of reduced prices. An extension of this supply scheme is pipeline supply where more than one customer is linked to the same plant. In Europe, currently there are around 4000 km of industrial gas pipeline linking a variety of industries to very large industrial gas production plants. This arrangement offers customers the economies of scale on the basic production facility.

2.2   Merchant Supply

Smaller customers are supplied with industrial gases in a liquid form which they may then vaporize for their own use, or in some cases use the inherent refrigeration in the liquefied gas. These supplies are either from 'Standalone' merchant plants or from the addition of liquefiers to tonnage plants in what is known as a 'piggyback'. This latter method is favoured by the industrial gas industry, since it offers significant economies in both capital and operating costs. Delivery is mainly by road tanker and contracts are of varying lengths but may be up to five years. The merchant business is governed by the economic radius of supply and is an intensely competitive market.

2.3   The Cylinder Supply

Very small customers are supplied by compressed gases in cylinders. Cylinders are supplied to the customer on a rental basis with the gas content being a small part of the overall cost. Contracts are generally short-term.

The one remaining method of industrial gas distribution is the sale or leasing of the plant for its production to the customer. Clearly the industrial gas companies would rather sell the gas, but this ability to sell their equipment to the market does demonstrate the competitiveness of their basic production facilities and it remains a vital area of activity for all industrial gas companies.

3   ATMOSPHERIC GASES

The core of the industrial gas business is formed by air-separation gases, the most important of which are oxygen, nitrogen, and argon.

3.1 Air

The composition of air varies slightly from region to region. The main variable component is water vapour, which ranges in content between about 4% v/v, in the tropics, to a negligible amount in cold dry regions. Other components such as carbon dioxide may vary due to natural phenomena or human activities. The principal components of dry air by volume are:

nitrogen – 78.03%
oxygen – 20.95%
argon – 0.93%
carbon dioxide – 0.04%

with the noble gases neon, helium, krypton, and xenon, together with a small amount of hydrogen making up the rest. The projected extraction of industrial gases in the year 2000 is about $2 \times 10^8 \, t \, y^{-1}$ on a world-wide basis. However, this seemingly large-scale extraction will have no perceptible effect on air composition for two reasons. Firstly, the total amount of the earth's atmosphere is about 5×10^{15} tonnes and secondly many of the processes which use industrial gases would otherwise use air.

3.2 Separation

3.2.1 Cryogenic Plants. There are many methods for the separation of air into its components, but it has been the development of cryogenic distillation which has allowed the large-scale manufacture of atmospheric gases. The process was invented at the beginning of the century, based on the work of Hampson (UK), Linde (Germany), and Claude (France). Linde and Hampson both recognized that the Joule–Thomson cooling effect, adiabatic expansion of a gas causing cooling, could be used to liquefy air using the process described in Figure 1. Air is compressed, the heat of compression is removed in an aftercooler and it then passes through a heat exchanger where it is cooled before being expanded adiabatically through a throttle valve. The cooled air then leaves through the heat exchanger where it cools the incoming air and is recycled through the compressor.

Successive cycles cause the air temperature to 'bootstrap' down until eventually air liquefaction temperatures are reached. This process requires high compression pressures and was thermodynamically inefficient. Claude recognized that the efficiency could be improved by replacing at least part of the Joule–Thomson effect by an isentropic expansion using an expansion engine. The mechanical work obtained in this engine would produce a much greater cooling effect than Joule–Thomson alone. The Claude cycle is shown in Figure 2 and some form of the Claude cycle is now used in all large air liquefaction process plants.

The development of the distillation systems for separating air into its basic components has no particular names associated with it, but there has been a gradual development over a number of years, with most of the major industrial companies playing a part. The principles are similar to those for the fractionation of other mixed liquids such as petroleum, but it necessarily operates at a very low

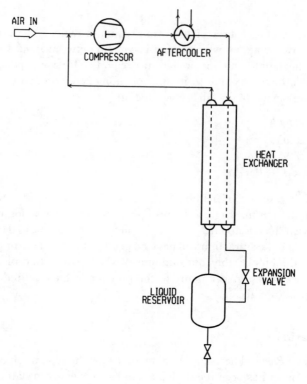

Figure 1   *Linde-cycle*

temperature. This means that special arrangements are required to provide reflux and paper boil-up for the process. The problem was solved by using the double column system shown in Figure 3. This makes use of the fact that oxygen boiling at low pressure is colder than nitrogen condensing at a higher pressure; hence the upper, low pressure, column condenses nitrogen in the lower, high pressure, column for reflux. At the same time the nitrogen in the high pressure column provides boil-up to the lower pressure column.

Air at its dew point temperature and at a pressure of approximately 6 bar (a) enters the high pressure column where it is partially separated. The more volatile nitrogen rises to the top of the column where it is condensed to provide reflux to the columns systems. The less volatile oxygen is stripped and accumulates at the base of the column as an enriched air stream containing about 40% oxygen. The oxygen-enriched stream from the bottom of the low pressure column then passes via a control valve to the upper column. It is separated into the more volatile, nitrogen, at the top of the column and the less volatile, oxygen, at the bottom. The pure liquid oxygen is cold enough to condense nitrogen in the high pressure column to provide reflux and, by suitable design, this reflux may be used for both the high and low pressure columns. Two other features should be noted: the noble gases and hydrogen are not condensed and accumulate in the reboiler–condenser at the top of the high pressure column.

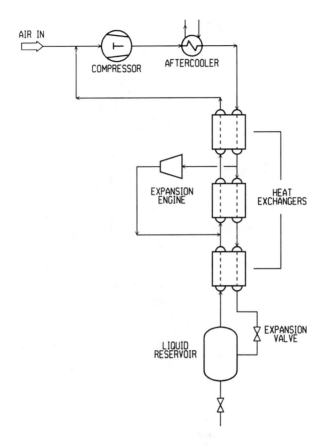

Figure 2 *Claude-cycle*

This stream is continuously purged to prevent blanketing of the reboiler–condenser by noncondensables. Neon may be recovered from this stream on some plants. Argon has a volatility between that of nitrogen and oxygen and therefore concentrates in the low pressure column approximately mid-way between the feed plate and the liquid oxygen in the sump. A stream may be removed for further processing and purification in a separate column if argon recovery is required. The remaining components of air, krypton and xenon, are less volatile than oxygen and accumulate in the sump of the low pressure column. It is possible to recover these gases by taking a side stream of liquid oxygen and processing it separately.

One other feature of the process which has not been mentioned up until now is the need to remove other trace components of air that would cause problems in the process. These components include water vapour and carbon dioxide, which would be solid at cryogenic temperatures and cause blockages, and trace hydrocarbons which would pose a threat to the safety of the plant if they came into significant contact with liquid oxygen.

Much of the history of the development of commercial-scale air separation plants has been concerned with the development of front-end equipment to

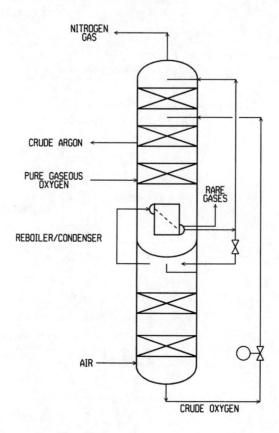

Figure 3   *Double Column Air Separation*

remove these trace components. In the early days chemical means, such as caustic soda, and absorption systems, such as silica gel, were used to remove contaminants. It was soon recognized that a more efficient way of removing these would be to provide a cold surface for them to condense on and from which they could be later evaporated. This led to the concept of the reversing cycle air separation plant. Each air separation plant was equipped with two large regenerator vessels, each consisting of a large pressure vessel filled with small coarse pebbles. During the process cycle one regenerator, which had been previously cooled, would receive the incoming air and cool it down to near air liquefaction temperatures, thereby removing the contaminants by condensation. At the same time, the other vessel which had previously been warmed, by the air passing through it, would be cooled down by the waste nitrogen from the air separation process and the contaminants would be evaporated. After a period of about 10 minutes the two vessels were changed over and the process repeated. This novel concept allowed the construction of much larger air separation plants, but since regenerators could still handle only one stream in each direction up to 40% of the air entering the plant had to be treated by conventional means to allow for heat exchange with the product streams.

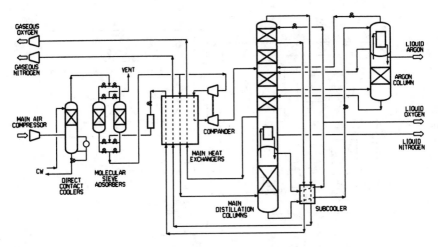

Figure 4   *Modern Air Separation Unit*

This problem was overcome by the development in the early 1960s of commercial plate-fin heat exchangers, which allowed as many as seven or eight streams per heat exchanger. Each heat exchanger was constructed in the form of an aluminium matrix by Brasing and contained about 100 plate-fin passages. The passages could be connected to provide any desired configuration. Through the availability of this equipment, the regenerators in the air separation plant were replaced by reversing heat exchangers. Air separation plants were then able to operate at lower pressures and to produce a wider range of product streams than before. Plate-fin heat exchangers were produced in modular form and, by manifolding them together, substantial plants could be built without a significant increase in risk. In the early 1970s there was a significant increase in the size of air separation plants, thus affording the economies of scale that many other industrial processes had been looking for, and a consequent increase in the demand for industrial gases.

In the late 1970s the development of zeolites caused a re-think in the design of air separation plants. It became possible, once more, to use front-end absorption systems economically, and most modern plants are now fitted with a 'molecular sieve' front-end.

The typical line diagram for a modern air separation unit for the production of oxygen and nitrogen is shown in Figure 4. After filtration to remove dust, ambient air is compressed in a high efficiency turbo-compressor to a pressure of about 6 bar (a). The air is then cooled in a direct contact after-cooler, sometimes using a chilled water supply, before passing to the molecular sieve adsorption system. In the adsorption system, carbon dioxide and water vapour are removed and the air then passes through plate-fin heat exchangers where it is cooled to near its dew point. The air is then passed to the distillation column system where it is separated into oxygen, nitrogen, and, if required, argon. Refrigeration for the process is provided by taking a side stream of the incoming air and passing it through a high efficiency expansion turbine, the output from which passes to the distillation

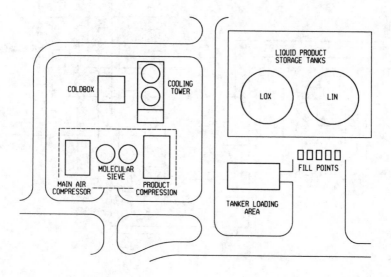

Figure 5 *Modern ASU Layout*

column. The effluent streams from the distillation columns pass back through the heat exchangers where they provide cooling to the incoming air stream. All of the products of the distillation process are available for use as industrial gases, with the exception of a small amount of nitrogen which is required for the regeneration of the molecular sieve vessels.

The whole system is designed for the lowest possible pressure drop since the main raw material for the production of atmospheric industrial gases is the energy consumed by the turbo air compressor. An air separation plant such as that in Figure 4 would form the basis of an industrial gas complex which would also include product compressors, nitrogen or oxygen liquefiers, argon purifiers, and possibly rare gas processing units.

Atmospheric gas plants on their present optimum economic scale of several thousand tonnes per day of oxygen and nitrogen products are large energy consumers. The theoretical minimum energy required to separate one normal cubic metre of oxygen from air at 1 bar is 0.074 kWh. However, in actual plants the energy required is currently around $0.38\,\mathrm{kWh\,Nm^{-3}}$, because of the many thermodynamically irreversible features of the real design. Similarly, the theoretical minimum energy for the liquefaction of nitrogen is $0.25\,\mathrm{kWh\,Nm^{-3}}$ whereas in practice a very efficient plant would achieve $0.45\,\mathrm{kWh\,Nm^{-3}}$. This performance has improved dramatically over the past 15 years when comparable numbers would have been $0.5\,\mathrm{kWh\,Nm^{-3}}$ for oxygen and $0.85\,\mathrm{kWh\,Nm^{-3}}$ for liquid nitrogen. A large air separation plant producing, say, $2000\,\mathrm{t\,d^{-1}}$ of gaseous and liquid products would have a power requirement in the order of 35 MW.

A typical layout for a commercial air separation plant is shown in Figure 5 and a picture of the same plant is shown in Figure 6.

Figure 6 *Modern Air Separation Unit*

3.2.2 Non-cryogenic Plants. Many methods for the separation of air other than by cryogenic distillation have been considered or tried, some of which are in commercial development. These have included a number of chemical processes in which unstable oxides or peroxides are formed and oxygen is recovered from them either by heat or pressure reduction, through complexes which form loose oxygen compounds in a manner similar to oxyhaemoglobin formation. All of these methods are regularly reviewed by the major industrial gas companies and a few of them still show some promise.

The most promising developments in non-cryogenic separation technologies have been the production of medium-purity nitrogen (0.5–5% oxygen) and oxygen (below 95% oxygen) through the use of adsorption technology and membrane technology.

It is estimated that currently these plants supply less than 5% of the nitrogen market, but it is projected that 30% of it will be non-cryogenic by the late 1990s. The other 70% of the nitrogen market is not readily amenable to non-cryogenic technologies for reasons of purity, pressure, use patterns, refrigeration, *etc.* Penetration of the oxygen market is even smaller, but it is projected that at the lower end of the market, particularly where purity requirements are less critical, as in combustion processes, significant market penetration will also occur through the use of adsorption processes. One consequence of the reduced cost of nitrogen and oxygen at low volumes from these technologies is that up to half the sales will be to applications that were not previously economic. The penetration of the industrial gas market by non-cryogenic technologies is currently limited by their ability to meet the product requirements of most of the customers whilst cryogenic technology can meet almost all of the requirements. A further factor that inhibits

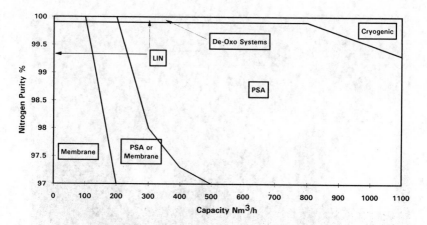

Figure 7 *Cost-effectiveness for N₂ Systems*

the pentration of the market by non-cryogenic technology is that most non-cryogenic plant is modular in form and does not benefit from the same economies of scale as cryogenic plant. In addition, contrary to many statements, cryogenic processes usually have lower specific energy than adsorption or membranes for gas separation. In general, they are only competitive when compared with the supply of merchant gases in a liquid form. Figure 7 shows the current status of the cost-effectiveness of different mechanisms for producing nitrogen. It can be seen that membranes and adsorption systems are competitive at some combination of low volumes and low purities. A similar chart could be produced for oxygen. Further development will undoubtedly raise both the purity and volume limits at which alternative processes are competitive and this has been taken into account in the projection of market share above.

Figure 8 shows the flow scheme for an adsorptive air separation plant. The adsorbant is chosen to have acceptable relative and total affinities for the adsorption of the components of the gas mixture, which are a complex function of pressure, temperature, and composition. Since the solid adsorbant is a stationary phase, the process must be operated in a cyclic manner using multiple beds to create a continuous supply of product gas. On a typical zeolitic adsorbant, the adsorption capacity is greater for nitrogen than oxygen and increases with partial pressure and decreasing temperature. It is possible to operate the separation process by either pressure-swing or by temperature-swing. However, the latter has a rather long cycle time and because the bed capacities are small would lead to extremely large beds.

Thus zeolitic air separation is usually accomplished by pressure swing of the adsorbent beds. Since nitrogen is preferentially adsorbed, an oxygen-rich gas is produced on the adsorption stage and nitrogen-rich gas at the desorption stage. It is possible to operate the process in one of two ways, either by feeding the system with air at an elevated level and desorbing at atmospheric pressure, *i.e.* pressure swing adsorption (PSA); or by feeding the process with air at atmospheric pressure and desorbing by use of a vacuum pump, *i.e.* vacuum swing adsorption

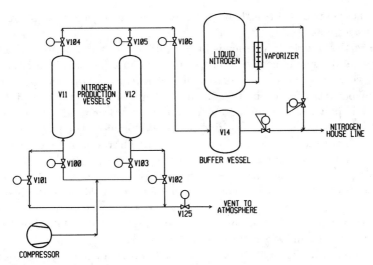

Figure 8 *Nitrogen PSA*

(VSA). Combinations are of course possible. In the system shown in Figure 8, which shows a typical three-bed VSA process, each bed undergoes the following steps:

- adsorption of nitrogen (with some oxygen);
- evacuation of the adsorbant;
- purge with oxygen product;
- pressure equalization; and
- repressurization.

Purge may be provided from stored product or directly from another adsorbant bed. Equalization may be between any pair of beds. Repressurization may be from the product and/or feed and/or an adsorbant feed. This process is currently capable of producing oxygen volumes at purities of between 91–94% and quantities up to $100\,t\,d^{-1}$ of oxygen. Further scale-up of the process is possible, but a number of problems caused by variations within the adsorbant bed will need extensive development, probably using computational fluid dynamics, in order to develop a better understanding of the process.

The production of nitrogen is carried out with a similar process but using carbon molecular sieve which has higher adsorptive capacity for oxygen over nitrogen. This process is run using pressure-swing adsorption and has the advantage of producing the nitrogen at nearly the feed pressure. It is generally run as a two-bed process, but still offers the opportunity to vary the adsorption by changes to the sequence and timing of the adsorption, depressurization, rest, purge, equalization, and repressurization steps.

Nitrogen production by adsorption has been practised commercially for about 15 years. But it is only in the past five years that the production of nitrogen by membranes has been commercially viable, and the economic production of oxygen has yet to be realized. However, the technology is young and is the subject of intensive research.

The separation process depends on the different transport rates of oxygen, nitrogen, and argon molecules in polymeric materials. The ideal material would exhibit very high selectivity and very high permeability. In practice there is a trade-off between these properties. For an economic process, a very thin film of the separating polymer is required (of the order of 0.1 mm) and thus must be supported to withstand the operating pressure difference. The polymer must be stable with time and capable of forming into fine fibres in order to achieve an economic surface to volume ratio. Membrane processes are attractive because of their simple continuous operation and correspondingly low capital cost. An increase of polymer selectivity above current levels would also offer the potential of low-energy requirements for medium-purity ($> - 1\%$ oxygen content) nitrogen. The process is simple: air is compressed, pre-cleaned to remove impurities harmful to the membrane, and separated in a single-stage membrane to produce nitrogen at the required purity and pressure. In order to match current cryogenic plant costs, membranes need a selectivity of about 8, although at lower purities selectivity could be as low as 5 to achieve parity. Present membrane module selectivities are in the range 4–6. Thus current nitrogen membrane processes operate at higher specific power than nitrogen PSA's but on a small scale they exhibit the benefits of a lower capital cost.

3.3 Oxygen

3.3.1 Production. Oxygen was first produced commercially towards the end of the last century by a number of now obsolete chemical processes. Air separation was revolutionized by the development of the cryogenic processes for its liquefaction and distillation as described previously, and most industrial oxygen is produced by this process today.

Some low-purity oxygen is currently produced by the adsorption processes already referred to, and an even smaller amount as a co-product in the electrolysis of water for hydrogen production.

Two significant developments led to the development of the current oxygen market. The first was the rapidly-increasing demand from the steel industry for the use of oxygen in converting iron into steel using modifications of the Bessemer process. The second was the innovation by Leonard Pool, founder of Air Products, of building the production unit next to the end-user and so developing the on-site concept.

Oxygen has grown from a few thousand tonnes per day in the early 1940s, through $50\,000\,t\,d^{-1}$ in the early 1970s, to a present world production of around $150\,000\,t\,d^{-1}$. Current forecasts suggest that this will more than double by the end of the century.

The price of oxygen to a user depends greatly on the method and scale of the distribution to him. In Europe currently, oxygen prices may range from about £3 per $100\,Nm^3$ for the largest customers on an on-site contract through to £6–£10 per $100\,Nm^3$ for merchant liquid customers to £20+ per $100\,Nm^3$ for small amounts delivered in cylinders.

3.3.2 Uses. Oxygen is an extremely reactive gas, combining with many substances at ordinary ambient conditions, and with many more at elevated pressure and temperature, and this affords the main use for oxygen. In general, oxygen use can be divided into two distinct parts; combustion uses and oxidation: reaction uses.

Over 60% of current oxygen production in Europe is used by steel and related industries. Initially this oxygen was used in the conversion of iron into steel in the basic oxygen furnace (BOF). The use of oxygen instead of air increases the speed of conversion and increases the productivity of the steel plant. This use started on a large scale in the late 1950s and was the basis for the rapid increase in industrial oxygen production for at least two decades. Even with the rationalization of the steel industry, oxygen use has continued to grow. Mini mills, which take a high proportion of scrap as feedstock, use oxygen in their electric arc furnaces. The large integrated mills, because of environmental pressures on the production of coke in coke ovens, have gradually been switching to coal injection processes and the use of oxygen enrichment in the blast furnace to enhance the flame temperature. The basic oxygen furnace uses about 0.07 tonnes oxygen/tonnes steel. Oxygen injection in blast furnaces, where practised, has levels as high as 0.25 tonnes oxygen/tonnes steel and offers the potential for the doubling of oxygen demand by the end of the century. Environmental pressures on coke ovens have led to the development of other processes for the direct reduction of iron using iron ore and coal. A typical process is the COREX[T] process of Voest Alpine which has an oxygen requirement of about 0.75 tonnes oxygen/tonne iron.

Oxygen also plays an important part in many other activities in the steel industry, from oxy-burner preheating to scarfing and cutting.

Oxygen is also used in many other furnace applications such as ferrous and non-ferrous melting and in industries such as glass manufacture where oxy-burner technology offers higher temperatures, greater productivity, reduced power consumption, and/or longer furnace life. The next largest user of oxygen, accounting for about 20% of the market, is the chemical industry. Oxygen is generally used in place of air in oxidation reactions, where it confers a number of benefits. These include faster reaction and greater reactor productivity, improved yield or energy efficiency by avoiding the dilution effects of the nitrogen in air, and possibly reduction in effluents due to the lack of inert venting.

Oxygen is used on a large scale for direct oxidation reactions:

- ethylene to ethylene oxide;
- propylene to propylene oxide;
- titanium dioxide by the chloride route;

as well as in the production of chlorinated hydrocarbons and in particular in the manufacture of vinyl chloride. Oxygen-using processes exist for the manufacture of vinyl acetate, acetylene, terephthalic acid, cyclohexanone, nitric acid, nitrous oxide, sulfuric acid, acetaldehyde, acetic acid, maleic anhydride, methylmethacrylate, and many other chemicals.

Large-scale oxygen use is developing in the production of synthesis gas (hydrogen and carbon monoxide) by partial oxidation (POX) of suitable hydrocarbons.

The greatest potential increase in oxygen use will come from current and future environmental legislation which limits effluents and emissions. Industry has two choices: either it changes the process to eliminate the problem or it develops methods for destroying and safely disposing of the hazardous substances. Many of the changes in the metal and chemical industries have been driven by environmental consideration, as discussed in Chapter 2.

Oxygen will play a key role in other sectors:

- gaseous effluents – oxy-burner NOX correction;
- substitution of air to reduce effluent nitrogen;
- liquids – BOD by oxygen sparging;
- oxygen/ozone sterilization;
- superciritical oxidation of slurries;
- solid effluents – oxy-burner vitrification;
- oxy-burner metal recovery systems;
- high-temperature incineration;
- energy;
- coal gasification combined cycle (CGCC);
- partial oxidation of residuals (POX).

In addition to the above, oxygen will continue to be used in its traditional role in cutting, welding, lancing, diving, and medical uses.

3.3.3 Nitrogen. Most nitrogen for industrial use is produced as a co-product with oxygen in air separation plants. Such plant can generally produce between 2–3 times as much nitrogen as oxygen, at a purity less than 5 p.p.m. of impurities. With the increase in oxygen demand from the late 1950s nitrogen has been available on a large scale. The world-wide production has thus increased from a few thousand tonnes per year in the early 1940s to approximately $150\,000\,t\,y^{-1}$ by 1990. Since nitrogen produced by this method is a co-product, general practice in the industry is to regard it as having zero cost at low pressure. The price of nitrogen to the customer depends on its state, liquid or gas, and in the case of gas the end-use pressure required. Current market prices for nitrogen vary from about £1–£2 per $100\,Nm^3$ for large customers, delivered by pipeline through about £7–£10 per $100\,Nm^3$ for large liquid nitrogen customers, to £20+ per $100\,Nm^3$ for cylinder nitrogen.

Nitrogen is also produced by small, on-purpose 'nitrogen generators'. These are small cryogenic air separation plants producing high purity, gaseous nitrogen, with volumes between $40–400\,t\,d^{-1}$. A typical plant is shown in Figure 9. Some nitrogen generators have been designed to produce 'ultra high purity' gaseous nitrogen for the electronics industry, where impurity levels in the p.p.b. range are required.

As mentioned previously, where nitrogen purity is less significant, pressure swing adsorption plants, similar to the type shown in Figure 8, are used to provide gaseous nitrogen in volumes generally from $1–50\,t\,d^{-1}$.

It is possible to produce low-purity nitrogen, mixed with about 10–15% CO_2 using inert gas generators. But this is not considered generally an industrial gas.

Figure 9 *Cryogenic Nitrogen Generator*

Nitrogen is a chemically insert gas, not reacting generally at ambient conditions, and showing only limited reactivity at high temperature. This inertness is generally the basis of many of its industrial applications.

Nitrogen is used in large volumes in metallurgical, chemical, and flammable material transportation processes, where the presence of air would present the opportunity for fire or explosion hazards, or would cause undesirable oxidation of the product. A major use of nitrogen occurs in petrochemical plants where it is used to provide an inert atmosphere before filling, or after discharging, flammable materials. The use of nitrogen allows safe changes in the process and prevents the formation of dangerous air/vapour mixtures. It is commonly used to purge vessels and equipment that have contained flammable materials to allow cutting and welding operations to take place. It should be noted that because nitrogen is an asphyxiant, precautions have to be taken to ensure a suitable level of oxygen exists before and during vessel entry. Nitrogen is also used in the petrochemical industries to blanket product storage, to prevent the oxidation and thereby discolouring of fibres, and as a diluent to moderate or inhibit reaction in reactors.

In the metallurgical industries nitrogen is used to provide an inert atmosphere to prevent formation of metal oxide scale on the surface of metals being treated, and it plays some part in reactive heat-treatment processes. It is seeing an increasing use in the transportation of flammable particulates such as powered coal, and in providing a gas seal for the charging operations for such particulates.

In recent years two new processes have developed which use large volumes of nitrogen. Firstly, it has found an increasing use in the oil and natural gas industries for enhanced recovery of their products. Because of the nature of the underground reservoirs it is possible to pump nitrogen at high pressure (200 bar +) into one end of a field, and thus to increase the production of oil or natural gas from the other end of the field without significant dilution occurring. The inertness of the nitrogen has many advantages over other potential methods of raising the pressure in the field. Not the least of these is that when nitrogen breakthrough does occur, it is relatively easy to remove it from natural gas by cryogenic distillation.

The natural gas industry is also responsible for the second large-scale use of nitrogen. Because in most of mainland Europe the natural gas systems have been developed around gas from the Groningen field in Holland, which has significant hydrogen content, gas from other sources would be too rich for the system. Large volumes of nitrogen are used to correct the calorific value of richer natural gas sources. Typically a single blending station may use $6000 \, t \, d^{-1}$ of high purity nitrogen.

A rapidly growing demand from the semi-conductor (silicon wafer) sector of the electronics industry for ultra-high purity nitrogen has led to the development of specialized plants for its production. The product specification for the nitrogen requires impurity level in the p.p.b. range and also precludes particulate matter. Such plants are based on standard cryogenic nitrogen generators but contain significant amounts of additional equipment to reduce impurity levels and to filter out particulate matter. Most of the piping in these plants consists of electro-polished stainless steel in order to maintain the required purity. Whilst the volumes of nitrogen used for this purpose are generally small, in the order of $50 \, t \, d^{-1}$ plant the value of the nitrogen product is extremely high.

Many other uses for nitrogen include the provision of an atmosphere to prevent the oxidation of the tin bath in the float-glass process; the packaging of processed foods, pharmaceuticals, and fibres to prevent spoilage by air; and the storage of vegetable oils, raw foods, paints, and perfumes. In fact, in any application where the presence of air could degrade the product. High-purity nitrogen has also found a limited application in one process for the production of ammonia.

3.3.4 Liquid Nitrogen. The second principal use of nitrogen is in its liquid form as a refrigerant, where its inertness makes it suitable for direct contact with most materials that are to be cooled. Initially the use of liquid nitrogen was limited to applications where extreme low temperatures were essential. In recent years, however, liquid nitrogen has been a competitive freezing medium for many other substances. Because its use entails a lower capital investment than mechanical refrigeration methods, it is competitive in systems that are non-continuous such as seasonal foods; and because liquid nitrogen freezing occurs rapidly, it is competitive with other systems where the quality of the frozen product is enhanced by more rapid freezing, for example the formation of smaller ice crystals in the freezing of products with a high water content.

Typically, liquid nitrogen is used in direct contact with the material by immersion or liquid spraying. This allows rapid heat transfer from the material to the refrigerant, and the effluent nitrogen may be used to pre-cool incoming material. Applications are in two classes: those which *require* a very low temperature such as:

- freeze grinding, for materials which are normally soft or resilient;
- shrink fitting, for the assembly of engineering components;
- hardening of soft materials to allow machining or deflashing;
- metal recovery systems where ferrous materials may be fragmented to allow them to be removed from non-ferrous materials;
- preservation of biological materials such as blood and semen;
- condensation of trace contaminants in effluent air systems;

and applications which do *not require* very low temperatures:

- food freezing, particularly of high value soft fruit, shrimp, *etc.*;
- pipe freezing, for stopping the flow of liquid in a pipe to allow maintenance;
- in transit refrigeration of foods;
- droplet dosing of containers in rapid filling processes such as beer canning.

3.3.5 Argon Most commercially produced argon is obtained by further distillation of the argon-rich fraction from the low-pressure column of an air separation plant. Some argon is obtained by its recovery from the purge gas system on ammonia plants, where it concentrates as the nitrogen is used in ammonia production. The concentrated argon fraction from an air separation plant typically contains 10–12% argon, the rest being mainly oxygen. This mixture is distilled in a column for which the reflux is obtained by by-passing part of the crude oxygen stream from the high-pressure column through an auxiliary re-boiler condenser. The column has a large number of stages and generally produces argon at about 97–98% purity, the remainder being oxygen. The bottom product of the argon column, oxygen, is returned to the low pressure column. With modern technology using packed columns it is possible to produce high purity argon by distillation; however, in most industrial plants the argon is further purified by removing the oxygen with hydrogen in a catalytic 'de-oxo' unit. After further purification to remove the hydrogen, a highly purified liquid or gaseous argon product is obtained.

In an ammonia-purge gas recovery plant, the argon in the air feed to the plant is concentrated, typically 9–11%, in the final purge gas stream, mixed with nitrogen, hydrogen, and other minor impurities. Argon is recovered by a complex system of fractional distillation. The cost of argon production by this method is significantly more than for argon from air separation, and is therefore only viable when argon supply from air separation is limited.

With the growth in oxygen production in the early 1950s and 1960s the potential supply of argon should have increased significantly, but it was only in the 1970s with the development of new applications, particularly in steel making, that production facilities were added to most air separation plants. European argon production has risen from less than $100\,t\,d^{-1}$ in 1940 to $1600\,t\,d^{-1}$ in 1990.

Argon pricing is not related to the cost of production. This is because in order to sustain competitive oxygen pricing, the industry has always given co-product credits from the argon to the oxygen. The oxygen prices given in the previous section have barely changed from the early 1970s, both for reasons of argon credits and improvement in production matters. Current argon pricing in Europe ranges from about £50–£70 per $100\,Nm^3$ for high volume liquid or gaseous argon users to £200 plus for $100\,Nm^3$ for cylinder gas.

Argon is used as an inert gas where in high-temperature applications the cheaper alternatives, nitrogen and carbon dioxide, would be reactive. Typically, this use occurs in high-temperature metallurgical processes. Argon finds its major use in stirring and blanketing applications in the manufacture of steel. Argon uses in steel making range between 1 and $2\,Nm^3\,t^{-1}$ steel. Current processes under investigation for the use of argon in the production of high quality steels and in the spray forming of sheet steels could lead to an order of magnitude increase in this demand. Argon is widely used as a shielding gas in electric arc welding, as an inert atmosphere in powder metallurgy and uranium fuel rod processing, and on a smaller scale for filling electric lamps and other electric discharge devices.

3.4   Neon, Krypton, and Xenon

Neon, krypton, and xenon can be recovered from air separation plants by further distillation processes. All of them have relatively limited uses, principally in specialist electronic lighting systems. In general, they are only recovered on a few very large air separation plants and their consumption is relatively small. Prices range from about £$1\,1^{-1}$ for neon, £$1.50\,1^{-1}$ for keypton, to £4 to £$6\,1^{-1}$ for xenon.

3.5   Helium

Whilst some helium can be recovered in the same stream as neon by air separation, most helium is produced by its recovery from natural gas. Large plants for this purpose exist in the USA, Eastern Europe, and Algeria. The helium content of natural gas varies significantly in different gas fields, ranging from less than 0.1% v/v to around 7%. Usually, the gas fields from which helium is extracted by cryogenic separation have contents of at least 1%.

Helium has a wide range of minor and specialized uses. Its low density without flammability leads to a use in blimps, balloons, and airships. It is also extensively used in mixtures with a few percent oxygen for deep-sea breathing gas.

The very low temperature of liquid helium gives it numerous uses in superconductivity activities.

Because all helium is produced as liquid and transported in bulk as liquid, the cost does not vary significantly for liquid or gas. High purity helium prices for most users range from £400 to £500 per 100 Nm^3.

4 HYDROGEN AND CARBON MONOXIDE

4.1 Production

Discussion of production of hydrogen and carbon monoxide as industrial gases excludes production of hydrogen in ammonia, methanol, or of its mixtures with carbon monoxide as a synthesis gas for the production of organic chemicals such as acetic acid. It has been estimated that industrial gas volumes of hydrogen in Europe represent less than 2% of the hydrogen generated for any purpose. However, hydrogen and carbon oxide produced by the industrial gas industry are generally of higher purity than those produced by end-users.

High-purity hydrogen may be recovered from brine electrolysis, water electrolysis, or from by-product streams in the petrochemical and refining industries. Similarly, carbon monoxide may be recovered from any process where incomplete combustion of carbon-containing materials has occurred, such as from blast furnace gas or BOF gas. However, the most common production method is by the reforming of hydrocarbons, where they are produced as co-products. It is for this reason that carbon monoxide has been included in this section.

There are three principal methods for the production of hydrogen and carbon monoxide from hydrocarbon feed stocks, namely: steam methane reforming (SMR); auto-thermal reforming (ATR); and partial oxidation (POX).

The production of hydrogen and carbon monoxide from hydrocarbon feed stocks has a similar basis to the old process for producing town gas from coke in the water–gas reaction, where the oxygen combined with the carbon to form carbon monoxide and liberated the hydrogen from the water molecule. Whilst in principal it is possible to use any hydrocarbon feed-stock to produce hydrogen and carbon monoxide, the availability of large quantities of natural gas which is readily transportable and easy to handle has meant that earlier processes which used naphtha as a feed-stock have largely been superseded.

The main cost elements in the production of hydrogen and carbon monoxide are the cost of the feed-stock and the capital cost of the plant used in their synthesis. The principal technical challenge in the process is to prevent undesirable side reactions from occurring, which range from the further oxidation of carbon monoxide to carbon dioxide, to the deposition of carbon. Since the process invariably runs with excess steam, reversion of hydrogen back to water is not a problem. The problem of carbon monoxide conversion can be overcome by recycling carbon dioxide to the process, thus pushing the equilibrium towards carbon monoxide. The deposition of carbon can be prevented by the selection of suitable catalysts and/or by running with an adequate excess of steam.

The principal difference between the three methods described is in the provision

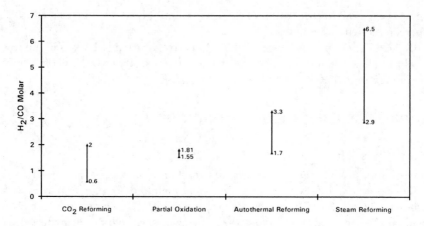

Figure 10   *Typical H$_2$:CO Ratios at Reformer Outlet*

of the source of heat and of the amount of catalyst required. Steam methane reformers use a number of tubes filled with catalyst which are heated to a high temperature by an external heat source. The auto-thermal reformer uses a small amount of oxygen in the feed, which reacts with a small amount of the feed-stock to produce the heat internally. The partial oxidation reactor uses a larger amount of oxygen and relies on a quenching mechanism to produce the correct mixture of hydrogen and carbon monoxide. The POX reactor contains no catalyst.

The choice of process route depends on the cost of feed-stock, the availability of capital, and on the hydrogen/CO split required. Figure 10 shows a typical range of hydrogen/CO ratios at the reformer outlet.

4.2   Steam Methane Reforming (SMR)

The SMR system shown in Figure 11 is typical of the most common hydrogen and CO production systems. Steam methane reforming on a nickel catalyst followed by water-gas shift reaction has long been used to produce hydrogen. The reactions are:

$$CH_4 + H_2O \rightarrow + 3H_2 \qquad \text{highly endothermic}$$
$$CO + H_2O \rightarrow CO_2 + H_2 \qquad \text{mildly exothermic}$$

The natural gas is hydro-desulfurized, if necessary, and then mixed with steam and preheated by the reformer effluents to 450–650 °C before entering the reformer, where the reactions are carried out in catalyst-filled tubes, up to 13 metres long, mounted vertically in a furnace fired by fuel/natural gas. Operating pressures typically range between 14 and 30 bar and outlet process gas temperatures range from 800 °C to 900 °C. Steam-to-carbon molar ratio varies between 3 and 5, and depends on product purity requirements and the type of purification process. The process conditions within the reformer are selected to prevent the following carbon formation reactions occurring, which would damage and deactivate the catalysts:

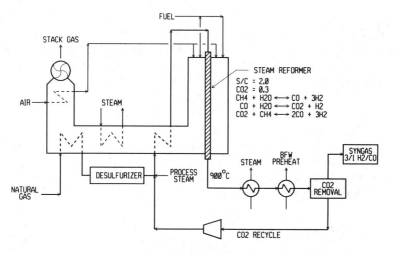

Figure 11 *Typical Steam Methane Reformer for Syngas Production*

$$CH_4 \rightarrow 2H_2 + C$$
$$H_2O + C \rightarrow H_2 + CO$$
$$2CO \rightarrow CO_2 + C$$

Of the three possible carbon-forming reactions, thermodynamic equilibrium supports carbon formation by the methane cracking reaction, occurring at temperatures below 700 °C. In this temperature range, however, thermodynamic equilibrium supports carbon removal by the other two reactions, and the kinetics mean that carbon is removed at a faster rate than it is formed. Carbon formation is a complex issue and is influenced by steam to carbon ratio, CO_2-to-carbon ratio, oxygen to carbon and hydrogen to carbon, heat flux across the tube and the catalyst, and reformer operating temperature and pressure. To achieve high CO contents in the reformer outlet gas, the steam-to-carbon ratio has to be reduced to the minimum consistent with avoiding carbon formation. Reforming must be carried out at the maximum possible temperature, and CO_2 must be recycled from the acid gas removal section to the reformer feed. To avoid carbon formation and to achieve reformer tube life at the higher operating temperature (900 °C) reformer operating pressures are limited to a maximum of 17 bar. With steam-to-carbon ratio of approximately 2.0 and total CO_2 recycle of the CO_2 generated in the process, a H_2/CO ratio of approximately 3.0 is possible.

Recycle of the CO_2 to the reformer promotes the following reaction:

$$CO_2 + CH_4 \rightarrow 2CO + 2H_2$$

If a source of CO_2 is available for import, hydrogen/CO ratios of below 1.3 are achievable.

The reformed synthesis gas mixture leaving the reformer then passes through a heat recovery system where it is used to raise steam for the process and for preheating boiler feed water. The optimization of the heat recovery systems is critical to the process economics.

There are now two potential process routes depending on the requirements for hydrogen and CO. Some of the synthesis gas may proceed via catalytic shift conversion, where the CO is 'shifted' to CO_2:

$$CO + H_2O \rightarrow CO_2 + H_2$$

The stream then passes to CO_2 removal, which can be done with most of the well-known acid gas removal systems, such as MEA. Depending on the purity specification, the hydrogen stream may be subject to methanation where the following reactions remove impurities:

$$CO + 3H_2 \rightarrow CH_4 + H_2O$$
$$CO_2 + 4H_2 \rightarrow CH_4 + 2H_2O$$

and may be further purified by, for example, pressure swing adsorption.

The alternative process route is for the recovery of CO and hydrogen from the synthesis gas stream, which will involve the removal of acid gas, the separation, and the purification of the product. As in the case of hydrogen, any of the well-known acid gas removal systems can be used, followed by cryogenic separation, chemical complexation, adsorption, or membrane separation, or any combination. The process configuration chosen is completely dependent on the purity requirements for the products and the value placed on the co-product. In general, large plants have tended to use a cryogenic separation route, but adsorption and membrane plant routes are developing rapidly.

4.3   Hydrogen

4.3.1   Production. Principal methods for the production of gaseous hydrogen are by the reforming of hydrocarbon feed stocks. The major production of hydrogen occurs in oil refineries where it is used by the producer. Small amounts of hydrogen are produced by the dissociation of ammonia, and by electrolysis.

Hydrogen is also produced in a liquid form, mainly as a consequence of its use as a rocket fuel in the early 1960s. The liquefaction of hydrogen presents no particular difficulties and is similar to that used for liquefying nitrogen, except that hydrogen is used as a refrigerant. One minor problem with hydrogen liquefaction is that the concentrations of *ortho*- and *para*-hydrogen change from 25% *para* at room temperature to 99.8% *para* at liquid hydrogen temperatures. The *ortho/para* transition is exothermic and it is therefore necessary to effect this transition before liquefaction, to avoid subsequent evaporation of the liquid hydrogen.

Volumes of hydrogen used have grown but are still relatively modest when compared with other industrial gases and the industrial gas market of about $100 \, t \, d^{-1}$ is a very small part of the total hydrogen produced. However, growth in hydrogen demand from the refinery sector, due to environmental legislation, will see a significant increase in this market over the next ten years with predicted volumes of more than $1000\,t\,d^{-1}$. Hydrogen prices vary significantly with volume and purity, ranging from as little as £3 per $100\,Nm^3$ for large, impure-hydrogen customers through £30 to £50 per $100\,Nm^3$ for large cylinder customers, and £60 to £80 per $100\,Nm^3$ for high purity liquid.

4.3.2 Uses. Nearly 80% of all hydrogen produced is used as chemical feed-stock for the production of ammonia or methanol. This is followed by its use for petroleum hydrogenation or desulfurization. Hydrogen for these applications has generally been generated by the owner of the down-stream process. More recently, environmental pressures on the petroleum refiners have lead to an increase in the demand for hydro-desulfurization and hydro-cracking. Participation of the industrial gas companies in these larger projects is becoming more usual.

Most of the marketed hydrogen is used where small volumes of high-purity hydrogen are required, particularly in cases where self-generation would be uneconomic. These applications tend to make use of hydrogen's strong reducing capability or its ability to saturate unsaturated hydrocarbons. Typical examples range from the production of hardened oils through pharmaceuticals to reducing atmospheres in metallurgy. Much work has also been carried out in recent years on the use of hydrogen as a potential transport fuel, either through burning in combustion engines or via use in fuel cells. The advance to this 'hydrogen economy' is being researched in many countries, particularly in Japan. As yet, apart from some work on city transportation in the Netherlands, which is heavily subsidized by the EC, little real progress is being made. To put the industrial gas market for hydrogen in perspective, one hydro-cracker project for an oil refinery uses about the same volume of hydrogen as the total merchant hydrogen market for the whole of Europe in 1990.

4.4 Carbon Monoxide

4.4.1 Production. Carbon monoxide is generally produced by the separation of a synthesis gas containing hydrogen and carbon monoxide as described in the previous section. Typical plant sizes range from $30\,t\,d^{-1}$ up to $500\,t\,d^{-1}$ of gaseous carbon monoxide. Because of its extreme toxicity, carbon monoxide is never produced as a liquid product.

The pricing of carbon monoxide is extremely variable, and depends on the co-product requirements for hydrogen and steam. At one extreme, large volumes of carbon monoxide, with good co-product credits, may cost as little as £15 per $100\,Nm^3$. Cylinders of high purity carbon monoxide may cost up to £70 per $100\,Nm^3$.

4.4.2 Uses. The largest quantities of carbon monoxide are used to react with methanol to form acetic acid, which is widely used in the production of a number of organic chemicals.

Phosgene, formed by reacting carbon monoxide with chlorine over an active carbon catalyst, can be reacted with the appropriate diamine to form the two isocyanates TDI and MDI, which are widely used in the production of flexible and rigid foam and micro-cellular plastics. Phosgene is also reacted with bisphenol A to

Figure 12   *A Large Steam Gas Facility*

form polycarbonate resins, used in the fast-growing engineering plastics market.

Other smaller commercial uses for carbon monoxide are in the production of versatic acids, which are used for surface coatings, vinyl stabilizers, in metal solvent-extraction, pharmaceuticals, and agrochemicals.

The development of further uses for carbon monoxide is being widely studied and in all cases the development depends on the availability of higher purities. That for acetic acid production need only be 98% pure, for polycarbonate production, there is a tendency to require purities at the p.p.m. level.

4.5   A Large Industrial Gas Facility

The industrial gas facility at Rozenburg in the Netherlands shown in Figure 12 is one of the largest production facilities in the world for mixed industrial gases. In the picture can be clearly seen the liquid hydrogen plant and its storage sphere, two hydrogen/carbon monoxide plants, two air separation plants, and a steam and co-generation system. Typically, the replacement value of such a plant would be several hundred million pounds.

CHAPTER 10

Production and Uses of Inorganic Boron Compounds

R. THOMPSON

Although much research has been done, over many years, into the preparation and applications of a wide and interesting range of inorganic boron compounds, those which make up the tonnage and value of the boron products industry, in 1995, are the chemically unspectacular substances boric acid and sodium tetraborate (borax). With total world-wide production in excess of two million tons annually, they are firmly in the class of commodity chemicals. As with many other chemically simple industrial inorganic compounds whose manufacture is described in this book, production from the naturally-occurring minerals and salts may be explained by a few simple equations. Similarly, the art of manufacture to the high purity specifications lies in optimizing the conditions for the primary reactions with regard to yield and elimination of impurities, as well as the avoidance or control of unacceptable by-products; and in doing so on a very large scale, with attention to energy and environmental factors as well as production cost issues.

Description of this important sector of the inorganic chemicals industry is therefore largely an account of mining and refining processes leading to borax and boric acid in various states of hydration, and to their uses; with lesser attention to other inorganic borates; and least to the chemically more interesting compounds which are usually fully described in standard inorganic chemistry textbooks. All aspects are dealt with comprehensively in Mellor.[1]

Inter-relationship of the bulk chemicals and their ores is shown schematically in Figure 1. One other boron compound might debatably find a place in the diagram, and that is sodium perborate. It accounts for a considerable proportion of the total output of borax, but for a variety of technical, commercial, and geographical reasons manufacture has traditionally been the preserve of the hydrogen peroxide manufacturing industry (described in Chapter 7). This and other applications for borax, and uses of borates in other forms, are discussed later in the present chapter.

For the most part, the user of these bulk inorganic boron-oxy compounds is

[1] 'Mellor's Comprehensive Treatise on Inorganic and Theoretical Chemistry', Vol. V, Part A, ed. R. Thompson, Longmans, Harlow, 1980.

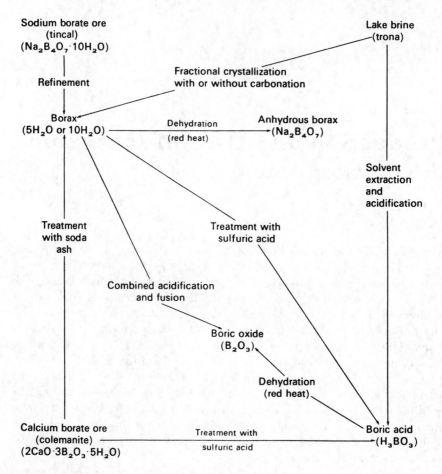

Figure 1

buying the contained B_2O_3 and this moiety is virtually the currency used. Where sodium is acceptable (or would otherwise be added to a formulation) it is common to use borax ($Na_2B_4O_7$, or $Na_2O.2B_2O_3$) in one of its hydration states. If alkalinity or an alkali metal cannot be tolerated, the more costly boric acid (H_3BO_3, or $3H_2O.B_2O_3$) or its anhydride B_2O_3 can be used.

1 OCCURRENCE AND MINING

The Western world's borax requirements are derived mainly from native sodium borate minerals (tincal, $Na_2B_4O_7.10H_2O$) and kernite, ($Na_2B_4O_7.4H_2O$) and lake brines. Solid deposits occur in sufficiently rich form for economic mining in California, Turkey, Chile, and Argentina; the brine lake is also in the same region of California. Most North American production of boric acid is also from these sources, but in almost all other areas, including Europe, it is often derived from the calcium ore, colemanite ($Ca_2B_6O_{11},5H_2O$) mined for this purpose mainly in

Turkey. Colemanite is also won in parts of California but is currently only used in upgraded form as a glass ingredient. In a few other places (Italy, Eastern Europe) this calcium mineral, again of Turkish origin, is subjected to decomposition by sodium sesquicarbonate to produce borax. Particular local circumstances render this route economically (or politically) viable, but it was abandoned in California when the rich sodium borate reserves were discovered over half-a-century ago.

The most important sodium borate deposit now being worked is the open pit mine of the US Borax Inc., which is situated near the town of Boron in the Mojave desert of Southern California. There is also a similar, though much smaller, mine operated by Borax Argentina at Tincalayu in Argentina. Both are wholly-owned subsidiaries of The RTZ Corporation Plc, a British company. At each location the orebodies are at sufficiently shallow depths to permit open-cast mining, which in the Kramer deposit at Boron was made possible by the removal of some 400 feet thickness of overburden over a pit area which is now about 3500 feet across. From its discovery until 1957 the deposit was worked by conventional underground mining methods and evidence of the original workings may still be seen.

Tincal predominates at both locations, but the Californian mine also contains substantial amounts of the tetrahydrate, kernite (in some textbooks erroneously called Rasorite after its discoverer). Mining at the Boron open-pit is by drilling and blasting selectively in benches. Electrically powered shovels transfer the ore via dump trucks to a primary crusher situated at the foot of the conveyor belt which then hauls it to the surface stockpile. From here it receives a secondary crushing before chemical processing.

The sodium borate deposit in Turkey is of more recent discovery and development. Owned by an agency of the Turkish government, not much has, as yet, been published about the operation but it can differ little from the already worked borax deposits.

2 PRODUCTION OF BORAX FROM SOLID ORES

As implied above, the overall chemical flow-sheet of borax production from its ore is simple: the crushed feed is dissolved in hot water, the siliceous gangue which accompanies it is removed by thickening and filtration, after which the product is cropped by crystallization and dried. In practice the chemistry is complicated by the combined facts that for borax yield, solubility, and water conservation reasons the process is operated in a cyclic manner. This leads to a build-up of, and the need to eliminate, certain impurities some of which are water-soluble. The ore is in fact dissolved in recycled weak hot mother liquor, the solution and suspended solids being passed through vibrating screens to remove the larger particles before the liquor is piped to the main gangue separation stage, a battery of six thickeners. Hot dissolution is necessary because of the moderately low solubility of borax in cold water [5 g $(100 g)^{-1}$ at 20 °C, rising to 191 g at 100 °C]. In order to avoid premature crystallization in pipe-lines and equipment, the liquors are maintained at around 100 °C during processing. This applies equally to the gangue separation step. Approximately 230 feet in diameter, when installed some 30 years ago, the thickeners were (and possibly still are) the world's largest covered type. They are operated in sequence, countercurrent, and thickener pulp or underflow is piped to sealed tailings ponds. The clarified liquor overflow is filtered and the borax recovered as either the decahydrate or pentahydrate in Struthers Wells vacuum crystallizers. Crystal is separated from mother-liquor by centrifuging, after which the product is dried in rotary dryers.

'Borax' by common usage implies sodium tetraborate decahydrate, $Na_2B_4O_7.10H_2O$. This contains 47% of water. Transport and storage charges are therefore potentially doubled compared with the anhydrous salt, but these costs have to be set against the energy expense of removing the water. From the 1950s the tendency was towards a greater proportion of total production being in the anhydrous form, but as fuel became both more expensive and at times scarce, the 1970s saw a reversal of this trend. Fortunately, the existence of a pentahydrate enables a compromise to be made, and by cropping the crystals at above the transition point of 60.80 °C half of the hydrate water is removed from the product at substantially zero energy cost. Water consumption is theoretically reduced also, but in all such processes some liquor bleed is necessary in order to eliminate impurities. Careful routing of make-up water from all points of introduction to final discharge via the thickener system maximizes utilization and reduces consumption. In spite of the increasing trend towards use of '5-mol' borax (also known by the trade name NEOBOR®) a considerable tonnage of the decahydrate is still required by US Borax and its customers, and the need to make varying quantities of the two hydrates according to market demand requires skilled production planning and control.

3 PRODUCTION OF BORAX FROM LAKE BRINES

A minor but significant proportion of North American borax production is from the brine of Searle's Lake, also situated in the Mojave desert. This apparently dry lake of about 34 mile² in area has saturated brine (*circa* 35% solids) to within a few inches of the surface, filling channels and voids which occupy 40–50% of the cross-section, effectively about 120 feet thick but in two main layers. The brine is rich in soda, potash, sulfate, and chloride but contains about $1\frac{1}{2}$–2% of B_2O_3 (equivalent to 5% $Na_2B_4O_7,10H_2O$). In the main process, operated by North American Chemical Co. at a location named Trona, upper-layer brine is pumped from wells and concentrated by evaporation. Rapid cooling precipitates potassium chloride, itself a valuable product. This leaves a liquor supersaturated with respect to borax, which is recovered above the transition temperature as the pentahydrate, and then recrystallized to the commercial product.

An alternative process, operated originally in 1925 at another part of the lake in the plant of the West End Chemical Company (formerly a division of Stauffer Chemical Company and now also part of NACC) but later adopted by American Potash, involves acidification of the alkaline (and rich in borax) lower-layer brine with flue-gas carbon dioxide. Sodium carbonate is thereby converted into sodium bicarbonate, which precipitates because of its lower water solubility. At the same time the $Na_2O : B_2O_3$ ratio of the liquors is altered from the 1 : 2 of borax (NB this ratio, even though borax is historically called the *tetra*borate) to the 1 : 5 of sodium *penta*borate, at which the water solubility with respect to B_2O_3 is much greater. When further quantities of lake brine are added, the pH and the ratio are restored to their original values after which the less-soluble borax is crystallized out upon refrigerated cooling.

Compositional and phase studies of the lake brine were classics of their time and were published for Teeple as an ACS monograph in 1929.[2]

4 MANUFACTURE OF BORIC ACID

Boric acid to both producers and consumers always means orthoboric acid, H_3BO_3. (It was formerly known as boracic acid.) Metaboric acid, HBO_2, is not manufactured on a significant scale and when needed by industry is usually made *in situ* (see below). There are now three important sources of the acid: solid sodium borates, lake brines, and colemanite. Although boric acid occurs naturally in the hot springs or fumaroles of Tuscany, an historical source, recovery directly therefrom is no longer practised.

Whether a sodium borate or colemanite is the raw material, boric acid is liberated by decomposition with a stronger acid, which implies any mineral acid. For economic reasons sulfuric acid is almost invariably used. This also has the production convenience with colemanite in that the precipitated calcium sulfate by-product of the metathetic reaction is insoluble and readily removed by filtration. Because of the relatively low solubility of boric acid in cold water (4.95 g per 100 g at 20 °C) the process is carried out at above 80 °C (solubility 23.6 g). This means that a large crop of boric acid crystals is readily obtained by cooling the liquor, from which it may readily be recovered on centrifugal hydro-extractors or similar equipment. A technical grade can be obtained directly, by merely introducing a washing cycle to the hydro operation before discharge, but for higher grades of product the options are total recrystallization or the removal of dissolved impurities (both anionic and cationic) by passing the liquors through ion-exchange columns. Crystallization used to be effected batch-wise in batteries of air-cooled or water-cooled granulators, but vacuum crystallizers are now preferred. After centrifuging, the moist crop is dried in continuous rotary dryers (or similar) to free-flowing granules (or, after grinding and sieving, to powder).

Production from borax involves a similar double-decomposition reaction with sulfuric acid. The water content of the process is conveniently reduced (and provision thereby made for later introduction of water for crystal washing) by starting with the pentahydrate. For solubility reasons this process is also operated hot. Boric acid is the first product to separate on cooling, and is removed on hydro-extractors. Further cooling precipitates the sodium sulfate by-product, which is anhydrous ('salt cake') if separated at above 33 °C. A process advantage of making boric acid from refined borax is that the minor impurities have already been removed.

[2] J.E. Teeple, 'The Industrial Development of Searles Lake Brines', American Chemical Society Monograph Series No. 49, 1929.

During the 1980s, US Borax brought on-stream at their plant at Boron, California, a process using the naturally-occurring tetrahydrate, kernite, $Na_2B_4O_7.4H_2O$. This mineral is found alongside tincal in the open-pit mine and being more slowly soluble in water was an inconvenient feedstock for the borax process and was avoided during mining. Kernite is however equally susceptible to decomposition with sulfuric acid and use in its own right has provided the opportunity to utilize more run-of mine ore and this effectively also increases the life of the deposit.

Building upon their expertise in brine processing, American Potash some years ago developed a supplementary process in which soluble borate in either lake brine or weak plant end-liquors is removed as an organic complex.[3] This is a solvent–extraction process in which the alkaline brine is first treated with 1,8-naphthalenediol or an aliphatic or aromatic polyol, alone or dissolved in a carrier such as kerosene or octanol. A borate complex is formed and may precipitate, in which case it is separated by filtration, dissolved in water, and treated with sulfuric acid in order to precipitate the diol. Boric acid is then recovered by an evaporation and crystallization cycle from the aqueous filtrate (which also contains sodium sulfate). With some extractants the borate complex does not precipitate but remains dissolved in the immiscible organic phase, from which after separation it is stripped with sulfuric acid, the boric acid recovered as before from the resultant aqueous phase, and the organic phase recycled. The preferred polyol solution is said to be 4-isooctyl-6-chlorosaligenin in kerosene.

5 ANHYDROUS BULK PRODUCTS

Freedom from combined water is a prerequisite for the borates used in certain sectors of the glass, pyrometallurgical, and chemical industries. Both borax and boric acid are readily dehydrated on an industrial scale and, although as stated above the practice is declining somewhat in the case of the former, these two 'fused products' are important items of commerce.

Dehydration of borax is usually effected in two stages: 'calcination' at relatively low temperatures to the approximate dihydrate, followed by fusion and heating to about 850 °C to eliminate remaining amounts of water. In the process currently operated by US Borax the first stage is accomplished in oil- or gas-fired rotary furnace and the second by feeding the powdery intermediate into large tank furnaces similar to the ones used in the glass industry. Anhydrous borax is in fact a simple glass and the fully dehydrated product is tapped continuously from the furnace as a vitreous stream, cooled on chill rolls, and finally crushed and sieved. American Potash described their somewhat different technique which employs a furnace of circular design, shaped like a giant tundish with a domed roof.[4] Gas burners are located in the dome and the partially dehydrated (calcined) feed enters through an annular space where the dome and the tundish meet. The design provides that the borax is finally dehydrated and fused on the crust of a bed of unfused feed, so avoiding direct contact with refractory brick. Other designs

[3] British Patent BR910541. US Patent US2969275 and US3111383.
[4] G.H. Bixler and D.L. Sawyer, *Ind. Rng. Chem.*, 1957, **49**, 330.

require the use of special refractories; molten borax is a powerful solvent for many refractory oxides, hence its extensive use in the glass industry.

Dehydration of boric acid to boric oxide, B_2O_3, is accomplished at lower temperatures. The reverberatory furnace equipment is again of generally similar design to the glass industry's melting furnaces, although much smaller for market reasons than the plant used for borax dehydration. Discharge, cooling, crushing, and sieving are done similarly. Both the lower fusion temperature and the lower corrosivity of boric oxide lessen the refractories problems.

6 OTHER INORGANIC BORATES

Although several other borates are produced commercially, none is made on a scale greater than a few thousand tons per annum and all are thus minor in comparison with borax and boric acid. Nevertheless, they play useful roles in industry.

6.1 Other Sodium Borates

The ready ability of the boron atom to form trigonal or tetrahedral bonds with oxygen renders possible a wide variety of stoichiometric combination ratios of B_2O_3, a metal oxide, and H_2O. This is displayed even with monovalent atoms like sodium, and the metaborate ($NaBO_2$, or $Na_2O.B_2O_3$) and pentaborate ($Na_2O.5B_2O_3$) are well-known in addition to borax itself (the $1:2$ compound). The existence in some cases of a number of hydrates and anhydrous crystalline or glassy products imposes a discipline on producer and consumer industries to restrict the commercial range. In practice, a $1:4$ borate and the $1:1$ sodium metaborate ($Na_2B_2O_4.4H_2O$ or $8H_2O$) are produced industrially at annual rates in the thousands of tons.

Reference was made in the lake brine context to the increased solubility of the $1:5$ borate compared with borax. Solubility is at about its highest with the $1:4$ $Na_2O:B_2O_3$ ratio, more than five times as much B_2O_3 entering solution in this form at $30\,°C$ compared with the dissolution of borax. Moreover, the resultant solution is of essentially neutral pH (7.3 at 15%), an advantage in many applications. There is no true compound of this formula, and hence a homogeneous solid cannot be made by crystallization. However, US Borax developed a process whereby a concentrated solution of the $1:4$ ratio is spray-dried to a stable granular tetrahydrate, and this product has been sold for many years under the trade-names POLYBOR®, TIMBOR®, and SOLUBOR®.

Sodium metaborate is made by the interaction of borax and sodium hydroxide, the product being crystallized readily from solution. It is more soluble than borax, as well as being more alkaline (a 1% solution at $20\,°C$ has pH $= 11.0$). Its principal use is *in situ* in the production of sodium perborate (see Chapter 7).

6.2 Potassium Borates

Both the tetraborate ($K_2B_4O_7.4H_2O$) and pentaborate ($K_2B_{10}O_{16}.8H_2O$) are in small scale commercial production. Each can be made either by the addition of the appropriate amount of potassium hydroxide to boric acid, or the interaction of borax and potassium chloride, with or without carbonation, *e.g.*

$$4KCl + 5Na_2B_4O_7.10H_2O + 6CO_2 \rightarrow 2K_2B_{10}O_{16}.8H_2O + 6NaHCO_3 + 4NaCl + 31H_2O$$

6.3 Ammonium Borates

The 1:2, usually referred to in this case as the **bi**borate, and 1:5 ratio compounds are made industrially from ammonia and boric acid. The pentaborate is much the more stable, with no odour of ammonia. Its principal use is as an electrolyte component in electrolytic condensers.

6.4 Calcium Borates

Hydrated synthetic calcium metaborate has been marketed, but not in significant quantities. Of far greater importance is the mineral colemanite $(2CaO.3B_2O_3.5H_2O)$ in dressed form. The main use is in glass fibre production (see below).

6.5 Barium Borate

Barium metaborate, $BaB_2O_4.4H_2O$, is produced on an industrial scale by the reaction of barium sulfide (derived from the carbothermic reduction of barytes) with borax. It is used as a waterborne paint pigment, primarily because of fungicidal properties although it does have corrosion inhibiting qualities. When prepared in a particular way, coated with silica or in dispersion with anatase or rutile, it is capable of controlled chalking, helping exterior paints to maintain a continually cleaned surface in climates of intermittent rainfall but which are sufficiently warm and humid to encourage mildew and fungal growth.[5]

6.6 Zinc Borates

Several grades of hydrated zinc borate are now on the market and used as a fire retardant pigment for plastics.[6] The precise ratio of zinc to boron is less important than the overall chemical properties, notably the amount of hydrate water and the temperature at which it is evolved. This temperature must be sufficiently high not to cause 'blowing' during fabrication or cure of the plastic article, but be released endothermically at the appropriate temperature during the build-up of the combustion process. The residual zinc borate then controls after-glow. A compound which has been found to be particularly suitable for incorporation in a wide range of plastics, notably halogenated polyesters and PVC, has the composition $2ZnO.3B_2O_3.3.5H_2O$ and is known in the trade as ZB 2335 or

[5] US Patent US3060049.
[6] W.G. Woods and J.G. Bower, 'Firebrake ZB, A New Fire Retardant Additive', SPI Reinforced Plastics/Composites Division, 25th Technical and Management Conference, Washington, USA, February 1970.

FIREBRAKE ZB®. It is made under carefully controlled conditions from zinc oxide and boric acid.[7] Zinc borates are now finding increasing use as anticorrosion pigments in primers for steel, aluminium and other metals, and are comparable in performance to zinc and strontium chromates.

6.7 Boron Phosphate

This material is no longer in regular production by the borate industry, users in the petrochemical field preferring to make it in-house. Its major application is as a heterogeneous acid catalyst, for which purpose proprietary variations are made in $B_2O_3 : P_2O_5$ ratio, surface area, and the presence of other solids in admixture. It is especially suitable for the hydration of alkenes and the dehydration of amides to nitriles. Boron phosphate exists as both low-temperature and high-temperature forms, although there is no sharp distinction. It is most readily made by intimately mixing one mole of concentrated phosphoric acid with one mole of fine boric acid powder and heating the resultant creamy slurry at a low temperature until an amorphous dry solid is obtained. The product at this stage is approximately BPO_4H_2O. Heating to higher temperatures (preferably after pulverizing) progressively eliminates water, until at about 400 °C all has been lost. At this stage it is still amorphous, but more slowly soluble in water. If finally heated to above 1100 °C it transforms to a white crystalline, insoluble, refractory product, BPO_4, which is isostructural with silica (crisobalite). This form has limited use as a ceramic flux in conjunction with pegmatite, and as an opacifier for glass utensils.

7 MAJOR APPLICATIONS OF B_2O_3 CONTAINING MATERIALS

The bulk-chemical boron compounds are employed both in fusion processes and in aqueous solution, making use of the specific qualities of the contained B_2O_3 in the former case and the overall chemical properties of the particular compounds in the latter. The glass and related industries are the main consumers in the one, and the general detergency area in the other. But there are few industries, including process industries, which do not use boric acid or borates in some form.

7.1 Glass and the Vitreous Industries

Borax and boric oxide are powerful solvents (or fluxes) for many metal oxides, including those commonly used in glass manufacture. They enable glass to be fused at lower temperatures, with consequent economy in fuel. Most important, they improve the quality of the product in a number of directions. B_2O_3 like SiO_2, is a 'network former' which readily enters into the main structure of the glass and alters its properties. Particularly important are the reduction of thermal expansion, thus improving the thermal shock resistance of kitchen and laboratory ware, and the increase in brilliance. These effects are evidenced in borosilicate glasses, *e.g.* Pyrex, which contain 10–13% B_2O_3, in craze-free pottery and tile

---

[7] US Patent US3549316 and US3718615.

glazes. In the vitreous industries (glass, glazes, enamels) it is often immaterial which borate is used, consistent with the qualitative and quantitative compatability of the accompanying cation with the formulation as a whole. Price per unit of contained B_2O_3, with due allowance for purity and water-content, is a primary factor influencing choice when such chemical inter-changeability is possible. Sodium tetraborate, notably the pentahydrate, is used for the production of Pyrex and the sodium borosilicate glasses, including glass wool for thermal insulation which contains 4–5% B_2O_3 and is a major consumer. However, glass fibre (known as 'E'-glass) for plastic reinforcement or textiles must be soda-free, in which case boric oxide or, more usually nowadays, specially upgraded colemanite ore are used. Textile glasses contain 6–8% B_2O_3.

7.2 Metallurgical Fluxes

The abilities of molten boric oxide and alkali metal borates to dissolve other metal oxides are exploited in a wide range of fluxing applications in the metallurgical and engineering industries. They are valuable cover fluxes in metal melting, especially of copper alloys. Borax is used as a flux in the assaying and refining of gold. In the steel industry it finds application for scale-free reheating of billets. Potassium pentaborate is principally used for welding and brazing stainless steel. Trimethyl borate, being volatile, is the active constituent of gas welding fluxes.

7.3 Bleaching and Detergency

A very large proportion of sodium borate production, notably the pentahydrate, is used in making sodium perborates (see Chapter 7). Regular bulk shipments are made from California to Europe and other parts of the world for this purpose. Tradition in European households had been to 'boil' clothes, or at least to heat to well above hand-washing temperatures in order to whiten domestic and personal linen. Sodium perborate is an increasingly effective bleach at temperatures above about 55 °C. As average washing temperatures have fallen over recent years, so has the sodium perborate content of the packaged washing powder increased to the present level of up to 25% (the other major ingredients being sodium tripolyphosphate and synthetic sulfonated detergents). Reduction in effective bleaching temperatures has been made possible by the introduction of 'activators' such as TAED (tetra-acetyl ethylene diamine) while the lower temperature permits the incorporation of enzymes for the so-called biological compositions. Sodium perborate production in Europe in 1992 was 700 000 tonnes (as the tetrahydrate), mostly for this end-use.

It has not been the practice in the USA to 'boil' household laundry, and in fact far fewer domestic washing machines were fitted with water heaters in that country. Consequently, chlorine-based bleaches (hypochlorites, chlorocyanurates, *etc.*) are often used, a practice not generally favoured by most European housewives. Nevertheless, it has been customary in the USA over several generations for borax to be added to the wash along with the proprietary detergent powder. Thus in the washwaters of both continents there is a presence of dissolved sodium borate which acts as a detergent builder of acceptably mild alkalinity (to hands and clothes). Borax solution is also valuable as a home laundry pre-soak, especially for diapers and the like, where its chemical buffering action and protein solubilizing properties are particularly valuable.

7.4 Biological Applications

Boric acid and its salts have low mammalian toxicity and are not surprisingly rather weak bacteriostats by modern standards. Formerly employed as food preservatives and medicinally (boric acid is still a favoured constituent of eye lotions), they have generally been supplanted by other materials in these areas. No role for boron has yet been identified unequivocally in animal metabolism, but there is developing evidence suggesting some human essentiality. Soluble inorganic borates are however toxic to certain forms of insect life, and for example they are specific stomach poisons for cockroaches. The compounds are simply administered by mixing either borax or boric acid with a bait such as sugar or starch and applied as powder to infested areas. The mixtures are harmless to domestic animals. Ant baits are formulated similarly.

Borates are also toxic to many wood-boring insects and the water-soluble 1:4 sodium borate (called TIMBOR® for this application) is widely used in timber preservation.[8] A process known as diffusion is operated by momentary immersion of newly-sawn or unseasoned timber in the hot and concentrated aqueous borate liquor and allowing it to drain; followed by diffusion of the residual solution on its surface into the depth of the wood during several weeks of storage and drying, leaving it preserved throughout (uniquely amongst commercial preservative systems) in a colourless and odourless manner, harmless to high life forms. Esters of boric acid (dioxaborinanes) are used to control fungus and microbial life in diesel and aviation jet fuels.[9]

Boron plays an interesting and unique role in plant life. Above certain levels (which vary widely according to the species) it is phytotoxic and borax, often in admixture with sodium chlorate, was used extensively for many years as a non-selective weed killer. More powerful herbicides have been developed but nothing can replace boron as a trace nutrient. Apart from the citrus family, most other fruit, fodder, oil, and beverage crops require boron as an essential element. Without it, apple and pear develop cork, yield of sugar beet is reduced, peanut quality is impaired, and alfalfa growth is stunted. The boron may be introduced along with the fertilizer, in soluble or slow-release form, or as a foliar feed. The highly water soluble 1:4 sodium borate which is also of substantially neutral pH, is sold for this use under the name of SOLUBOR®.

7.5 Uses of Boric Acid

One of the most important single uses for boric acid is in the nylon industry. It is employed to modify the course and selectivity of the liquid phase oxidation of

[8] L.H. Williams, 'The Chemistry of Wood Preservation', Ed. R. Thompson, Special Publication No. 98, The Royal Society of Chemistry, 1991, pp. 34–52.
[9] R. Thompson, 'Use of Boron Compounds for Microbial Control in Fuel and Machine Oils', The Institute of Petroleum Quarterly Journal of Technical Papers, London, April–June 1986, pp. 52–65.

cyclohexane by molecular oxygen to cyclohexanol and cyclohexanone, thus enabling a better yield of the co-products required for nylon 66 manufacture to be achieved. The function of boric acid in this industrial extension of the Bashkirov reaction[10] is a stoichiometric participant and not merely as a catalyst. The acid, dehydrated *in situ* at 160 °C to metaboric, esterifies cyclohexanol (which also helps in separating it from the cyclohexanone), and is partially recovered during the subsequent hydrolysis of the ester. A similar controlled oxidation of n-paraffins can be used to produce a detergent alkylate for making biodegradable 'soft' syndets.

7.6   Nuclear Applications

The boron atom has two isotopes, ^{10}B and ^{11}B, the former of which is a neutron absorber and is present up to 18% in naturally-occurring boron compounds. For some military and medical applications suitable compounds are made in isotopically enriched form, but normally the cheaper non-enriched form is used. Relatively large tonnages of high-purity boric acid are used in pressurized water reactors (PWR) and boron carbide is an important control rod material (see below).

7.7   Miscellaneous Applications of Borax and Boric Acid

Both borax and boric acid have been used for many years in flameproofing treatments for combustible building materials and textiles, particularly where the substrate is of a cellulosic nature. The optimum $Na_2O:B_2O_3$ ratio for fire retardancy is close to the maximum water solubility within the system, at approximately 1:4. The spray dried product POLYBOR® is frequently used for the purpose. Such solutions also diffuse readily into veneers used in plywood manufacture. Wood particle board (sometimes called chipboard) is best flameproofed with boric acid alone, which is mixed with urea–formaldehyde resin and sprayed onto the wood chips before they are compacted in a heated platten press or between heated rolls. The boric acid, which complexes with hydroxyl groups on the cellulose, also acts as the catalyst in heat setting the resin. Wood particle board flameproofed with boric acid has the added advantage of producing very little smoke under fire conditions, and 'after-glow' is prevented. Matchsticks are also glow-proofed in this way. The largest single outlet for boric acid in this area however is in flameproofing cotton mattresses (mandatory for public institutions in parts of the USA). For this application the cotton linters and dry boric acid powder are milled intimately together. A mixture of shredded waste newsprint and borax or boric acid found extensive use as loose-fill cellulosic insulation for homes in the USA during the oil crisis years and is still used for loft and dry cavity wall insulation.

Borax, with its mild alkalinity and high buffering capacity, is used extensively as a corrosion inhibitor for ferrous metals. Its high solubility in ethylene glycol combines to make it especially useful in car antifreeze formulations (*e.g.* British Standard 3152). Aqueous solutions of borax have replaced chromates in railroad

[10] A.N. Bashkirov *et al.*, 'The Oxidation of Hydrocarbons in the Liquid Phase', Ed. N.M. Emanuel, Pergamon Press, Oxford, p. 183, 1965.

and other diesel coolants. In the dry drawing of steel wire, borax is used to neutralize residual acid from the pickle and the deposit of the salt remaining on the wire is valuable as a lubricant carrier.

Boric acid and its soluble salts and esters find major or minor applications in textile dyeing, dyestuffs manufacture, leather tanning, the photographic industry, laundry starches, starch-based adhesives for corrugated cardboard manufacture, anodizing, electrolytic capacitor production, plaster and concrete set retardants, paper sizing, and in many other industries.

8 OTHER BORON COMPOUNDS USED INDUSTRIALLY

The non-oxygenated inorganic boron compounds for which commercial application has been found may conveniently be divided into two classes: those having boron–halogen and boron–hydrogen bonds; and the binary interstitial solids usually of a hard or refractory nature, to which class elemental boron itself may be considered to belong.

8.1 Boron Trifluoride

BF_3, a colourless gas, boiling point $-100.4\,^\circ C$, is produced in tonnage quantities by relatively conventional chemical process methods. It is the only boron halide possessing any hydrolytic stability and may be made simply by the action of sulfuric acid on a mixture of borax and fluorspar:

$$Na_2B_4O_7 + 6CaF_2 + 8H_2SO_4 \rightarrow 4BF_3 + 2NaHSO_4 + 6CaSO_4 + 7H_2O$$

Sulfuric acid in excess of the stoichiometic amount acts as a dessicant. The evolved gas, separated from spray, may be collected and stored in cylinders as the pure compound; or more usually is absorbed in media such as ether, phenol, or acetic acid with which, being a strongly electrophilic molecule, it readily forms addition complexes. Other processes involve the reaction of boric acid or borax with fluorine sources such as HF, NH_4HF_2, or HSO_3F; each has relative merits in BF_3 purity and, compared with the fluorspar route, economy in sulfuric acid usage. BF_3 is used either alone or in the form of one of the above or similar adducts as a catalyst for a wide variety of alkylation, polymerization, and other reactions in the petrochemical industry. Its dimethyl etherate finds some use as the boron source for the vapour fractionation method of ^{10}B isotope enrichment or separation.

For a more detailed account of boron trifluoride and fluoroboric compounds, see Chapter 8.

8.2 Fluoroboric Acid and Metal Fluoroborates

HBF_4 is made either by dissolving boric acid in hydrofluoric acid, or by treating boric acid and fluorspar with sulfuric acid. It is sold as an approximately 50% aqueous solution, but production is not large and uses are limited to metal cleaning and as an organic reaction catalyst. Sodium, potassium, and ammonium

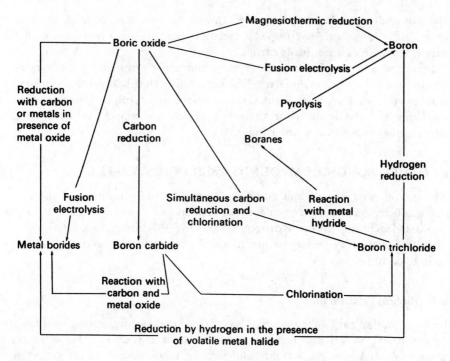

Figure 2

fluoroborates are all available commercially and can be made by neutralizing fluorobic acid with the hydroxide or carbonate of the desired cation. Alternatively, HF (*e.g.* waste from other processes) can be absorbed in borax solution to produce $NaBF_4$ directly; KBF_4 can then be made from this by the addition of KOH. Fluoroborate uses are in the electroplating industry, with the sodium and potassium salts also having an important outlet as grain refining agents for aluminium (along with fluorotitanates they react in the molten metal to form titanium diboride nuclei).

8.3 Other Boron Halides

BCl_3, BBr_3, and BI_3 are all prepared under high temperature anhydrous conditions, usually from boron carbide and the halogen. While the tribromide and triiodide are made only in laboratory amounts, for research or electronics applications, boron trichloride is produced in tonnage quantities. The preferred process is the exothermic chlorination of boron carbide at above 700 °C (Figure 2). Production is accomplished by passing chlorine upwards through a column packed with B_4C granules. Reaction is initiated by heating those at the base of the column to redness, after which it is necessary to control the flow of chlorine such that the reaction zones rises through the bed at a suitable rate. The boron trichloride gas (b.p. 12 °C) is condensed, along with the halides of any impurity elements (Al, Si, Fe) which may be present in the boron carbide, and then purified by fractional distillation. The process is usually run batch-wise, but continuous operation (feeding new boron carbide granules at the top and removing the spent ones from the bottom of the column) is practicable according to product demand. BCl_3 is a colourless mobile liquid, rapidly (and even violently) hydrolysed by water.

8.4 Sodium Borohydride

Metal borohydrides (sometimes called, pedantically and misleadingly, 'hydroborates') are the only B—H bonded compounds in industrial tonnage production. By far the most common is $NaBH_4$, a white, crystalline, and water-soluble solid. It is usually made from sodium hydride and trimethyl borate:

$$4NaH + B(OMe)_3 \rightarrow NaBH_4 + 3NaOMe$$

The process is carried out in two stages: hydrogenation of a very fine dispersion of sodium in mineral oil, followed by reaction with the borate ester at 250–270 °C. The product may be separated by extraction with liquid ammonia, or converted into potassium borohydride by precipitation with potassium hydroxide.[11] The extraction stage is expensive and, as ultimate use is often in aqueous solution, it can be cost-effective to ship and apply as a 12% solution. This is sold under the trade name of BOROL®. An alternative method was developed but no longer used by Farbenfabriden Bayer AG.[12] This is a two-stage process in which anhydrous borax is fused with silica and the resulting sodium borosilicate treated with hydrogen and molten sodium at 450–500 °C and 3 atm pressure:

$$Na_2B_4O_7 + 7SiO_2 + 16Na + 8H_2 \rightarrow 4NaBH_4 + 7Na_2SiO_3$$

The product is extracted with liquid ammonia.

Sodium borohydride is a powerful and selective reducing agent, employed in both inorganic and organic systems. Its ability to reduce carbonyl groups irreversibly is of value both for organic syntheses and in pulp bleaching, but by far the most important use is for sodium hydrosulfite regeneration (or replacement of toxic hydrosulfite) in connection with the latter. Borohydrides are also able to reduce heavy metal ions in solution to the elemental form, a quality useful in electroless plating, hydrometallurgical winning, and effluent treatment.

8.5 Boranes

The boron hydrides, or boranes, are not produced commercially in other than what may be termed laboratory quantities. From the 1940s, they had been made by US government contractors on an industrial scale for military purpose (high-energy fuels for missiles) and it is likely that stockpiles still exist. The lower boranes are hazardous to handle, being highly toxic as well as spontaneously inflammable and explosive. For these reasons together with their very high cost it is

[11] W.S. Fedor, M.D. Banus, and D.P. Ingalls, *Ind. Eng. Chem.*, 1957, **49**, 1664.
[12] Anon., *Chemical Week*, 10 June 1961, **63**.

unlikely that they will find serious industrial use outside the electronics and fine chemicals industries. It would be as illogical to consider boranes for commercial fuels as it would be to make petrol from limestone. One of the closest approaches to commercialization has been with a carborane–siloxane high-temperature resistant polymer sold under the trade name DEXSIL®.[13]

8.6   Boron

Elemental boron is not made on a large scale. World-wide, probably less than 10 tons are produced annually and most of this is impure material required for flares and pyrotechnics. Such material is made by magnesiothermic reduction of boric oxide, in much the same way as Moissan prepared it originally. The product is capable of being upgraded by vacuum melting after the acid-soluble impurities have been removed, and a small amount of material is so treated for electronics and research purposes. An interesting new use of amorphous boron is as the initiator in air bags used as a safety feature in cars.

The greatest advances in elemental boron technology in earlier years were in connection with the production of reinforcement filament for plastics and metal composites such as helicopter blades and aircraft wing sections. This use is based on the high specific stiffness of the boron filament; it resembles carbon fibre in its attributes for these special applications, and has an advantage in being a less reactive reinforcement for aluminium, especially under moist conditions. Boron filament is made continuously by chemical vapour deposition onto a fine tungsten wire (similar to the filament of electric light bulbs) which is heated resistively while being passed through a boron trichloride–hydrogen atmosphere. The tungsten core remains in the product, adding marginally to its density. This process represents by far the largest use for boron trichloride.

8.7   Boron Carbide

Boron carbide, B_4C, is produced at a rate of several thousand tons per annum, mainly for abrasives purposes and similar uses. The bulk is made by the electrothermal reduction of a source of boric oxide with petroleum coke, the cooled product being a clinker which is mechanically fractured into granules and size-classified according to its final destiny. Gravel-size material is chosen for boron trichloride production, less coarse product being used as an abrasive grit, or pulverized even finer into polishing powder and hot-pressing grades.

Like the metal borides (see below) boron carbide is a hard, refractory solid. Its uses are mainly based on these qualities, but the properties at the same time demand special techniques other than casting in order to fabricate usable objects: melting point is around 2250 °C, a temperature which no conventional mould material can withstand. Resort is therefore made to hot-pressing, a high-temperature powder metallurgical technique which in this case is practised in graphite dies at around 2000 °C and pressure of about 2 tons inch^{-2}. Because of its

---

[13] D.J. Mangold, *Appl. Polym. Symp.*, 1969, **11**, 157.

extreme hardness (over $9\frac{1}{2}$ on Moh's scale, or approximately 4000 kg mm^{-2} VPN) it is imperative to press to as near as possible the shape and size required in order to minimize the amount of diamond machining necessary. Currently the major non-military uses for pressings are shot-blast nozzles, which out-last tungsten carbide and are only about one-sixth of the weight, and as mechanical seals. This low specific gravity (2.5 g cm^{-3}), coupled with high acoustic impedance, led to the development of boron carbide as lightweight ceramic armour. Scores of thousands of suits were used to protect battle personnel during the Vietnam war and the need greatly accelerated the development of hot-pressing technology.[14] Boron carbide is now the leading ceramic armour for use where weight reduction is a priority and it is used extensively in crews' seats in military aircraft and for the protection of vital components such as avionics and helicopter gearboxes.

One other important application of boron carbide makes use of neither hardness nor refractoriness, but the neutron-absorbing properties of the boron present. One-fifth of this is of the ^{10}B isotope, and with 78% total boron, B_4C is much more cost-effective than the free element. The cost difference, appreciable even in the powder form, is greatly magnified by the relative ease of fabrication of boron carbide, especially for nuclear reactor control rods. The most spectacular reported neutron-absorbing use was the dropping of 40 tonnes onto the damaged Chernobyl reactor after the 1986 disaster.[15]

8.8   Metal Borides

These compounds, like boron carbide, are interstitial solids which collectively fall into the class of substances known as 'refractory hard metals'. The inter-relationship and methods of manufacture are summarized in Figure 2. Titanium diboride, TiB_2, is commercially the most important. Possessing melting points from just under $2000\,°C$ in the cases of chromium mono- and diborides to over $3000\,°C$ for the titanium and zirconium diborides, fabrication is by hot-pressing in a similar manner to boron carbide. The products are only marginally less hard (in the region of 3000 kg mm^{-2} VPN) and hence diamond machining is also necessary for the final shaping of components. Most metals form at least one boride within the series MB, MB_2, MB_4, MB_6, and MB_{12} and combination in other ratios is common. Small amounts of hexaborides of rare earth elements, LnB_6, are used in the electronics industry. The chemistry, fabrication, and applications of metal borides, boron carbide, and boron nitride have been reviewed fully elsewhere.[16,17] For this brief account of the contemporary industrial significance of borides it is appropriate simply to mention that the diborides are produced by carbothermic reduction of the metal oxide in the presence of boric oxide, *e.g.*

[14] J.V.E. Hansen, *Research/Development*, Technical Publishing Company, Barrington, Illinois, 1968, **26**.
[15] 'Summary Report on the Post-Accident Review Meeting on the Chernobyl Accident', Safety Series No. 75-INSAG-1 International Atomic Energy Agency, Vienna, 1986, p. 46.
[16] R. Thompson, 'Borides, their Chemistry and Applications', The Royal Institute of Chemistry, Lecture Series, Vol. 5, 1965.
[17] R. Thompson and P. Sindlhauser, 'Borides – The Unusual Ceramics', Ceramic Technology, '93, Sterling Publications Ltd., 1993.

$$TiO_2 + B_2O_3 + 5C \rightarrow TiB_2 + 5CO$$

The reaction is endothermic and effected by heating in vented electric furnaces at 2000 °C briquettes made from an intimate mixture of the finely powdered reactants. Chromium monoboride is made by the exothermic reduction of the parent oxides with aluminium powder. Titanium diboride, which is a good electrical conductor (the conductivity is similar to that of iron) and is chemically resistant to attack by most non-ferrous metals, including even boiling aluminium, finds its main outlet as the standard crucible material used in the vacuum metallizing industry. For this application the overall conductivity of the boat-shaped crucible is reduced (to facilitate resistive heating during use) by mixing the titanium diboride powder with boron nitride (hexagonal) powder prior to hot-pressing. The principal use for the chromium borides is as ingredients for welding alloys, especially those employed for hard facing. A crude calcium boride has been used for deoxygenating copper.

8.9   Boron Alloys

These are stoichiometrically indefinite but invariably metal-rich compositions of commercial importance formed between boron and metals such as copper, manganese, and the iron group. Production of ferroboron (10–25% B), manganese–boron (15–20% B), and nickel–boron (15–18% B) alloys is by the reduction, usually aluminothermic, of boric oxide in the presence of the alloying metal or its oxide. Ferroboron is used to introduce small amounts (0.002%) of boron into steel in order to improve hardenability, or larger amounts (up to about 7%) for nuclear shielding. The other boron alloys are used in hard facing and the production of special alloy steels.

Ternary compositions containing iron and a rare-earth element are coming into use as magnetic alloys, for example $Fe_{14}Nd_2B$. They are employed principally in the starter and ancillary motors in automobiles where there is a weight and volume saving compared with cobalt–samarium magnet alloys of similar performance. Iron–boron alloys have also found use as amorphous metals, which are produced in tape or ribbon form by quenching from the melt at the rate of a million degrees per second. Uses range from tyre reinforcement to electrical transformer laminates.

8.10   Boron Nitride

Isoelectronic with carbon, boron nitride is its total analogue in existing as a soft hexagonal form and an intensely hard cubic form. Hexagonal BN is a white solid with a graphitic feel and crystal structure, but unlike graphite it is a non-conductor of electricity. This property and its refractoriness (it begins to decompose at 1700 °C) account for its industrial applications. One such outlet (as a constituent of vacuum metallizing crucibles) has already been mentioned; its others are largely in the electrical and related industries, where its high thermal conductivity as well as its high resistivity is of especial value. Boron nitride is usually made as a fine powder by heating a mixture of boric acid and melamine:

$$2H_3BO_3 + C_3N_6H_6 \rightarrow 2BN + 3CO + 2N_2 + 3H_2O$$

followed by milling and classification.

It is fabricated by hot-pressing, including hot isostatic pressing, as in the cases of the other binary refractory compounds of boron. Being much softer than either the metal borides or boron carbide, it is customary to press large blocks and cut to shape using normal engineering tools. The powder itself has a growing number of applications ranging from a filler for electrical industry plastics, where its thermal conductivity coupled with electrical insulation uniquely come into play, to abradable seals in aircraft jet engines and as a cosmetics facepowder ingredient, utilizing its soft lubricious nature.

Cubic BN has the same crystal structure as, and is practically as hard as, diamond but is slightly more oxidation resistant. It is made from the hexagonal form in the same type of high-temperature, high-pressure equipment as is used for synthetic diamond production. It is sold as boart and as tool tip forms, which are especially useful in the machining of ferrous metals.

8.11 Boric Acid Esters

Most alcohols can be condensed with boric acid, but few such esters have found industrial use, largely because of the hydrolytic instability. Trimethyl borate is used as a volatile flux in gas welding, and as an ingredient in manufacturing paint driers. High alcohol esters are used to control organic growth in hydrocarbon fuels.[9]

Production and Use of Aluminium Compounds

K. A. EVANS

1 INTRODUCTION

Aluminium compounds have been of service to men for over 4000 years. A fragment of leather, now in Cairo Museum, taken from an Egyptian tomb provides the earliest evidence of the use of an aluminium compound, in this instance as a mordant for dyes. The skill of the Egyptian dyers became renowned and large quantities of alum were exported from Egypt; Pliny in AD70 described the skills of the Egyptian dyers and the colourfastness of their wares. The Ebers' papyrus of *ca.* 1550 BC is believed to contain hieroglyphic signs referring to alum. The Greeks and Romans continued the use of alum for dyeing and further applications quickly spread to leather tanning, flame proofing, metal production, glass and ceramic making, and for medicinal uses.[1]

Today aluminium compounds find their way into almost every sphere of human activity and several are manufactured on a vast industrial scale. This article will deal with the production and use of a number of the more important commercially available compounds, namely the hydroxide, oxide, sulfate, chlorides, nitrate, phosphate, fluoride, fluoroaluminates, bromide, alkoxides, alkyls, carboxylates, and aluminates of sodium and calcium.

2 OCCURRENCE OF ALUMINIUM COMPOUNDS

Aluminium is the third most abundant element in the Earth's crust at 8.8 wt.%, though it does not occur in the metallic state because of its strong affinity for oxygen. It is of widespread occurrence in minerals and rocks, particularly in chemical combination with the oxides of silicon, *e.g.* feldspars, mica, and clays. The mineral corundum, Al_2O_3, containing 52.9% aluminium has the highest known aluminium content. A number of other minerals consisting of oxides of aluminium and other metals arise as alumina spinels but their occurrence is rare.

---

[1] C. Singer, 'The Earliest Chemical Industry', The Folio Society, London, 1948.

Aluminium is also present in a large variety of other ores and minerals, for example in hydroxide, oxide–hydroxide, phosphate, fluoride, and sulfate.

The alumina content of primary igneous rocks averages 15%, ranging from 5% in peridotite to 32% in anorthosite. The most abundant of the aluminium silicate minerals are the feldspars with an alumina content ranging from 18.4% in orthoclase to 36.7% in anorthite.[2]

Exposure of primary igneous rocks to weathering leads to the leaching out of the more soluble constituents. Feldspars, feldspathoids, and other aluminium silicates are leached of most of their sodium, potassium, calcium, and magnesium and some of their silicon. The material remaining consists of mixtures of oxides, hydroxides, and hydrated silicates of aluminium, iron, and other elements together with insoluble primary minerals such as quartz, rutile, zircon, ilmenite, and magnetite. In temperate climates the residual products consist mainly of the minerals kaolinite, halloysite, and allophone: the process is termed kaolinization.

Under certain conditions the process of weathering continues beyond the kaolinization process. The silicates are broken down further and silica is leached out to give a material containing hydroxides of aluminium with oxides and hydroxides of iron, manganese, titanium, and other elements and some hydrated silicates. These materials are termed bauxites when they contain a level of aluminium hydroxide which is of economic interest as a source of alumina or other aluminium containing compound.

The name bauxite derives from the location of the deposit near Les Baux in Provence, France which was discovered in 1821.

The aluminium hydroxides in the bauxite occur in three main forms: gibbsite (alumina content 65.4%), boehmite (alumina content 85.0%), and diaspore (alumina content 85.0%). The major impurities present in bauxite are iron oxides (as haematite, goethite, or magnetite) 2–20%, titania (2–8%), and silica (0.5–10%). Other impurities present in minor quantities are the oxides of calcium, magnesium, gallium, chromium, manganese, and phosphorus.[3]

The bauxites high in gibbsite content are the most readily processed by the Bayer process and occur in extensive deposits in Jamaica, Brazil (Trombetas), Australia (Weipa, Gove), Guinea (Boke, Fria), Guyana, Surinam, India, USA, *etc.* Those deposits high in boehmite require more aggressive extraction conditions and are commonly found in European bauxites, *e.g.* France, Greece, former Yugoslavia, CIS, Hungary. Diaspore occurs predominantly in China, CIS, Romania, and Greece: these bauxites are particularly suitable for use in refractory compositions.

3 ALUMINIUM HYDROXIDE

World-wide production of aluminium hydroxide exceeds 55 million tonnes year^{-1}. Some 90% of this aluminium hydroxide is then calcined to aluminium oxide and used in electrolytic cells for the production of aluminium metal.

[2] S. Bracewell, 'Bauxite and Aluminas', HMSO, London, 1962.
[3] T. G. Pearson, 'Chemical Background of the Aluminium Industry', Royal Institute of Chemistry, London, 1955.

Approximately 3.6 million tonnes of aluminium hydroxide is used annually for uses other than metal production.

Uses of aluminium hydroxide and the oxides derived therefrom range from fire retardants to refractories, cement to ceramics, paint to paper, abrasives to catalysts, and water treatment feedstocks to toothpaste.

Crucial to the development of this range of uses has been the very wide range of properties that arise as you move from gelatinous boehmites to crystalline hydroxides to activated alumina to α-alumina. The properties range from relatively soft materials and compounds with high solubility in acids and alkalies, to an extremely inert and hard material.

The growth of these uses has been considerably helped by the large scale manufacture of aluminium hydroxide which has kept the costs to a minimum.

3.1 Nomenclature of Aluminium Hydroxides

Aluminium forms a wide range of hydroxides; some of these are well characterized crystalline compounds whilst other are ill-defined amorphous substance. Often the terms alumina hydrate, alumina trihydrate, or ATH are used for aluminium hydroxide but these names incorrectly suggest that they are forms of alumina with different amounts of water of crystallization. The crystalline form of the aluminium hydroxides can be divided into the trihydroxides $Al(OH)_3$, which exists in three forms: gibbsite (γ-aluminium trihydroxide), bayerite (α-aluminium trihydroxide), and nordstrandite; and the oxide hydroxides, which exists as boehmite (γ-aluminium oxide hydroxide) or diaspore (α-aluminium oxide hydroxide). All these are found in nature although bayerite is found only rarely. A less well documented form of an aluminium oxide hydroxide is tohdite which has the composition $5Al_2O_3.H_2O$.[4]

Commercially the most important of the hydroxides is gibbsite although bayerite and boehmite are also manufactured on a small industrial scale. Non-crystalline forms of the hydroxides encompass gelatinous pseudo-boehmites and aluminium hydroxides with amorphous X-ray patterns.

3.2 Properties of Aluminium Hydroxides

As described earlier gibbsite is a major constituent of tropical bauxites which are normally tertiary in age; it is sometimes referred to as hydrargillite.

It consists of close-packed sheets of hydroxyl ions with aluminium ions in two-thirds of the octahedral holes between close-packed layers, which are themselves in open-packing with neighbouring pairs, the whole approximating to hexagonal symmetry. Gibbsite is invariably monoclinic though a triclinic form has been reported. The crystal structure of gibbsite is depicted in Figure 1.

Gibbsite cleaves in the (001) plane, is translucent or white depending upon its crystallite size, and has a Mohs' hardness of $2\frac{1}{2}$–$3\frac{1}{2}$. Its density is 2.42 g cm^{-3} and its

[4] K. Wefers, in 'Alumina Chemicals', ed. L.D. Hart, American Ceramic Society, Westerville, USA, 1990, p. 13.

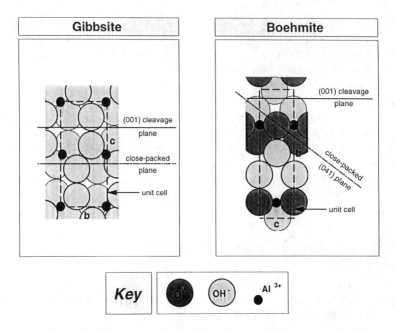

Figure 1 *Crystal Structure of Gibbsite and Boehmite*

ΔH_f is $-1293.2\,\mathrm{kJ\,mol^{-1}}$. It is virtually insoluble at neutral pH but it becomes readily soluble at pHs greater than 9 and less than 5.

3.3 Production of Aluminium Hydroxides

With the exception of small tonnages of boehmite and bayerite, the aluminium hydroxide normally produced is gibbsite, the γ-aluminium trihydroxide. The most important commercial route for the production of gibbsite is the Bayer process starting from bauxite.[3]

In the Bayer process, crushed bauxite is treated with caustic aluminate solution containing $100\text{--}300\,\mathrm{g\,dm^{-3}}$ of soda. The dissolution reaction is generally carried out under pressure at elevated temperatures ranging from 140 °C to 280 °C. The caustic solution reacts only with the alumina so that the impurities can be separated by settling and filtration leaving a clear solution.

To recover the dissolved aluminium hydroxide the solution is diluted, if necessary, to $100\text{--}150\,\mathrm{g\,dm^{-3}}$ soda and cooled to 50 °C to 70 °C. The solution is then mixed with large quantities of recycled seed particles of gibbsite (up to four times by weight of the amount dissolved) and agitated in large crystallizers for periods of up to three days. The product slurry is then passed through a classification system in which the coarse fraction is separated and passed forward to the next step in the process while the finer fraction is recycled as seed for subsequent crystallization. Filtered solution, depleted of about half of its alumina, goes back to the extraction system via the evaporators. Figure 2 shows the various stages of the Bayer process.

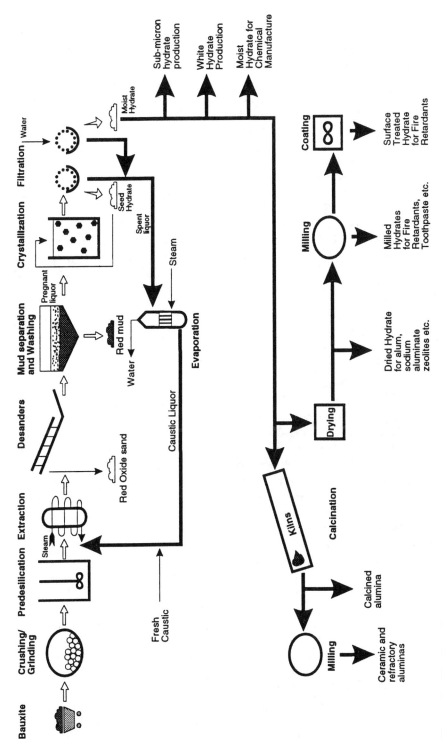

Figure 2 *The Bayer Process*

It is essential to control the process so that the product obtained is of the chemical and physical character required and the seed surface area is maintained. The addition of previously crystallized gibbsite as seed provides a measure of control over the particle size distribution of the product and is in fact the basis of the original Bayer patent.[5] The temperature at which crystallization takes place, the degree of agitation, and the quantity of seed added also influence the particle size distribution and purity of the gibbsite obtained.

During crystallization, the particle size of the original seed undergoes changes due to crystal growth, production of new crystals by secondary nucleation, and cementing of particles into consolidated agglomerates/aggregates.[6,7] The extent of particle size enlargement that takes place depends upon the starting alumina supersaturation of the solution, the temperature, and the quantity of seed added. Figure 1 shows a typical particle of gibbsite produced in the Bayer process.

Under the conditions of temperature and soda concentration employed in the Bayer process, the product from the crystallizers is always gibbsite irrespective of the nature of the alumina present in the starting bauxite. At lower temperatures, however, bayerite starts to be formed. Autoprecipitation, in the absence of seed crystals at 20 °C, or neutralization of sodium aluminate solution with carbon dioxide gas, leads preferentially to the formation of bayerite. However, because of the relatively rapid rate of crystallization under these conditions, the product particles are usually of colloidal size and low crystalline order.[8]

The smallest plants currently in operation typically have an annual capacity of 100 000 tonnes; the output of these is invariably devoted to higher purity chemical or 'specials' usage whilst the newer plants are up to three million tonnes per annum in size and devoted principally to providing feedstock for calcination to alumina for use in aluminium metal production. The particle size of agglomerates produced by this route is normally in the range 30–150 μm but crystals in the range 0.3–200 μm can be produced by controlling seed concentration, seed size, seed activity, temperature, and degree of supersaturation.

When producing aluminium hydroxide from a solution of pH < 7, the initial product is usually a gel which however will transform to crystalline trihydroxide on ageing if the hydroxyl ion concentration is subsequently kept sufficiently high. In the absence of alkali ions, the product is bayerite. The first X-ray crystalline form to occur in the ageing sequence is usually boehmite. Gelatinous boehmite itself is an important technical product.[8]

3.4 Uses of Aluminium Hydroxides

Whilst the direct application of aluminium hydroxide (more accurately aluminium trihydroxide) in paper, paint, glass, ceramic glazes, pharmaceuticals, toothpaste, and as a fire retardant accounts for several hundreds of thousands of tonnes, the use

[5] K. J. Bayer, UK Patent No. 10093, 1887.
[6] N. Brown, *J. Cryst. Growth*, 1972, **12**, 39.
[7] N. Brown, *J. Cryst. Growth*, 1972, **16**, 163.
[8] K. Wefers and G.M. Bell, 'Oxides and Hydroxides of Aluminum', Alcoa Technical Paper No. 19, Alcoa Aluminum Company of America, 1972.

in manufacture of other chemicals is even larger. Largest of all is the manufacture of zeolites and of 'iron-free' aluminium sulfate for use in paper making and water treatment as a coagulant. Other very large uses are: manufacture of aluminium fluoride, aluminium nitrate, poly(aluminium chloride) (PAC), poly(aluminium silicate sulfate) (PASS®–Handy Chemicals Ltd.), sodium aluminate, catalysts, and titania pigment coating.[9]

3.4.1 Fire Retardancy. Interest in the use of aluminium compounds for imparting fire resistance to materials stretches back a long way. The annals of Claudius record that in the attack on Piraeus in 83 BC the storming towers were protected against fire by treatment with a solution of alum. The use of aluminium hydroxide as a fire retardant, however, has more recent origins and is now one hundred years old. In the last decade of the Nineteenth century, Sir William Henry Perkin was investigating the flammability of cotton flannelette. Aluminium hydroxide was found to be the best of the 25 potential flame retardants for cellulose that he investigated.[10] An 1896 patent covered the use of aluminium hydroxide as a fire retardant for wood fibre. The next reference to the use of aluminium hydroxide in a flame retardant composition was in 1921 when its use, together with antimony chloride, was patented for flame retarding rubber.[11] There is, however, little evidence of this mixture being adopted commercially at that time.

Several patents appeared in the mid 1950s extolling the benefits of aluminium hydroxide in polyester resin, neoprene, butyl rubber, and epoxies to improve the arc resistance of electrical components. The usage of aluminium hydroxide as a filler in rubbers and plastics, however, remained small throughout the fifties and sixties.

Stimulation to the usage of aluminium hydroxide as a fire retardant was provided by the work of Connolly and Thornton in 1965[12] which showed the improvements that could be obtained in the limiting oxygen index and reduced burning rate of polyester resins incorporating aluminium hydroxide. The trigger to large tonnage usage of aluminium hydroxide came a few years later after several fires in the USA where carpets were highlighted as responsible for spreading fires. In one particular case in Ohio, USA, 32 people died in a modern old people's home. A Senate enquiry attributed the spread of fire along a corridor to a foam backed nylon pile carpet; the report stressed that the fatalities were due not to burns but to asphyxiation. The outcome was that legislation was introduced in the USA in 1971 covering mandatory Federal standards for the surface spread of flame on carpets and rugs when subjected to a small ignition source. Aluminium hydroxide was found to be the ideal fire retardant to meet this legislative requirement. It could be incorporated into either the foam rubber backing or the adhesive rubber anchor coat which bonds the carpet fibres. High levels of addition could be used without adversely affecting the performance of the rubber and the

[9] K. A. Evans and N. Brown, in 'Speciality Inorganic Chemicals', ed. R. Thompson, Special Publication No. 40, The Royal Society of Chemistry, London, 1981, p. 164.
[10] J. W. Lyons, 'The Chemistry and Uses of Fire Retardants', Wiley, New York, 1970.
[11] Sovereign Mills, UK Patent No. 183922, 1922.
[12] W. J. Connolly and A. M. Thornton, *Mod. Plast.*, 1965, **43**, 154, 156, 202.

necessary degree of fire resistance could be obtained. Many thousands of tonnes of aluminium hydroxide are used for this application annually.

During the 1970s, the use of aluminium hydroxide as a fire retardant in plastics grew from a few hundred tonnes to many tens of thousands of tonnes. The largest use was in glass reinforced unsaturated polyester in a wide variety of applications including electrical switch boxes, building panels, machine housings, and automotive parts. Significant markets also emerged in PVC, especially for conveyor belting, polyurethane foam, and epoxies. Considerable tonnages, particularly of specially produced white grades of aluminium hydroxide, are now used in the manufacture of acrylic based (Corian®-DuPont) or polyester based compositions for use in kitchen work surfaces, bathroom vanity units, surrounds, sinks, and shower trays. Very high levels of aluminium hydroxide are utilized, 60–70%, to provide good fire retardancy and good stain resistance together with a translucent appearance which may be likened to marble.

The current annual tonnage of aluminium hydroxide used as a fire retardant world-wide is estimated at over 200 000 tonnes. The USA and Canada account for some 120 000–127 000 tonnes of this and Europe 40 000–50 000 tonnes. Japan consumes about 20 000–30 000 tonnes annum^{-1}.

Combustion of a polymeric material can be divided into a number of stages: heating, decomposition of the polymer to give combustible gases, ignition, heat release to the surrounding material, and propagation.

Aluminium hydroxide operates in the first two stages of the combustion process, namely heating and decomposition. On heating to above 200 °C, aluminium hydroxide decomposes to give alumina and water vapour; this reaction is strongly endothermic absorbing $1.97 \, kJ \, g^{-1}$ of aluminium hydroxide. The heating of the polymeric material is thereby slowed down and the onset and rate of decomposition of the polymer also reduced. Additionally the water vapour that is evolved from the decomposing aluminium hydroxide dilutes any combustible gases evolved and hinders access of oxygen to the surface of the polymer thereby further suppressing ignition.

For an endothermic material to be efficient as a fire retardant, it is important that its endothermic decomposition process occurs over the temperature range at which the polymer decomposes.

The smoke suppressing property of aluminium hydroxide is not fully understood but it is believed that the heat dissipation that occurs in the burning polymer favours cross-linking reactions over pyrolysis. This leads to the formation of a char in preference to soot particles.

The physical and chemical characteristics of aluminium hydroxide which combine to make aluminium hydroxide such an effective fire retardant are summarized in Table 1.

The limitations to the use of aluminium hydroxide as a fire retardant are its unsuitability for polymers processed at temperatures above 200 °C and the high loadings necessary to impart a sufficient degree of fire retardancy.

Considerable development work has been undertaken to overcome these problem areas and to satisfy the growing use of polymer applications. Products in the particle size range of 100 μm to a few tenths of a micrometre have been

Table 1 *Characteristics of Aluminium Hydroxide Which Make it an Effective Fire Retardant*

(1)	Endotherm occurs at an optimum temperature range for many plastics and rubbers.
(2)	It is an effective smoke suppressant for many polymer systems.
(3)	It does not evolve any corrosive or toxic product on decomposition.
(4)	It presents no health hazard when handled.
(5)	It is virtually insoluble in water, so will not leach out of a filled polymer.
(6)	Grades with purities of 99.6% are readily obtained.
(7)	Its electrical properties make it an ideal filler for insulators.
(8)	It is non-volatile so will not exude out of the polymer on ageing.
(9)	It is relatively cheap compared to other flame retardants and to the polymer which it replaces.
(10)	The price of the basic hydrate is likely to remain stable as it is made in very large tonnages being an intermediate in the production of aluminium.
(11)	It is a crystalline material and hence the absorption capacity of its surface is relatively low.
(12)	It is available in a wide variety of particle sizes (a few tenths of a μm to $100\,\mu$m).
(13)	It can be produced with a high level of whiteness and is unaffected by ultraviolet light.
(14)	It has a Mohs' hardness of $2\frac{1}{2}$ so does not cause excessive wear on equipment.

developed with a range of particle size distributions in order to achieve maximum filler loading at minium viscosity in resins.

Sub-micron grades of aluminium hydroxide are extensively used in cable sheathing compounds. Loading levels up to 65% are commonplace in order to achieve high levels of fire retardancy and low smoke evolution whilst providing a cable which is free from halogens so obviating the problem of hydrochloric acid fume evolution during a fire. For these applications, grades of aluminium hydroxide have been specially developed with very low levels of soluble soda levels such that they have the minimum adverse effect on the cable when it is immersed in water over a prolonged period. The finest grades are found to impart a reinforcing effect in several elastomer systems.

The polymeric systems where aluminium hydroxide can be employed as a fire retardant/smoke suppressant are very wide-ranging[13,14] and are summarized in Table 2. Aluminium hydroxide is compatible with many other fire retardants and blends are frequently employed in order to impart the necessary degree of fire retardance/smoke suppression at minimum cost.

The use of organic compounds and organo-metallic compounds has been widely investigated in order to improve the compatibility between the surface of the aluminium hydroxide and the polymer. Stearates, plasticizers, silanes, titanates, zirco-aluminates, and aluminium organics have all been employed with varying degrees of success. The appropriate choice of substituent on the metal can result in easier processing or improved mechanical properties (especially retention of mechanical properties after immersion in water).

[13] S. C. Brown and M. J. Herbert, 'New Developments in ATH Technology and Applications', in 'Flame Retardants', Plastics and Rubber Institute, Elsevier Applied Science, London, 1992, p. 100.
[14] L. A. Musselman, in 'Alumina Chemicals', ed. L. D. Hart, American Ceramic Society, Westerville, USA, 1990, p. 195.

Table 2   *Application Areas for Aluminium Hydroxide as a Flame Retardant/Smoke Suppressor*

Thermosetting polymers
Polyester – spray
            – hand lay up
            – SMC (Sheet Moulding Compounds)
            – DMC (Dough Moulding Compounds)
            – Foam
            – Cast (*e.g.* Synthetic marble)
Epoxy
Phenolic
Polyurethane
Methacrylic resins [*e.g.* Corian® (DuPont), Avron® (ICI)]

Thermoplastics
PVC – rigid
      – plasticized
Polyethylene
Polypropylene
Polystyrene

Elastomeric
Natural rubber
Styrene butadiene rubber
Neoprene
Nitrile rubber
Butyl rubber
Silicone rubber
EPDM   – ethylene propylene diene monomer
EPR/EPM   – ethylene propylene rubber
Vamac®   – ethylene vinyl acetate
Hypalon®   – chlorosulfonated polyethylene
Vinylacetate – vinyl chloride copolymers

Cellulosics
Loft insulation
Chipboard or particle board
Hardboard

3.4.2 Toothpaste. The usage of aluminium hydroxide as an abrasive in toothpaste formulations grew dramatically in the 1970s and 1980s to a level of several tens of thousands of tonnes. The growth was driven by the trend towards therapeutic/orthodontic pastes and the compatibility of aluminium hydroxide with the majority of compounds which are now frequently incorporated into a toothpaste in order to bestow a therapeutic or prophylactic property.

Additives which are added include:

- fluoride to minimize dental caries;
- bacteriostats to destroy mouth bacteria, *e.g.* chlorohexidine;
- zinc salts to inhibit plaque formation, *e.g.* zinc citrate;
- enzyme inhibitors to prevent or reduce acid production; and
- enzymes to accelerate the breakdown of protein, starch, and lipids.

A typical toothpaste might contain approximately 50% abrasive filler, 20% water, 25% humectant such as glycerin or sorbitol, 1–2% surfactant, *e.g.* sodium lauryl sulfate, 1% thickening agent, *e.g.* sodium carboxymethyl cellulose, and 1% flavour, together with active ingredients to reduce dental caries (fluorides), minimize plaque formation (*e.g.* zinc citrate), or prevent gum disorders. The abrasive filler used should be sufficiently abrasive so that material adhering to the teeth is removed without causing damage to the tooth enamel or any exposed dentine. A particular advantage imparted by aluminium hydroxide, when incorporated into a dentrifice, is that the toothpaste has very good cleansing properties but causes less damage to tooth enamel than most of the other commonly used abrasive fillers such as calcium carbonate or dicalcium phosphate.[15] Aluminium hydroxide has a Mohs' hardness of between $2\frac{1}{2}$ to $3\frac{1}{2}$, considerably below that of tooth enamel, dentine, or cementum; its abrasivity falls between that of chalk and dicalcium phosphate. The degree of abrasivity and cleaning properties can to some extent be controlled by the particle size and particle size distribution of the aluminium hydroxide used in the formulation. Grades with median particle sizes of between $30\,\mu m$ and $1\,\mu m$ have been used to produce dentrifices with different characteristics. Grades of aluminium hydroxide with median particle sizes of $1\,\mu m$ have the minimum abrasivity and grades of this size have been used since the 1950s to increase the viscosity of toothpastes, especially under low shear conditions. The utilization of grades containing particles with a narrow distribution of sizes will result in more cost-effective toothpastes as the higher viscosity that they impart allows higher levels of water to be utilized.

The effect of the filler on the active ingredients is important, especially with respect to fluoride additions. The utilization of calcium carbonate has declined sharply because it can react with the fluorides added and so obviate their effectiveness as anticaries additives. Sodium fluoride, stannous fluoride, and sodium monofluorophosphate are all frequently used as fluoride sources in toothpastes. Tests have shown good compatibility between aluminium hydroxide and sodium monofluorophosphate and toothpastes containing them have been found by clinical trials to remain effective at reducing dental caries for relatively long periods.[16,17] Elevated levels of soluble soda adversely affect the effectiveness of the fluoride-containing pastes. Co-milling the aluminium hydroxide with long chain fatty carboxylic acids can improve the effective utilization of the fluoride.[18] Utilization of grades of aluminium hydroxide with low surface activity also enhances the effectiveness of the fluoride,[19] zinc citrate,[20] and other additives.

3.4.3 Paper. Sub-micron, normally high whiteness precipitated grades of aluminium hydroxide have been used since the early 1950s in paper manufacture but its use is now declining. Initially it was used as a filler replacing some of the titania and substituting for calcium carbonate and china clay. Aluminium

[15] Unilever, UK Patent No. 1188353, 1968.
[16] R. J. Andlaw and G. J. Tucker, *Br. Dent. J.*, 1975, **138**, 462.
[17] P. M. C. James *et al.*, *Com. Dat. Ord. Epid.*, 1977, **5**, 67.
[18] Colgate Palmolive, US Patent No. 4098878, 1978.
[19] Alcan, Eur Patent No. 328407, 1987.
[20] Unilever, US Patent No. 4988498, 1991.

hydroxide has a lower refractive index than titania but does not adversely affect the functioning of some of the optical whiteners used in paper. It has very high retention and dispersibility characteristics during paper manufacture and improves the retention behaviour of other fillers and pigments. It produces a higher opacity than china clay and results in an improvement in printing performance.[14,21,22]

The use of aluminium hydroxides in paper coating dates back to the 1960s as a replacement for china clay.[23] The aluminium hydroxide imparts high brightness, high opacity, high gloss, and good ink receptivity.

3.4.4   Paint. Developments have been in progress since the 1960s partially to replace the titania in paint. Williams[24] showed that a 25% replacement of the titania with a sub-micron precipitated grade of aluminium hydroxide could be made in a vinyl latex coating formulation without loss of opacity but at lower cost. The effectiveness of the aluminium hydroxide was attributed partly to the fact that it is close to the optimum size for providing the ideal spacing of titania particles for light reflection and scattering.

Some use of aluminium hydroxide as a titania substitute is made in acrylic and alkyl paint systems but the scale is currently small. Small quantities are also used in fire retardant paints, mastics, intumescent coatings, and ablative coatings for its fire and smoke inhibiting characteristics.

3.4.5   Pharmaceuticals. Significant tonnage of amorphous aluminium hydroxide gel is used in antacid preparations and medicines for the control of phosphate levels. A high degree of purity is required and the product is designed to have a high reactivity in order to readily react with excess stomach acids. Compositions containing magnesium hydroxide in addition to aluminium hydroxide gel are sometimes employed.

4   ALUMINIUM OXIDE

4.1   Nomenclature and Properties of Aluminium Oxide

There are many forms of aluminium oxide (α, χ, β, η, δ, κ, θ, γ, ρ, ι) but α-aluminium oxide is the only thermodynamically stable form. The other forms are frequently termed transition aluminas and arise during the thermal decomposition of aluminium trihydroxides and oxide–hydroxides under varying conditions. β-Alumina, when first reported, was thought to be another form of aluminium oxide but subsequent investigations showed the presence of sodium oxide. The most common form of β-alumina is $Na_2O.11Al_2O_3$ although several other forms exist with differing ratios and incorporating other alkali or alkaline earth oxides.

A number of sub-oxides of aluminium have been reported to exist in the gas phase: AlO, Al_2O, and Al_2O_2.

[21] J.J. Koenig, *Tappi*, 1964, **48(12)**, 123A.
[22] W.H. Bureau, in 'Pulp to Printing', Graphic Arts Publications, Chicago, 1968.
[23] H.H. Murray, 'Paper Coating Pigments', Monography Series No. 30, New York, 1963.
[24] J.E. Williams, Jr., *Paint Varn. Prod.*, 1967, **57**, 54.

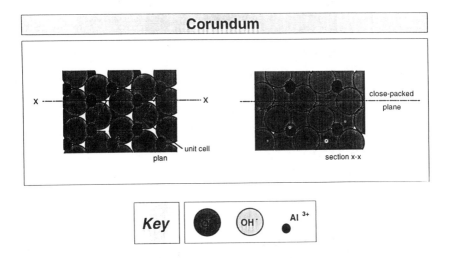

Figure 3 *Crystal Structure of Corundum*

α-Aluminium oxide or alumina is composed of hexagonal close-packed layers of oxygen ions with two-thirds of the octahedral holes occupied statistically by aluminium ions. Bragg and Bragg[25] investigated the structure of α-alumina showing it to be hexagonal–rhombohedral. The crystal structure of corundum is shown in Figure 3. α-Alumina is extremely hard with a Mohs' hardness of 9; this is surpassed by only a few other materials. It has a density of $3.98\,\mathrm{g\,cm^{-3}}$.

Corundum is a common constituent in many igneous and metamorphic rocks. Corundum of gemstone quality is found in alluvial deposits in Burma, Sri Lanka, Kashmir, Australia, and East Africa. The presence of metal ions gives rise to a number of precious gemstones. Traces of Cr^{3+} give ruby and of Fe^{2+}, Fe^{3+}, or Ti^{4+} give blue sapphire. Synthetic rubies, and blue and white sapphires are produced in significant quantities. Lower quality corundum is a major component of emery, which is used as an abrasive. It is mined in South Africa and, the Greek island, Naxos.

α-Alumina is the thermodynamically most stable form of the compounds formed between aluminium and oxygen and is the final product from thermal or dehydroxylation treatments of all the hydroxides or oxides. The dehydroxylation of gibbsite to the various aluminium oxides is shown in Figure 4.

4.2 Industrial Grades of Aluminium Oxide

Industrial grades of aluminium oxide are often divided into: smelter, activated, catalytic, calcined, low-soda, reactive, tabular, fused, and high purity.

These differ in their particle size, morphology, α-alumina content, and impurities, especially the soda level. With the exception of very small quantities they are produced by calcination of Bayer derived gibbsite.

[25] W. H. Bragg and W. L. Bragg, 'X-Rays and Crystal Structure', Bell, London, 1916.

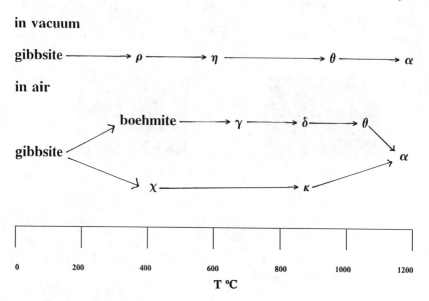

in vacuum

gibbsite ⟶ ρ ⟶ η ⟶ θ ⟶ α

in air

boehmite ⟶ γ ⟶ δ ⟶ θ

gibbsite

χ ⟶ κ

⟶ α

T °C

Figure 4 *Thermal Transformation Sequences for Gibbsite to Alumina*

4.2.1 Smelter Grade Alumina. Smelter grade alumina is the name given to alumina utilized in the manufacture of aluminium metal. Historically it was manufactured using rotary kilns but is now generally produced in fluid bed or fluid flash calciners. In fluid flash processes the aluminium hydroxide is fed into a counter-current stream of hot air obtained by burning fuel oil or gas. The first effect is that of removing the free water and the next is removal of the chemically combined water; this occurs over a range of temperatures between 180–600 °C. The dehydrated alumina is principally in the form of activated alumina and the surface area gradually decreases as the temperature rises towards 1000 °C. Further calcination at temperatures > 1000 °C converts this into the non-absorbent α-form. Smelter grade aluminas generally retain the physical nature of the starting aluminium hydroxide. The conversion into the α-form is typically of the order of 25% and the specific surface area is relatively high at $> 50 \, m^2 g^{-1}$ due to the presence of transition aluminas. The nature and percentage of the transition aluminas depend on the atmosphere within the calciner, the amount of moisture present, and the heating rate. The alumina most commonly used for smelting today is termed 'sandy' alumina as it has good flowing properties. The older, more traditional smelters required an alumina with a low bulk density, high α-content, and with poor flowing characteristics so that when heaped up in the cell it tended to stay in position. This alumina is traditionally called 'floury' alumina. Fluid flash calciners are much more efficient than rotary kilns and reduce fuel costs by about 30–40%. The higher performance speciality aluminas, however, are generally manufactured using rotary or static kilns.

4.2.2 Calcined Alumina. Nearly four million tonnes out of the total of 40 million tonnes of alumina made annually is utilized for applications other than aluminium production.

The markets for speciality calcined alumina is approximately divided into:

refractories	50%
abrasives	20%
whitewares and spark plugs	15%
ceramics	10%
others	5%

Calcined aluminas are generally manufactured in rotary kilns with the aluminium hydroxide going through the phases described earlier. In the rotary kiln mineralizers are frequently added to catalyse the reaction and bring down the temperature at which the α-alumina forms. Fluorides are the most widely used type of mineralizer; they have a major effect on the morphology of the crystals formed, giving platey crystals.

Soda is the major impurity and can rise as high as 0.6 wt. %; the other major impurities are SiO_2 at 0.02–0.05 wt. % and Fe_2O_3 at 0.02–0.04 wt. %. The α-alumina content is normally in the range 75–100% and the surface areas will consequently be much lower than those for smelter grade alumina – 0.5 to 25 $m^2 g^{-1}$. The agglomerates produced during rotary calcination are typically 40 to 150 μm in size; these agglomerates consist of crystals 0.5 to 15 μm in size. For most applications these agglomerates are milled or micronized to release the fine crystals. On dehydroxylation the gibbsite converts to transition aluminas with partially disordered structures. As the calcination temperature increases, the structures become more ordered until the transformation to α-alumina occurs. As demonstrated in 'An Atlas of Alumina',[26] the alumina aggregates are pseudomorphs of the original gibbsite with little particle size change on calcination.

4.2.3 Low Soda Aluminas. Low soda aluminas are defined generally as aluminas with a soda content of <0.1 wt. %. These can be manufactured from gibbsite by a number of methods:

- use of a low soda gibbsite prepared using modified decomposition conditions;
- addition of chlorine gas to the kiln;
- addition of boron additives to the gibbsite prior to calcination;
- washing of activated or high surface area alumina prior to final calcination; or
- addition of compounds, such as silica sand, to the gibbsite prior to passage through the kiln: the soda reacts with the silica to form sodium silicate which is then removed by sieving.

'Medium' or 'intermediate' soda aluminas is the name given to aluminas with soda contents of 0.15–0.25 wt. %. Some micrographs of aluminas displaying varying morphologies are shown in Figure 5.

'Reactive' aluminas is the term normally given to aluminas of relatively high purity and small crystal size (<1 μm) which sinter to a fully dense body at lower

[26] B A Chemicals, 'An Atlas of Alumina', 1969.

(a)

Figure 5   *Micrographs of Aluminas Displaying Different Morphologies. (a) Fluoride Mineralized Alumina and (b) Boron Mineralized Alumina*

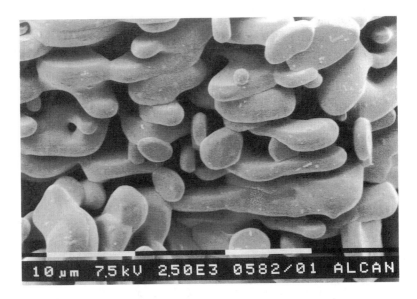

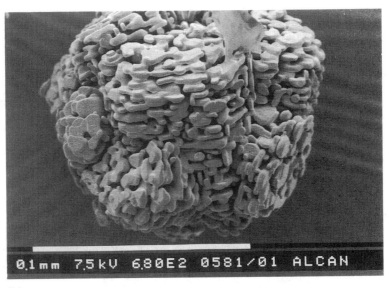

(b)

temperatures than low soda, medium-soda, or ordinary-soda aluminas. They are normally supplied after intensive ball-milling which breaks up the agglomerates produced after calcination. They are utilized where exceptional strength, wear resistance, temperature resistance, surface finish, or chemical inertness are required.

4.2.4 Tabular Alumina. Tabular alumina is recrystallized or sintered α-alumina, so called because its morphology consists of large, 50–500 μm, flat tablet-shaped crystals of corundum. It is produced by pelletizing, extruding, or pressing calcined alumina into shapes and then heating these shapes to a temperature just under their fusion point, 1700–1850 °C; shaft kilns have been found to be especially effective for the production of tabular aluminas.

After calcination the spheres or shapes of sintered alumina can be used as they are for some applications or crushed, screened, and ground to produce a wide range of sizes. As the material has been sintered it has an especially low porosity, high density, low permeability, good chemical inertness, high refractoriness, and is especially good for refractory applications.

4.2.5 High Purity Aluminas. High purity aluminas are normally classified as those with a purity of 99.9% and can be manufactured by routes starting from Bayer hydrate using successive activations and washings or via a chloride to achieve the necessary degree of purity. The higher purities are manufactured by calcining ammonium aluminium sulfate or from aluminium metal. In the case of the route via ammonium aluminium sulfate, the necessary degree of purity is obtained by successive recrystallizations. Especially high purities can be made from aluminium by reacting the metal with an alcohol, purifying the aluminium alkoxide by distillation, hydrolysing, and then calcination.

Applications include the manufacture of synthetic gem stones such as rubies and yttrium aluminium garnets for lasers, and sapphires for instrument windows and lasers. A use that has grown rapidly in recent years has been in 'non-scratch' watch glass faces.

4.3 Uses of Aluminium Oxide

4.3.1 Alumina in Technical Ceramics. Alumina ceramics have outstanding resistance to heat, wear, and chemicals which has lead to their multiplicity of applications and their use in almost every industry. These alumina ceramics can be manufactured from almost pure aluminium oxide or with other compounds to form a glassy phase and allow components to be fabricated at lower temperatures and sometimes with more reproducibility. Other constituents may be added to give enhanced mechanical properties, *e.g.* other oxide ceramics such as zirconia or platelets, fibres, or whiskers of silicon carbide. Recent years have witnessed an explosion of literature about alumina composites which are derived directly from aluminium metal. In this process, the Lanxide™ (Lanxide Corporation) process, very complex shapes can be manufactured and oxide can be made to grow through a wide variety of ceramic materials.[27,28]

[27] M. S. Newkirk, A. W. Urquhart, H. R. Zwicker, and E. Breval, *J. Mater. Res.*, 1986, **1**, 81.
[28] M. S. Newkirk, H. D. Lesher, D. R. White, C. R. Kennedy, A. W. Urquhart, and T. D. Claar, *Ceram. Eng. Sci. Proc.*, 1987, **8**, 879.

Table 3 *Properties of High Alumina Ceramic Bodies*

High tensile strength.
High compression strength.
Hard and abrasive.
Resistant to abrasion.
Resistant to chemical attack by a wide range of chemicals even at elevated temperatures.
High thermal conductivity.
Resistant to thermal shock.
High degree of refractoriness.
High dielectric strength.
High electrical resistivity even at elevated temperatures.
Transparent to radio frequencies.
Low neutron capture area.
Raw material readily available and price not subject to violent fluctuation.

Work on alumina ceramics dates back to the last century and a patent was filed in 1907[29] covering the production of ceramic alumina. It was initially used for laboratory equipment and then for spark plugs. Uses now extend to electronic applications, bioceramics, armour, grinding media, abrasion resistant tiles, cutting tools, and wear parts for the textile and paper making industries. The properties of alumina ceramics are summarized in Table 3.

Laboratory ware. Ceramics for use in high temperature applications such as laboratory crucibles and thermocouple sheaths are manufactured with the minimum glassy component and with an alumina content of $>99\%$. The ceramics are produced in such a way as to give large grain sizes after sintering.[30]

Spark plugs. One of the most familiar examples of a high alumina ceramic is the sparking plug insulator which typically contains 88–95% alumina. High-alumina content insulators were developed during the World War II to meet the needs of high compression aero-engines for which the existing mullite plugs (*ca.* 55% Al_2O_3) were inadequate.[31] The high silica content of these mullite insulators made them susceptible to attack from the lead in the fuel and other additives in the lubricants. Alumina was able to meet all the necessary properties: ability to withstand attack by lead tetraethyl and the other compounds in the fuel or lubricating oils; ability to withstand temperatures from sub-zero to over 1000 °C; reasonably high thermal conductivity in order to conduct heat from the nose of the plug to the engine block and thereby avoid pre-ignition; ability to withstand voltages of the order of 20 000 V at high temperature; and high strength. Generally, low soda aluminas (Na_2O $<0.05\%$) are used for the manufacture of sparking plug insulators but in some instances a medium soda alumina (Na_2O level $<0.25\%$) in a slightly lower alumina content plug is required.

Electronic/electrical applications. These applications include thin and thick film substrates for integrated circuits, electronic packages, vacuum tube envelopes, r.f.

[29] German Patent No. 220394, 1907.
[30] R. Morrell, 'Handbook of Properties of Technical and Engineering Ceramics. Part 1. An Introduction for the Engineer and Designer', HMSO, London, 1985.
[31] J. S. Owens, *Ceram. Bull.*, 1977, **56**, 437.

windows, rectifier housings, sodium vapour lamp tubes, and high voltage insulators.

Initially alumina-based ceramics were used for the packaging of electronic devices so as to provide environmental protection.[32] Progressively their use has been expanded such that the ceramic substrate became a functional part of the system. In 1965 IBM introduced a solid logic technology (SLT) module. The sheet of 96% alumina was metallized and resistors screened onto these substrates; silicon device chips were then connected. These devices have become progressively more sophisticated with more and more layers and more and more chips whilst becoming smaller.[33] The relatively high thermal conductivity of the alumina conducts heat away and reduces thermal gradients. The strength of the alumina substrate enables the devices to withstand both the thermal and mechanical stresses. Multilayer devices with 36 layers of ceramic substrate are now commonplace. The layers of alumina are made by making a slurry of alumina in a polymer binder [*e.g.* poly(vinyl chloride acetate)] and then spreading it out as a thin film using a doctor blade. The binders are then carefully burnt out in order to control the shrinkage. Low-soda, thermally reactive aluminas are frequently utilized in these applications. As such thin films (<1 μm) are applied to the substrate very smooth finishes (<0.1 μm roughness) are required.

High purity aluminas ($>99.99\%$ Al_2O_3) find use in the manufacture of high-pressure sodium vapour lights. Normally, alum derived aluminas are utilized so that the fired polycrystalline alumina tubes are translucent and allow the transmission of $>96\%$ of the light whilst still providing excellent resistance to the sodium vapour.

Alumina has been used in the less sophisticated area of electrical porcelain since at least 1921 when it was used to replace flint.[34] Alumina additions of up to 40% are used together with ball clay, feldspar, and china clay.

Bioceramics. Whilst the quantities are small compared to other uses the applications of alumina in bioceramics are well established and present some particularly challenging problems. The main applications are as bone substitutes in joints and dental implants. Alumina is used because of its excellent wear characteristics and resistance to attack by body fluids. Alumina ceramics containing 99.99% Al_2O_3 are employed in order to achieve the necessary chemical inertness. Particularly exacting production conditions are required to manufacture components with small grain size (<4 μm), high strength, and close size tolerance. Chemical polishing using fluxes is sometimes employed to reduce the coefficient of friction and improve wear resistance.

Alumina ceramic hip joints have been employed for many years. The alumina femoral head (ball) and acetabulum (cup) in either alumina or ultra high molecular weight polyethylene (UHMWPE) have been found to be far superior to the conventional titanium metal–UHMWPE joint, especially with respect to wear characteristics. This considerably extends the time before a patient needs a replacement joint. It has been found that the surface roughness of the metal to

[32] B. Schwartz, *J. Phys. Chem. Solids*, 1984, **45**, 1051.
[33] B. Schwartz, in 'Alumina Chemicals', ed. L. D. Hart, American Ceramic Society, Westerville, USA, 1990, p. 299.
[34] R. Twells, Jr. and C. C. Lin, *J. Am. Ceram. Soc.*, 1921, **4**, 195.

UHMWPE joint increased as time progressed whilst the roughness for a ceramic system became lower.[35]

Other uses for alumina ceramics in the body include maxillary reconstructions, middle ear ossicular replacements, and dental implants. Dental implants include single crystal alumina inserts that are threaded into the alveolar bone – these can serve as anchors for bridges or can be capped with crowns of polycrystalline alumina.[36] In addition to its high strength and high corrosion resistance, no calculus plaque or concrements adhere to the alumina surface.[37]

Some interesting lines of research have involved developments with respect to encouraging the bonding of the alumina ceramic with bone or body tissue. It has been shown that connective tissue infiltration with initiation of bone growth to a depth of 100 μm can occur if porous alumina with a pore size of 75 to 100 μm is used under certain conditions.[38]

Later developments have included alumina poly(tetrafluoroethylene) composites which enable large pore size ranges (50 to 400 μm) to be produced which promote rapid growth of tissue.[39] Excellent bone attachment has been reported with $MgAl_2O_4$ based ceramics composite.[40] This ceramic composite, or osteoceramic, is composed of $MgAl_2O_4$ and α-$Ca_3(PO_4)_2$ in equal molar quantities. The spinel provides the skeleton and provides the strength whilst the tricalcium phosphate which fills the skeletal pores is bioresorbable and is progressively replaced by bone.[41,42]

Armour applications. High alumina ceramics have been used in armour materials since the early 1960s. The main areas where alumina ceramics have been used are for bullet-proof flak jackets for police and military personnel and in strategic parts of vehicles especially helicopters. The mechanism by which ceramic materials such as alumina work as armour materials is very different from metals. Metals absorb the energy of a projectile by a plastic deformation mechanism whilst in a ceramic the kinetic energy of the projectile is absorbed by a fracture energy mechanism. The ceramic must therefore have a high hardness, a high compressive strength, and a high Young's Modulus. On encountering a ceramic surface the tip of a projectile becomes blunted so its effective cross-section increases. This leads to a diminution in the compressive stress. A compressive shock wave occurs at the point of impact, travels through the material resulting in star shaped radial cracks and circumferential cracks.[43] The microstructure of the alumina ceramic plays an important role in the energy dissipation process.[44]

[35] H. Dörree, D. Bentler, and Geduldig, *Arch. Orthop. Unfall. Chir.*, 1975, **83**, 269.
[36] J. W. Boretos, in 'Alumina Chemicals', ed. L. D. Hart, American Ceramic Society, Westerville, USA, 1990, p. 337.
[37] E. Brinkmann, 'Das Keramik – ~Anker Implantat, nach E. Mutschelknauss', Zähnarzte Prax 29, 1978, p. 148.
[38] S. F. Hulbert, F. A. Young, R. S. Mathews, J. J. Klawitter, C. D. Talbert, and F. H. Stelling, *J. Biomed. Mater. Res.*, 1970, **4**, 433.
[39] R. L. Westfall, C. A. Howay, and J. N. Dent, *J. Oral Maxillofacial Surg.*, 1982, **40**, 771.
[40] G. G. Niederauer, T. D. McGee, and R. K. Kudej, *Am. Ceram. Soc. Bull.*, 1991, **70**, 1010.
[41] D. Cutright, S. Bhaskar, M. Brady, L. Getter, and W. Posey, *Oral Surg.*, 1972, **33**, 850.
[42] L. L. Hench and J. Wilson, *MRS Bull.*, 1991, **16**, 62.
[43] V. D. Frechette and C. F. Cline, *Am. Ceram. Soc. Bull.*, 1970, **49**, 994.
[44] S.-K. Chung, *Am. Ceram. Soc. Bull.*, 1990, **69**, 358.

A backing material with a high fracture toughness is required such as high strength steel or precipitation hardened aluminium alloys on fibre reinforced plastics. There is no 'ideal' armour ceramic for all applications. The optimum is very dependant upon what specific ballistic threat the armour is required to resist. As the projectiles become more sophisticated (*e.g.* kinetic energy penetrators and shaped charges) more use has been made of ceramic matrix composites such as silicon carbide whisker filled alumina ceramics and the Lanxide™ (Lanxide Corp., Newark) alumina/aluminium based system containing particulates or whiskers. Composites offer integrity after impact and demonstrate multi hit performance better than pure ceramics.[45]

Ceramic tableware. Some alumina is present in ceramic tableware from the alumina that occurs as a constituent in the kaolins, ball clays, feldspars, nepheline syenite, and talcs that are used during production. Milled calcined alumina is frequently added as a replacement for flint or quartz at a level of 10–20% to increase the fired strength of the tableware giving better mechanical strength and thermal shock resistance.[46,47] In some cooking ware an addition of up to 50% Al_2O_3 is made. This is widely used for 'hotelware' intended for canteens, restaurants, and hospitals where the crockery is subjected to particularly heavy use. Additional benefits that accrue from the incorporation of milled alumina include extension of the firing range, increased whiteness, and fewer flaws,[48] with a more defined size distribution.

Wear resistant components. The extreme hardness and chemical inertness of alumina ceramics make it the ideal material to solve many wear problems. The excellent wear characteristics have been attributed[49] to the absorption behaviour of the surface of alumina ceramics. The outer ionic layers consist of oxygen ions which possess a surface charge; water or long-chain carboxylic acids are attracted by this charge and are adsorbed into the surface by van der Waals forces. This layer of adsorbed molecules forms a protective layer which reduces wear.

One of the earliest large scale uses of alumina ceramic components in wear applications was for thread guides in the textile industry. Synthetic fibres travelling at high speeds were found to be very abrasive and hardened steel, glass, or glazed porcelain components were found to last for only a few days. High alumina ceramics could be fabricated to last for up to ten years. Additionally the very low surface roughness that could be achieved with alumina ceramics meant that little damage was caused to the thread.

Other wear components for which alumina ceramics find use are: wire drawing pulleys for steel and copper wire fabrication, bearings for pumps, joint seals, tap washer seals, linings for coal or mineral handling equipment, cyclone liners, grinding media, ball-mill linings, tines, and seed drills in agricultural equipment.

[45] D. J. Viechnicki, M. J. Slavin, and M. I. Kliman, *Am. Ceram. Soc. Bull.*, 1991, **70**, 1035.
[46] C. R. Austin, H. Z. Schofield, and N. L. Haldy, *J. Am. Ceram. Soc.*, 1946, **29**, 341.
[47] W. E. Blodgett, *Am. Ceram. Soc. Bull.*, 1961, **40**, 74.
[48] R. J. Beals, in 'Alumina Chemicals', ed. L. D. Hart, American Ceramic Society, Westerville, USA, 1990, p. 323.
[49] E. Dörre and H. Hübner, 'Alumina: Processing, Properties and Applications', Springer-Verlag, Berlin, 1984.

In paper making equipment it is used as the support for the wire which supports the web of paper in the fourdrinier section of the equipment.

Cutting tools. Cutting tools inserts composed of high alumina ceramics allow very high cutting speeds to be achieved, up to $2000 \, \text{m} \, \text{min}^{-1}$. This compares to approximately $10 \, \text{m} \, \text{min}^{-1}$ for carbon steels, $50 \, \text{m} \, \text{min}^{-1}$ for high-speed steels containing 25% carbide, and approximately $200 \, \text{m} \, \text{min}^{-1}$ for tungsten or titanium carbide sintered materials. The requirements for an alumina ceramic to meet the stringent demands in cutting tools are a high wear resistance; high mechanical strength at elevated temperature; and high impact strength. The key to producing a good alumina cutting tool is to maximize the impact strength, which is relatively low compared to metals. A number of routes have been developed to obtain improved impact strength:

- fabrication of the ceramic with a small grain size;[50]
- addition of zirconium oxide;[51]
- addition of titanium carbide to increase the thermal conductivity and hence improve the thermal shock resistance;[51]
- incorporation of silicon carbide whiskers;[52]
- addition of titanium nitride, boron nitride or Sialon® (Trade mark of Lucas Cookson).

A common way to reduce the grain size of the alumina ceramic is to add magnesia to inhibit exaggerated grain growth;[53,54] an addition here of 250 p.p.m. is typical and the target grain size is approximately $1-5 \, \mu\text{m}$.

High alumina ceramic cutting tools are especially effective for cast iron and they are used with particular economic advantage for brake discs, brake drums, and cylinder liners. In addition to the higher cutting speeds, high-purity aluminas produce a superior surface finish compared to other cutting tools which reduces the subsequent time and effort required for further finishing operations.

Zirconia toughened alumina. As discussed previously, one characteristic of alumina ceramic materials which has limited its usage has been poor fracture toughness. In 1976 Claussen[55] excited great interest by incorporating zirconia into an alumina matrix to give dramatically higher toughness values and substantially higher fracture strengths.

Zirconia may exist in one of four polymorphic forms: monoclinic, tetragonal, cubic, and orthorhombic. The orthorhombic form only exists under high pressure conditions. For pure zirconia, monoclinic is the form stable at ambient temperature. On heating to $1172 \, °\text{C}$ the monoclinic form transforms to tetragonal which on further heating transforms to the cubic form at approximately $2370 \, °\text{C}$. On cooling back to ambient temperature the monoclinic form is again achieved. The

[50] C. H. Kim, W. Roper, D. P. H. Hasselman, and G. E. Kane, *Am. ceram. Soc. Bull.*, 1975, **54**, 589.
[51] E. D. Whitney, *Powder Metall. Int.*, 1978, **10**, 16, 18–21.
[52] G. Fisher, in 'Alumina Chemicals', ed. L. D. Hart, American Ceramic Society, Westerville, USA, 1990, p. 353.
[53] H. P. Cahoon and C. J. Christensen, *J. Am. Ceram. Soc.*, 1956, **39**, 337.
[54] R. J. Brook, in 'Ceramic Fabrication Processes. Treatise on Materials Science and Technology', ed. F. F. Y. Wang, Academic Press, New York, Vol. 9, 1976, p. 331.
[55] N. Claussen, *J. Am. Ceram. Soc.*, 1976, **59**, 49.

transformation on cooling from the tetragonal form to the monoclinic form results in a volume increase of 3 to 5%. The addition of certain oxides such as Y_2O_3, MgO, certain rare earth oxides, and CaO can stabilize first the tetragonal phase and as increasing amounts are added, the cubic form.[56]

If zirconia particles are incorporated into an alumina matrix then the volume increase on cooling through the tetragonal–monoclinic can generate microcracks in the alumina matrix. These microcracks dissipate the energy of the propagating crack and considerably increase the toughness of the material. The size of the zirconia particles must be large enough to transform but small enough such that they cause only limited microcracking. If the zirconia particles are less than a certain critical size then the tetragonal–monoclinic transformation does not occur. Thus if very small size zirconia particles (less than 0.6 μm) are incorporated into the alumina then on cooling the zirconia remains in the metastable tetragonal form. In the vicinity of a propagating crack tip the tensile stresses release the matrix constraints on the metastable tetragonal zirconia particles which are then able to transform to the monoclinic form. The volume expansion and compressive strain thereby generated opposes the tensile forces opening the crack and means that additional work is required to propagate the crack further: this results in increased toughness and strength.

Typically zirconia is incorporated at levels of 10–15 vol. % but the optimum value will be dictated by the zirconia particle size, the presence or otherwise of stabilizers in the zirconia, and the balance between the fracture toughness and strength required. Fracture toughness values of three times the levels for conventional aluminas, some in excess of 12 MPa m$^{\frac{1}{2}}$, have been reported by several groups of workers. To date the main growth area for zirconia-toughened ceramics are wear components and cutting tool tips.

Ceramic matrix composites. Considerable interest has been aroused by the incorporation of fibres, whiskers, and platelets into alumina ceramics in order to improve specific properties, especially toughness.

The addition of 20 vol. % silicon carbide whiskers to alumina has given material with a fracture toughness value of 7.8 MPa m$^{\frac{1}{2}}$ and a flexural strength of 700 MPa. The addition of single crystal silicon carbide fibres (diameter 0.6 μm, length 10–80 μm) raised the fracture toughness from 4.6 MPa m$^{\frac{1}{2}}$ with no addition to 8.7 MPa m$^{\frac{1}{2}}$ (crack plane parallel to pressing direction) and 5.4 MPa m$^{\frac{1}{2}}$ (crack plane perpendicular to pressing direction).

β-Aluminas. β-Alumina has the idealized chemical formula $Na_2O.11Al_2O_3$ having a layer structure where the sodium ions are present in discrete layers separate blocks of aluminium ions and oxygen ions in the form of a spinel structure. There are two main subgroups, designated β'- and β"-alumina, which have different stacking sequences. β"-Alumina has the approximate chemical formula $Na_2O.5.33Al_2O_3$; it has similar chemical properties to β-Al_2O_3 but is a lower resistivity material and a contiguous mixture of the two is now the preferred material for use as an ionic conductor in electrochemical cells.[57] The β"-alumina structure is stabilized by the addition of magnesium or lithium ions: final levels of

[56] R. Stevens, 'An Introduction to Zirconia', Magnesium Elektron Ltd., Manchester, UK, 1986.
[57] W. T. Bakker, in 'Alumina Chemicals', ed. L. D. Hart, American Ceramic Society, Westerville, USA, 1990, p. 309.

magnesia in the product may be 2.5–4% and final levels of lithia between 0.2–0.8%. The presence of CaO, SiO_2, and K_2O are especially critical as they adversely affect the electrical resistivity and electrochemical behaviour of β''-alumina. A preferred method, patented by Ducan *et al.*,[58] which gives β''-aluminas with improved purity and excellent conductivity utilizes crystalline boehmite (CERA® Hydrate, Alcan Chemicals Ltd.) as the alumina source. High levels of lithia (0.80% by mass) enable tubes for electrochemical cell separators to be made containing substantially 100% β''-alumina.

The addition of 5–15% zirconia to the β''-alumina aimed at transformation toughening the material inhibits the formation of large crystals and enhances the performance of the cell.[57] In practice small amounts of zirconia (less than 1%) are added to achieve micro-structural control, preventing secondary grain growth. The zirconia also has the beneficial effect of acting as a 'getter' for impurities such as Ca^{2+} thereby improving grain boundary conductivity.

The goal of a relatively light, cheap battery which could provide sufficient energy density to power electric vehicles has been a focus of researchers for decades. The development of the sodium–sulfur battery in the late 1960s was a major step in achieving that objective. These batteries had a much higher energy density than lead acid batteries (100–150 W h kg^{-1} compared to 40–48 W h kg^{-1}) and enabled an electric vehicle to travel more than twice as far for the same weight of battery. At present a range of 150–200 miles has been achieved in test vehicles.

The sodium–sulfur battery comprises metallic sodium contained in a β-alumina tube as the negative electrode and sulfur impregnated with carbon fibre as the positive electrode. The β-alumina keeps the sodium and sulfur separate but allows the conduction of sodium ions. The operating temperature of the battery is about 350 °C which is necessary to achieve the ionic conductivity required for practical usage. This is one of the major disadvantages as precautions must be taken to ensure heat insulation and adequate containment in the event of an accident. Considerable opportunities exist for sodium–sulfur batteries for relatively low cost power load levelling systems. This would enable power plants to run at maximum efficiency. An electrically powered vehicle would require approximately 500 sodium–sulfur batteries whilst a load levelling storage battery would require 10 000 or more cells. Sodium–sulfur battery powered vehicles have been running on an experimental basis in Manchester (UK) for some years.

Considerable effort has been expended in enhancing the performance of sodium–sulfur batteries. A key area has been the improvement in consistency and number of charging/deep discharge cycles prior to failure. Individual test cells have achieved more than 5000 such cycles, well in excess of commercial requirements. Recent development with systems based on sodium aluminium chloride show particular promise and should one day enable a relatively cheap vehicle to be produced which is free from air pollution and conserves liquid fuels.

Alumina in glass. The alumina content of natural glasses (mainly rhyolitic obsidious) is 12–15 wt. %. The substitution of a silicon ion by an aluminium ion in a silicate glass structure has a significant affect on the stability of the three-dimensional structure. Binary Al_2O_3—SiO_2 glasses are only stable up to

[58] J. H. Duncan, P. Barrow, A. Van Zyl, and A. I. Kingon, 'Making β''-alumina', British Patent No. 2175582.

about 7 wt. % Al_2O_3; above this level a liquid phase, high in alumina, separates out.

Low level additions of alumina to soda–lime–silica glasses significantly reduces the tendency for devitrification or phase separation. Alumina also reduces the liquidus temperature when added to binary silicate glasses in amounts up to 10 wt. %.

Alumina improves the durability of certain glasses especially sodium borosilicate glasses used for scientific equipment, pharmaceutical products, or for body fluid containment where low alkali extraction rates are required. The acid resistance of Na_2O–CaO–SiO_2 glasses is improved by alumina additions. The addition of up to 3 wt. % Al_2O_3 is used in binary alkaline earth silicate and Pyrex® (Corning Inc.) type borosilicate glasses; without the alumina the borosilicate may exhibit immiscibility problems on cooling resulting in poor durability and unpredictable properties.[59]

Alumina has a strong network–forming effect in ternary aluminosilicate glasses, thereby increasing their viscosity. The alkaline earth alumina silicate glasses have high melting temperatures and high viscosities; this, however, allows strengthening via physical tempering and use at high temperature without deformation or loss of stress. This effect has been used in Correlle® (Cornings Inc.) laminated dinner ware and a similar composition used for 'E'-glass fibres and tungsten–halogen lamp envelopes.

Calcined alumina is used in optical lens and other special glasses requiring a high degree of purity in raw materials particularly with respect to impurities such as iron and chromium that can affect the colour and light transmission of the glass. High purity aluminas are used in some photochromic glasses where control of Al_2O_3 levels and impurities is critical.

4.3.2 Glass-ceramics. Glass-ceramics have been available since the 1950s.[60] Alumina is added to control the crystallization of glasses in order to produce a ceramic with a fine, uniform crystal size. One of the earliest materials developments to have practical use was cordierite, $2MgO.2Al_2O_3.5SiO_2$; the alumina content is about 35 wt. %. In this magnesium aluminosilicate, titania is added as a nucleating agent. Cordierite is still widely used because of its excellent thermal shock resistance, strength, toughness, and low thermal expression. Applications include catalyst substrates and radar-transmitting missile nose cones. Considerable interest exists in replacing alumina electronic substrates with glass-ceramics such as cordierite. They have a lower dielectric constant and a lower sintering temperature than alumina. This allows multi-layer substrates coated with Cu, Ag, or Au conductor patterns to be cofired and higher circuit densities to be obtained.

Another commercially significant area of glass-ceramics are the lithium aluminosilicates. A wide range of compositions, between $Li_2O.Al_2O_3.4SiO_2$ and $Li_2O.Al_2O_3.10SiO_2$, all have similar structures and properties. β-Spodumene $(Li_2O.Al_2O_2.4SiO_2)$ can be nucleated within a lithium alumino-silicate to produce a microcrystalline glass ceramic with a thermal expansion coefficient close to zero. This has lead to its utilization in cooking dishes. These

[59] P. F. Becher and G. C. Wei, *J. Am. Ceram. Soc.*, 1984, **67**, C-267.
[60] S. D. Stookey, *Ind. Eng. Chem.*, 1959, **51**, 805.

commercial lithium aluminates have a typical alumina content of 18 wt. %.

The incorporation of titania or zirconia nucleating agents into lithium alumino-silicates enables β-quartz glass-ceramics to be produced which have a crystallite size $<0.1\,\mu$m. This type of material is used for transparent cookware (*e.g.* Visions® – Corning Inc.), wood-stove windows, and electric cooker tops.

Glass-ceramics based on mullite, $3Al_2O_3.2SiO_2$, with alumina contents above 35 wt. % can be produced and exhibit good high temperature properties, excellent dielectric properties, and low thermal expansion. Mullite fibres have been produced by the sol–gel process and sold under the designation Nextel® (3M Company).

Nepheline based glass-ceramics form the basis of highly durable tableware. They consist in the main of $Na_2O.Al_2O_3.2SiO_2$ (nephelines) and $BaO.Al_2O_3.SiO_2$ (celsian) utilizing titania as the nucleating agent and have an alumina content of about 30 wt. %.

Glass-ceramics based on fluorophlogpite $(KMg_3AlSi_3O_{10}F_2)$ were developed in the 1970s. They contain 10 to 20 wt. % Al_2O_3. They can be made to have excellent dielectric properties, dimensional stability, zero porosity, and the capability of being machined by ordinary metalworking tools.[61] This latter feature is possible because randomly oriented tiny flakes of mica allow only local fracturing to occur. A commercially available material, Macor® (Corning Inc.), is used for windows for microwave tube parts, precision electric insulators, seismograph bobbing, and boundary retainers for the space shuttle.

4.3.3 Refractories. Alumina is widely used in refractory products to enhance their chemical resistance and improve thermal properties. High alumina refractory bricks and castables are inert to most corrosive fumes and slags and will resist both reducing and oxidizing atmospheres. In addition they exhibit increased load-bearing ability and resistance to spalling, thermal shock, and flame impingement. Calcined, fused, and tabular alumina are all used in the production of refractories for the ferrous, non-ferrous, cement, glass, chemical, petroleum, ceramics, and mineral processing industries. The usage of alumina in refractories falls into two areas which are almost equal in size: refractory bricks and monolithic refractories.

Refractory bricks. Alumina-silica refractory materials are often classified into low-alumina ($<45\%$ Al_2O_3) and high-alumina materials. The low-alumina refractories are made from fireclays, that is clays with an alumina content of $>20\%$. High-alumina refractories utilize the aluminas listed above, bauxite or sillimanite, kyanite or andalusite.

The choice of the composition of a refractory brick is dictated by its ability to withstand:

- chemical attack;
- erosion;
- thermal cycling;
- thermal shock; and
- transfer of heat.

[61] G. H. Beall, in 'Advances in Nucleation and Crystallisation of Glasses', ed. L. L. Hench and S. W. Frieman, American Ceramic Society Special Publication, Westerville, USA, 1972.

Whilst fully dense ceramics have many of the required properties they do not normally have sufficient resistance to the thermal shock witnessed in furnaces. The key is therefore to combine the necessary components to give a material which has the required degree of chemical resistance but has a controlled microporosity which then imparts the necessary hot strength.

Alumina in monolithic refractories. Monolithic refractories can be sub-divided into four main types: castables, mouldables (or plastic refractories), ramming mixes, and gunning mixes. They were initially introduced for repair of furnaces but are now used for whole furnace linings. Monolithic refractories have advantages over refractory bricks when kilns or furnaces of complex shapes are involved and are frequently easier and quicker to install than refractory bricks. Monolithic refractories comprise an aggregate and a binder. The aggregate in alumina-based monolithic refractories can be: calcined alumina, reactive alumina, fused alumina, tabular alumina, calcined bauxite, kyanite (a naturally occurring alumino-silicate from the USA, Sweden, Spain, or Brazil with an Al_2O_3 content of 56–60%), sillimanite (an alumino-silicate formed under higher temperature and pressure than andalusite, found in South Africa and India), andalusite (an alumino-silicate containing theoretically 63.2% Al_2O_3 mined in Spain, South Africa, France, China, and Brazil with an Al_2O_3 content of 55–61%), calcined kaolin pyrophyllite (theoretically $Al_2O_3.4SiO_2.H_2O$ mined principally in North Carolina), mullite ($3Al_2O_3.2SiO_2$), or calcined fire clay. Normally precalcined materials are used in order to minimize volume change on firing. The aggregates are incorporated to a level of between 50 to 80%.

There are two distinct ways of effecting the bonding in monolithic refractories. Depending upon the system, they are supplied in a moist form which develop their initial strength on drying or reaction with oxygen, or in a dried mix which require the addition of water on installation to achieve the necessary degree of hydration.

Plastic refractories which utilize clay to introduce plasticity have been commercially available since 1914. The clay provides a low level of strength on drying out and them imparts full strength on firing by ceramic bond formation.

Phosphates, especially monoaluminium phosphate and phosphoric acid, are widely used as binders. They provide excellent hot strength and abrasion resistance to the refractory.

Aluminium chlorophosphate hydrate, urea phosphate, chromium aluminium phosphate, and alkali metaphosphates have all been used as binders. Aluminium monophosphate is the preferred binder as direct use of phosphoric acid can cause bloating under certain circumstances. The ultimate heat treatment and aggregates present in the mix are particularly important in achieving the desired properties. Aluminium hydroxychloride, alumina sols, and aluminium sulfate are used as air-bonding binders in plastic refractories. The aluminium hydroxychloride and sulfate give products with good performance but evolve corrosive decomposition products on baking out.

Ramming mixes are forced into position by mechanical action and fired *in situ* in the furnace. They are similar to plastic refractories but contain less bonding agent.

Castable refractories generally utilize calcium aluminate as the binder. Calcium aluminate cements with different levels of alumina content (36–47%, 48–62%, and 70–80% Al_2O_3) can be added to provide progressively higher hot strength properties.

In order to improve service temperatures under certain conditions development work has focused on cement free castables and both aluminium and ρ-alumina additions have been found to be successful.[62]

Gunning mixes may comprise of the other types of formulation but are generally closest to castables. Water is added to the mix immediately prior to use and it is sprayed into position by compressed air. The size of the aggregate used is carefully chosen in order to have the appropriate viscosity and flow characteristics. This is a particularly rapid method of applying a furnace lining. The higher the alumina content in the monolithic refractory the higher the service temperature. The service temperature rises from 1540–1600 °C for a 40–50% Al_2O_3 containing plastic refractory to 1800–1900 °C for a 90% Al_2O_3 containing refractory. The choice of the alumina-containing aggregate or binder is however critical in ensuring that the correct mineralogy is obtained in the material after installation and heating. Higher alumina contents will result in refractories with higher thermal conductivities and thermal expansion.

Spinel. The period of the late 1980s and early 1990s has seen very significant growth in the usage of the mineral spinel, magnesium aluminate ($MgAl_2O_4$: theoretical Al_2O_3 content 71.7%). Spinel–magnesia refractories are resistant to both acidic and basic slags but have better thermal shock characteristics than do basic magnesia refractories. Spinel melts at 2135 °C and has a theoretical density of $3.59\,g\,cm^{-3}$. This combination of good chemical resistance to alkalis and suitable thermal properties make it the preferred choice in situations such as the hot zone of a cement kiln. As cement clinker is high in calcia, it will react with alumino-silicate-based refractories. During cement production the clinker forms a continuous coating inside the kiln. At periodic intervals this clinker breaks away exposing the refractory to a very high thermal shock. That the shock is too great for dolomite-based refractories led to the development of direct-bonded chrome–magnesite bricks. Concerns about the disposal of chrome-based refractories has led the industry subsequently to search for alternatives. Periclase–spinel bricks provide a good solution to the problem by virtue of the following characteristics:

(i)   they are resistant to the alkali environments exhibited in a cement kiln;
(ii)   they have a lower thermal conductivity than do magnesia bricks so the kiln shell runs at a lower temperature and less heat is wasted;
(iii) the density is lower than that of magnesia, leading to better reliability of the kiln;
(iv)   they have much better resistance to thermal shock; and
(v)   they have a lower thermal expansion.

Spinels rich in either alumina or magnesia have been developed.

[62] Y. Hongo, Y. Tsuzuki, and M. Miyawaki, 'Refractory Compositions', US Patent No. 4331733, 1982.

Recent years have also witnessed considerable growth in the use of spinel in refractories for the Japanese steel industry. The addition of spinel to alumina improves the slag and spalling resistance of castable refractories for the ladle linings of basic oxygen furnaces (BOFs). Bricks with a chemical composition between 80% MgO and 80% Al_2O_3 can be produced to give not only a range of slag and temperature shock resistances, but also a range of creep behaviour. These materials find application in sliding gate plates, continuous casting ceramics, and zone-lining in torpedo cars.

Calcium aluminate. Calcium aluminate cements of high purity may be manufactured from high purity limestone and calcined alumina. The primary hydraulic setting phase is $CaO.Al_2O_3$ but the secondary phases $CaO.2Al_2O_3$ and $12CaO.7Al_2O_3$ are also important. A particular advantage of calcium aluminate cement over Portland cement is its rapid strength development. High purity calcium refractory castables to form linings up to 1 m thick. They are extensively used with aggregates such as fire clays, kaolin, andalusite, kyanite, sillimanite, mullite, pyrophyllite, and refractory grade bauxites.

Ceramic fibres. Amorphous alumina-based ceramic fibres were developed in the 1940s. The concern about the use of asbestos led to renewed interest and they are now employed in a wide range of refractory insulation systems. They are more prone to chemical attack than is asbestos but have better thermal stability; aluminosilicate fibres can be utilized at temperatures up to 1400 °C. Standard grades contain 45–50% Al_2O_3 but compositions richer in alumina are made for use under higher service temperatures. They are made from natural alumina-containing minerals or from calcined alumina, depending on the final composition required. Minor quantities of ZrO_2, Cr_2O_3, or B_2O_3 are added to control various characteristics. The fibres are manufactured by melting the raw materials in a de Bussey-type furnace and a stream of molten material is then allowed to flow onto a series of rotating wheels, or the melt is subjected to a high-velocity air-jet. High-alumina fibres can be made by utilizing a solution of sol–gel method. Saffil® (ICI Ltd.) is a 95% Al_2O_3/5% SiO_2 fibre manufactured from an aluminium oxide precursor such as basic aluminium chloride.

4.3.4 Polishing and Grinding. Naturally-occurring alumina in the form of corundum or emery (alumina mixed with magnetite) has been used as an abrasive for over two-and-a-half thousand years. Corundum has a Mohs' hardness of 9 and there are only a few materials harder, *viz.* diamond, boron carbide, silicon carbide. Alumina in an appropriate form has been used for polishing a wide variety of materials: rocks, stainless steel, non-ferrous metals, plastic spectacle lenses, plastic aircraft windows, jewellery, or car paint finishes.

The alumina is sometimes used in an oil- or water-based slurry or can be mixed with oil or wax to form a paste or block. The critical alumina properties which affect its performance are its particle size, particle shape, particle size distribution, α-alumina content, and internal porosity. For most of these applications purity is not critical and 'ordinary' or 'normal' soda products (*i.e.* 0.1 to 0.6% soda) are used. Coarse aluminas (particle size 3–30 μm) are used when a high cutting rate is required whilst aluminas with a particle size of 2 μm or less are used when a polishing action is required. Control of particle size and particle size distribution is especially important for use in polishing plastic lenses and aircraft windows.

Alumina is added to toothpastes especially formulated for use by smokers to aid the removal of nicotine deposits from teeth. Here the avoidance of coarse particles which might cause scratching of the enamel is important.

Very fine aluminas with a narrow particle size distribution are sometimes added to magnetic tape at a low level (*e.g.* 1% addition) to prevent the build-up of iron oxide deposits on the recording/playback mechanisms.

4.4 Activated Alumina

As described earlier, various transition aluminas are obtained when aluminium hydroxide or aluminium oxide–hydroxide is heated at temperatures below that necessary to form α-aluminium oxide. These transition aluminas are frequently termed 'activated' or 'active' aluminas. Ulrich designated the aluminas obtained by dehydroxylation of aluminium hydroxide as γ-aluminas. This term persisted and came to be used for all the aluminas obtained by calcination at temperatures below 1000 °C. The generally agreed forms of transition aluminas are γ, δ, η, θ, χ, κ, ρ, ι. ρ-Alumina is amorphous but the other forms have reasonably well defined X-ray diffraction patterns. The transition phases that occur on heating aluminium trihydroxide (gibbsite) in vacuum are ρ-alumina (100–400 °C), η-alumina (270–500 °C), θ-alumina (870–1150 °C), and finally α-alumina. In air two distinct routes may be followed: either via boehmite (180–300 °C), γ-alumina (500–850 °C), δ-alumina (850–1050 °C), and θ-alumina to α-alumina; or via χ-alumina (200–500 °C) and κ-alumina (900–1000 °C) to α-alumina. The sequences of transformation are shown in Figure 4. The route followed is determined by a number of factors, in particular the particle size of the gibbsite and the hydrothermal conditions in the vicinity of the particles; other factors that have an influence on the path include the heating rate, moisture level in the feed, pressure, bed depth, and soda content. Thus, for coarse gibbsite particles (>60 μm) the route via boehmite is favoured as the water of dehydroxylation cannot readily escape. In the case of sub-micron gibbsite particles, however, hardly any boehmite may be produced. The amorphous form, ρ-alumina, can be produced by heating gibbsite at temperatures above 700 °C in a fast air stream or in a vacuum.

As for gibbsite, the dehydroxylation sequence of bayerite may follow two path-ways. Coarse particles (>100 μm), high pressures, fast heating rates, and moist conditions tend to favour the route via boehmite, γ-alumina, δ-alumina, and θ-alumina to α-alumina in the same way as with gibbsite. Fine bayerites transform under conditions of dry air and low heating rates via the following route: η-alumina (250–500 °C) $\rightarrow$ θ-alumina (850–1150 °C) $\rightarrow$ α-alumina.[63] The dehydroxylation sequence of nordstrandite is similar to that of bayerite.

[63] J. B. Barnett, US Patent No. 1868869, 1932.

4.4.1   Production of Activated Alumina. World-wide production of activated alumina is of the order of 100 000 tonnes annum^{-1}, principally for adsorption and catalysis. Historically it was produced by the dehydroxylation of gibbsite which is sometimes deposited as a scale at the bottom of the decomposition (precipitation) vessels during the Bayer process. Dehydroxylation was undertaken in a stream of air at 400 °C to give a material, principally γ/η-alumina, with some χ-alumina and boehmite with a surface area of *ca.* 250 m^2 g^{-1} and a broad distribution of pore-sizes up to 500 nm diameter.

Higher surface area material, 250–375 m^2 g^{-1}, can be made by heating aluminium hydroxide to temperatures between 400 to 800 °C to give predominantly ρ-alumina which is amorphous. Pelletization of this material allows spheres to be made which can be rehydrated and activated to give the desired strength and pore volume. On rehydration pseudoboehmite and bayerite is produced which transforms to η-alumina on dehydration. Higher initial dehydration temperatures lead to very amorphous products with improved rehydration characteristics and final surface areas of 320–380 m^2 g^{-1}. Activated aluminas of this type are widely used as desiccants, selective adsorbents, catalysts, and catalyst supports.

An especially favoured route for the production of aluminas for catalyst manufacture is via alumina gels to produce pseudoboehmites. There is considerably scope via this approach to modify the pore volume distribution of the product.

A considerable body of patent literature exists covering different routes to make alumina gels but the most common routes are based on: neutralization of aluminium sulfate or ammonium alum by ammonia; or neutralization of sodium aluminate with carbon dioxide, sodium bicarbonate, or aluminium sulfate. The mixing conditions in order to obtain the desired purity and morphological characteristics are critical. The precipitates obtained are frequently gelatinous and difficult to wash free from impurities. The filter cake may then be formed into the desired shape or spray-dried to give spheres. These materials are then activated at a temperature of 400–600 °C to give products with a surface area of 300–600 m^2 g^{-1}, a pore size of approximately 4 nm and which are mixtures of γ- and η-alumina.

Another route which is widely used for the production of high surface area activated alumina is via hydrolysis of aluminium alkoxides. The Ziegler process involves reacting aluminium metal with ethylene and hydrogen to form aluminium triethyl; controlled addition of more ethylene leads to polymerization of the ethylene on each branch of the aluminium ethyl arm to give the desired aluminium trialkyl which may then be partially oxidized to form the alkoxide. Hydrolysis of the alkoxides to produced results in linear alcohols and a high purity pseudoboehmite. The main impurities are titania and traces of carbon. The product is widely used as a wash coat for automotive emission catalysts. It can be extruded to various shapes and can be peptized with various acids to give a high green strength to the extruded body. The extruded material may be activated at 500–600 °C to give γ-alumina with a surface area of 185–250 m^2 g^{-1}.[64]

[64] C. Misra, 'Industrial Alumina Chemicals', American Chemical Society, Monograph 184, Washington, DC, 1986.

4.4.2 Uses of Activated Alumina. Adsorbents. The use of activated alumina as an adsorbent goes back to the 1930s; large tonnages are currently used for gas and liquid drying.

Even though the surface of an activated alumina has a strong affinity for water, making it is very effective for desiccant use, the process can be reversed readily at above 250 °C under low relative humidity conditions. The adsorption–desorption cycle can be repeated many hundreds of times without significant deterioration of efficacy. At a relative humidity of 50%, activated alumina can adsorb up to 20% of its own weight of water. The mechanism proposed for water bonding assumes oxide ions on the outermost surface layer and an incompletely co-ordinated aluminium ion in the next lower layer. This exposed cation is located in a 'hole' that is electron-deficient and acts as a Lewis acid site. The rehydroxylation of the dehydroxylated alumina surface on exposure to water vapour is accompanied by considerable heat evolution.[64]

Activated alumina can be used for removing water from a very wide range of compounds including: acetylene, air, alkanes, alkenes, ammonia, argon, benzene, chlorinated hydrocarbons, chlorine, hydrocarbons, natural gas, oxygen, petroleum fuels, sulfur dioxide, and transformer oils.[65] Activated alumina can dry a gas to a lower water content than any other commercially available desiccant.

In addition to water removal, activated alumina can be used to selectively adsorb certain other chemical species from gaseous or liquid streams. Polar molecules such as fluorides or chlorides are readily adsorbed, so in petroleum refining activated alumina is used to adsorb HCl from reformer hydrogen gas and organic fluorides from hydrocarbons produced by the HF alkylation process. Other important adsorption processes include:[66,67]

- removal of trace chloride and iron contaminants from ethylene dichloride prior to PVC manufacture;
- carbonyl sulfide removal from propylene prior to polypropylene manufacture;
- removal of fluoride from waters with excessively high natural fluoride levels;
- colour and odour removal from industrial effluents; and
- removal of degradation products from transformer oils.

Chromatographic uses for activated alumina in the laboratory have been common since the 1930s but large scale chromatography is now finding increasing application for example in the manufacture of flavours and fragrances, vitamins, steroids, proteins, and the separation of the *meta-* and *ortho-*isomers of xylene. Activated alumina has better resistance to moderately higher pHs than silica, so is the preferred choice for high pH systems. Activated alumina has a surface with both Lewis and Brønsted acidic and basic sites.

Acidity is derived from the Al^{3+} ions and protonated hydroxyls, whilst basicity is due to basic hydroxyls and O^{2-} anion vacancies.

Catalytic uses. Activated aluminas find widespread application as both catalysts

[65] R. D. Woosley, in 'Alumina Chemicals', ed. L. D. Hart, American ceramic Society, Westerville, USA, 1990, p. 241.
[66] H. L. Fleming and K. P. Goodboy, in 'Alumina Chemicals', ed. L.D. Hart, American Ceramic Society, Westerville, USA, 1990, p. 251.
[67] H. L. Fleming, in 'Alumina Chemicals', ed. L. D. Hart, American Ceramic Society, Westerville, USA, 1990, p. 263.

in their own right and as catalyst substrates. Some of the more significant applications are:

(i)   Claus catalyst – removal of hydrogen sulfide from natural gas processing, petroleum refining and coal treatments; sulfur is obtained as a by-product.

(ii)   Alcohol dehydration – to give olefins or ethers.

(iii)   Hydrotreating – to remove oxygen, sulfur, nitrogen, and metal (V,Ni) impurities from petroleum feed stocks, and to increase the H/C ratio.

(iv)   Reforming catalysts – Pt and Re catalysts on a γ-alumina substrate are used to raise the octane-number of petrol.

(v)   Automotive exhaust catalysts – used to oxidize the hydrocarbons and carbon monoxide to water and carbon dioxide. The alumina, as a γ-alumina slurry, pseudoboehmite, or boehmite sol is coated onto a cordierite or metal honeycomb. The surface is then activated by heating and platinum, palladium, or rhodium coated onto the alumina.

η- and γ- are the preferred forms of activated alumina used in catalytic processes. η-Alumina is preferred in acid-catalysed reactions because of the higher number of Lewis and Brønsted acid sites. γ-Alumina is preferred for hydrotreating applications because of its superior thermal and hydrothermal stability.

In addition to being able to control pore size, pore size distribution, and pore structure, increasing attention is now devoted to modifying the alumina surface so that it is optimized for different catalyst or adsorption uses.

4.5   Fused Alumina

Several hundreds of thousands of tonnes of alumina are utilized annually in the production of fused alumina which is used for the manufacture of grinding wheels. In this process a 'normal' soda grade of alumina is melted in an electric arc furnace using graphite electrodes. The melt is allowed to solidify, and the block of alumina is then broken up into a grain of various sizes. This grain is mixed with a resin or glass bonding agent and formed into shapes to produce grinding wheels. Small additions of the oxides of titanium, chromium, or zirconium are sometimes incorporated into the melt to give improved performance. Particles of fused alumina of different sizes are also utilized to make sanding or abrasive papers and belts.

Abrasive materials with exceptionally good performance can be manufactured by mixing colloidal aluminas with a magnesium compound and then calcining the mixture. Material with an alumina core and envelope of magnesium spinel is produced which has especially good cutting action giving an abrasive with a longer performance.

5   ANHYDROUS ALUMINIUM CHLORIDE

5.1   Properties of Anhydrous Aluminium Chloride

In the pure form, anhydrous aluminium chloride exists as a white solid but commercial grades vary in colour and are most often pale yellow. It sublimes at

180 °C as the dimer which dissociates to monomer above 300 °C. Liquid aluminium chloride only exists at elevated temperature and pressure; the triple point is 192.5 °C at 233 kPa. Its density is 2.46 g cm^{-3} and its heat of formation is 705.6 kJ mol^{-1}.

Anhydrous aluminium chloride reacts violently with water to give hydrogen chloride, steam, and fine particulate aluminium oxide. It reacts vigorously with most alcohols and forms complexes with polar aprotic solvents such as acetonitriles, ethers, nitromethane, and nitrobenzene.

5.2 Production of Anhydrous Aluminium Chloride

Anhydrous aluminium chloride can either be prepared from aluminium metal or from aluminous materials such as clay or bauxite.[68,69] Its production from aluminium metal involves bubbling gaseous chlorine through molten aluminium and condensing the aluminium chloride vapour.

The reaction is exothermic and therefore requires careful control of the chlorine feedback to maintain the reactor temperature of 600–750 °C. To lower the melting temperature, small amounts of copper or magnesium are frequently added. The aluminium chloride then collects in a condensing chamber before being removed, crushed, sized, and packaged under an atmosphere of dry air or nitrogen. The overall purity of aluminium chloride produced is >98% with the main impurities being Cl_2, Al, Fe, Si, Ti, and Ca. The presence of free chlorine of ferric chloride gives rise to a yellow coloration whilst a grey or green colour indicates aluminium.

Chlorination of aluminous ores has been extensively investigated principally as a route to produce aluminium by subsequent electrolysis of aluminium chloride.[70,71] This route to aluminium has a lower energy requirement than the Hall–Héroult process (by approximately one-third). It should be environmentally cleaner, but there are severe corrosion difficulties to overcome. The early patents involve feeding bauxite calcined at ~800 °C into a reactor and introducing, at the required rate, chlorine and oxygen. The aluminium chloride vapour was condensed as described above (product purity ~95%). The Gulf Oil Company produced aluminium chloride from calcined bauxite and coke for some 40 years but stopped in the 1960s.

More recently, the production of anhydrous aluminium chloride from non-bauxite ores has received considerable attention. The Toth process[72] involves the calcination at 700 °C of a mixture of kaolin, sulfur, and carbon which has been ground and pelletized. The mixture is then chlorinated at 800 °C and aluminium chloride is evolved. Silicon tetrachloride and titanium tetrachloride are removed by distillation and the anhydrous aluminium chloride purified further by resublimation.

Other routes that have been utilized involve the chlorination of γ-alumina in a

[68] A. Thomas, 'Anhydrous Aluminium Chloride in Organic Chemistry', Rheinhold, 1961.
[69] C. M. Marstiller, 'Aluminium Chloride', in 'Kirk–Othmer Encyclopaedia of Chemical Technology', 3rd edn., Vol. 2, Wiley, New York, 1978.
[70] J. Hille and W. Durrwachter, *Angew. Chem.*, 1960, **72(22)**, 850.
[71] Du Pont, US Patent No. 4214913, 1980.
[72] Toth Aluminium, US Patent No. 4695436, 1987.

fluidized bed. A mixture of chlorine and carbon monoxide is passed over a beechwood charcoal catalyst and then passed through a bed of γ-alumina. Small amounts of sodium chloride are added to catalyse the reaction.

Another route that involves alumina as the starting material is the Alcoa process.[73] Chlorine is reacted at approximately 600 °C in a fluidized bed reactor with a mixture of Bayer alumina ($\sim 80\%$), carbon ($\sim 20\%$), plus a sodium compound ($\sim 1\%$).

5.2.1 Applications of Anhydrous Aluminium Chloride. The catalytic properties of anhydrous aluminium chloride were discovered by Friedel and Crafts in 1877 and are now employed in a wide variety of reactions in many sectors of industry. Reactions in which it is used include dehydrogenation, decarboxylation, oxidation, alkylation, acylation, desulfurization, amination, and polymerization of many types. In volume terms the largest uses are for the production of ethylbenzene (for styrene production), ethyl chloride (used in tetraethyl lead and ethylcellulose production), cumene, aluminium alkyls, ibuprofen, phenylethyl alcohol, and the polymerization of isoprene, isobutylene, and ketones.

Aluminium chloride is also utilized as a nucleating agent in the manufacture of titanium dioxide. It is mixed with titanium tetrachloride and encourages the formation of the rutile form of titanium dioxide.

Other uses include the removal of dissolved gases and magnesium from aluminium alloys during secondary smelting, as a component of the flux used in arc welding titanium alloys,[69] as a water repellent agent for the impregnation of cotton textiles, as a binding agent for fire-resistant ceramic products, and as a hardening agent for photographic fixing chemicals.

The market for anhydrous aluminium chloride is over 100 000 tonnes annum^{-1}. The US market is estimated at 30 000 tonnes annum^{-1}, that for Europe 40 000 tonnes annum^{-1} whilst Japan's production is 10–12 000 tonnes annum^{-1}. The requirement for anhydrous aluminium chloride in catalysis is essentially static: some sectors of the market are growing, but the reduction in usage of lead as an anti-knock additive in petrol/gasoline has had a dramatic effect. The demand for anhydrous aluminium chloride for use during titanium dioxide production is growing as the chloride route is increasingly preferred to the sulfate route because it produces less solid waste so is environmentally more acceptable.

6 HYDRATED ALUMINIUM CHLORIDE

6.1 Production of Hydrated Aluminium Chloride

Aluminium chloride hexahydrate, $AlCl_3.6H_2O$, is a deliquescent powder which is soluble in both alcohol and water. It is commercially available as a 28% by weight solution and as a crystalline solid.

Aluminium chloride hexahydrate, $AlCl_3.6H_2O$, is usually made by dissolving aluminium hydroxide in concentrated hydrochloric acid at atmospheric boiling.[74] When the acid is used up, aluminium chloride hexahydrate is crystallized from the

---

[73] Alcoa, US Patent No. 3929975, 1975.
[74] C. W. Grams, 'Aluminium Halides and Aluminium Nitrate' in 'Kirk–Othmer Encyclopaedia of Chemical Technology', 4th edn., Vol. 2, Wiley, New York, 1991, p. 281.

solution by first of all cooling to 0 °C and then sparging with HCl gas. Any impurities from the starting aluminium hydroxide remain in solution so that a purified $AlCl_3.6H_2O$ crystallizes out. The crystals are readily filtered from the liquor and can be washed with ethyl ether before drying. Hydrated aluminium chloride can also be made by the hydrolysis of anhydrous aluminium chloride with dilute cold hydrochloric acid.

6.2 Uses of Hydrated Aluminium Chloride

Aluminium chloride hexahydrate has a large number of applications in industry, being used in textile finishing to impart crease-resistance and non-yellowing properties to cotton fabrics, antistatic properties to polyester, polyamide, and acrylic fibres, and improved flammability resistance of nylon.[74]

6.3 Basic Aluminium Chlorides

Basic aluminium chlorides are used in deodorants, anti-perspirant, fungicidal preparations, as water treatment agents, thickeners in paint, earth stabilization to prevent water permeation during oil drilling and as the starting material for the production of alumina fibres. They have the general formula $Al_2(OH)_{6-n}Cl_n$ (where n is 1 to 6). There are several preparative routes:

- reacting an excess of aluminium with 5–15% hydrochloric acid at 65–100 °C;
- the hydrolysis of aluminium alkoxides with hydrochloric acid;
- electrolysis of dilute aluminium chloride solutions; or
- the reaction between aluminium trihydroxide and hydrochloric acid under conditions of high pressure and temperature.

7 POLYMERIC ALUMINIUM COMPOUNDS

Aluminium compounds are used extensively in water treatment. The critical first stage is the hydrolysis of the aluminium ion and the subsequent polymerization to multinuclear hydrolysis products. The hydrolysed species are then adsorbed on the surface of extraneous colloidal particles present in the water and destabilize them. Van der Waal's forces then cause the destabilized particles to aggregate together and form flocs which can be readily removed, thereby clarifying the water. With a chemical such as aluminium sulfate the hydrolysis reaction requires the presence of natural alkalinity in the water, *e.g.* calcium bicarbonate or carbonate. The process can, however, be greatly improved by the use of pre-polymerized aluminium compounds, *e.g.* PAC, PASS® (trademark of Handy Chemicals Ltd.).

7.1 PAC

PAC or poly(aluminium chloride) was developed by the Taki Fertilizer Company in the 1960s and is now widely used as a water coagulant; world-wide usage is of the order of $200\,000\,\mathrm{t\,a^{-1}}$. It is normally manufactured by reacting a mixture of concentrated hydrochloric acid and sulfuric acid with aluminium hydroxide followed by the addition of a base such as calcium hydroxide.[89] If calcium hydroxide is utilized calcium sulfate is precipitated resulting in a product with the approximate formula: $Al_2(OH)_3Cl_{2.5}(SO_4)_{0.25}$. The waste calcium sulfate generated however presents a considerable disposal problem, but may under certain circumstances be used for the manufacture of building panels. Other routes to manufacture PAC type products include manufacture from scrap aluminium, spent aluminium chloride catalyst waste, and aluminium sulfate. The ratio of the hydroxyl/aluminium/polyvalent anion is critical to the performance of the PAC.

7.2 PASS®

PASS® or poly(aluminium silicate sulfate) is manufactured by a patented route from aluminium sulfate, sodium aluminate, and sodium silicate and has the empirical formula:[90,91]

$$Al(OH)_b(SO_4)_c(SiO_x)_d$$

The molecular weight, as determined by ultrafiltration and electrophoresis, is of the order of 300 000 Daltons. Products with alumina contents between 8–10% are normally manufactured.

Polyaluminium salts contain a mixture of monomeric, oligomeric, and polymeric cations. It is often assumed that 50–60% are in the polymeric form, a typical species might be $Al_{13}O_4(OH)_{26}(H_2O)_{12}{}^{5+}$. There polymeric aluminium compounds are lower in acidity than aluminium sulfate and are typically 50% neutralized or basic. The decrease in pH following a given aluminium dosage is less for polyaluminium salts. Whilst the total number of available positive charges of the flocculant is decreased following partial neutralization, the presence of polymeric units bearing high positive charge leads to stronger electrostatic forces between these polymers and negatively charged species (dissolved or colloidal). Additionally, the prehydrolysed aluminium species present in basic-aluminium salts may achieve complete hydrolysis faster than the initial Al^{3+} species. This is particularly so at low temperatures and polymeric aluminium compounds therefore have much better coagulation/flocculation characteristics in the cold than simple aluminium coagulants. The high molecular weight polymeric species can also temporarily act as a flocculant aid during the flocculation process, improving floc growth and floc resistance to shear.

These characteristics of polymeric aluminium compounds give rise to much improved behaviour in practice and may be summarized:

- produce faster forming flocs, especially at low temperatures;
- treat a wider range of waters than simple inorganic coagulants;
- reduce or eliminate the need for additional pH correction chemicals;
- produce larger and faster settling flocs reducing or eliminating the need for coagulation aids; and
- reduce residual aluminium concentrations.

These characteristics offer many benefits in terms of better performance and increased plant throughput.

8 SODIUM ALUMINATE

8.1 Production of Sodium Aluminate

The use of sodium aluminate, $NaAlO_2$, as a commercial product began in the 1920s. It is now being used increasingly in the water treatment and paper industries amongst others. The commercial product is available as either a liquid containing about 40% by weight of sodium aluminate or as a solid product which is essentially anhydrous.

Sodium aluminate is of course the circulating liquid medium of the Bayer process for the production of alumina from bauxite. The process liquor, however, tends to be too contaminated with impurities so that commercial quantities of sodium aluminate are normally made by dissolving aluminium trihydroxide from the Bayer process in sodium hydroxide solution at atmospheric boiling. To obtain a colourless solution or to obtain a white product completely soluble in water and free from colour, it is important to start with a low-iron aluminium trihydroxide ($<0.010\%$ Fe_2O_3). Even so, colloidal iron associated with certain bauxites, *e.g.* Guyana, can, via the trihydroxide, impart a brown coloration to the sodium aluminate solution. This however, can be removed with a small amount of active carbon.

Commercial grades of sodium aluminate generally contain some water of hydration and an excess of soda. In solutions, the high pH retards the reversion of sodium aluminate to the insoluble aluminium hydroxide; liquid sodium aluminate generally requires the higher $Na_2O:Al_2O_3$ ratios (up to $1.5:1$) for good stability and storage properties. Grades with particularly long stability have been developed; these use patented stabilizing agents, *e.g.* Hanfloc-Handy Chemicals Ltd. Commercial sodium aluminates are therefore not accurately represented by any of the commonly used formulae $NaAlO_2$, $Na_2O.Al_2O_3$, or $Na_2Al_2O_4$.

There are many other processes known for preparing sodium aluminate that do not require the addition of water.[75] For example, sodium aluminate can be prepared as a hygroscopic white crystalline solid by fusing equimolar amounts of sodium carbonate and aluminium acetate at a temperature of $800\,^\circ C$.[56]

8.2 Applications of Sodium Aluminate

Sodium aluminate has a large number of uses as a coagulating agent. It is also used in conjunction with other coagulating agents such as aluminium sulfate, ferric salts, clays, and polyelectrolytes.[76] In the paper industry, sodium aluminate improves sizing and filler retention. When added to titania paint pigment, the

[75] W. R. Busler, 'Aluminates' in 'Kirk–Othmer Encyclopaedia of Chemical Technology', 3rd edn., Vol. 2, Wiley, New York, 1978.
[76] Nalco Chemical Company, Canadian Patent No. 964808, 1975.

non-chalking behaviour of outdoor paints is improved. In the processing of acrylic and polyester synthetic fibres, the use of sodium aluminate enhances dyeing, antipiling, and antistatic properties of the fibres. Sodium aluminate is often used in the preparation of alumina based catalysts. The reaction of sodium aluminate with silica or silicates can produce porous crystalline alumino-silicates which are used as absorbents and catalyst supports.

9   ALUMINIUM NITRATE

Aluminium nitrate nonahydrate, $Al(NO_3)_3.9H_2O$, is a stable white crystalline material, has a melting point of 73.5 °C, and is soluble in cold water, alcohols, and acetone. It is used primarily as a salting-out agent in the extraction of actinides. It is also used a source of alumina used in the preparation of insulating papers, in transformer core laminates, and in cathode-ray tube heating elements.

   In general, aluminium nitrate nonahydrate is prepared by dissolving aluminium trihydroxide in dilute nitric acid and crystallizing from the resulting aqueous solution. Other methods which have been patented involve treating bauxite or calcined clay with nitric acid. Liquor purification generally requires the removal of dissolved iron from the leach liquor before crystallization takes place. This is generally achieved by solvent extraction using, for example, di-(2 ethyl hexyl) phosphoric acid.[77] There is negligible co-extraction of aluminium.

10   ALUMINIUM PHOSPHATES

Aluminium forms a number of phosphates: the orthophosphate, $AlPO_4$; the primary and secondary orthophosphates, $Al(H_2PO_4)_3$ and $Al_2(HPO_4)_3$; the pyrophosphate, $Al_4(P_2O_7)_3$; the acid pyrophosphate $Al_2H_6(P_2O_7)_3$; and the metaphosphate, $Al(PO_3)_3$.[78]

   The orthophosphate can be prepared by dissolving aluminium hydroxide in a stoichiometric amount of phosphoric acid. If an excess of phosphoric acid is used than the acid orthophosphates will be formed. An alternative route to produce the orthophosphate is to mix a salt of lower-boiling acid, such as sulfate, chloride, or nitrate, with the appropriate proportion of phosphoric acid.[79]

   The principal use of aluminium phosphates is as high temperature binding agents. The aluminium poly(orthophosphates) form viscous, adhesive solutions in water at room temperature which are highly effective in permanently bonding refractories even at high temperatures.[80] The aluminium poly(orthophosphates) and metaphosphates are employed in glasses, ceramics, and catalysts.[79] They are incorporated into special glasses when particularly high chemical resistance is required.

[77] Arthur D. Little, UK Patent No. 1311614, 1973.
[78] J. W. Mellor, 'A Comprehensive Treatise on Inorganic and Theoretical Chemistry', Longmans, Green & Co., London, 1929.
[79] J. R. Van Wazer, 'Phosphoric Acids and Phosphates', in 'Kirk–Othmer Encyclopaedia of Chemical Technology', 2nd edn., Vol. 15, Wiley, New York, 1968.
[80] W. D. Kingery, *J. Am. Chem. Soc.*, 1952, **35**, 61.

11 ALUMINIUM CARBOXYLATES

The aluminium salts of carboxylic acids are known as aluminium carboxylates. They are derived from aluminium hydroxide by successive replacement of hydroxyl by carboxylate anions.

The aluminium formate series would, for example, comprise of the following:

Normal aluminium formate	$Al(OOCH)_3$
Monobasic aluminium formate	$(HO)Al(OOCH)_2$
Dibasic aluminium formate	$(HO)_2Al(OOCH)$

The general methods for preparing mono- and dibasic aluminium carboxylates include:

(i) direct reaction of aluminium as the catalytic electrode with an aqueous solution of the acid;[81,82]

(ii) reaction of the metal with the acid in the present of mercury or mercuric chloride in catalytic amounts;[83]

(iii) reaction by double displacement between aluminium alkoxide and the sodium salt of the organic acid.[84]

Normal aluminium carboxylates are prepared by the direct reaction of the acid with aluminium chloride in an organic solvent, or by a double displacement reaction using a soluble aluminium salt of the organic acid.[85,86]

About twenty of the aluminium carboxylates are of industrial importance and have commercial applications. The applications fall into four general areas:

(i) finishing agents for the water-proofing of cloth;

(ii) as mordants in the dyeing of textiles;

(iii) pharmaceutical preparations because of their antiseptic, astringent, and basic properties; and

(iv) the manufacture of cosmetics due to the gelling properties of some of the higher molecular weight aluminium carboxylates.

The most commonly used aluminium carboxylates for water-proofing textiles and as mordants in dyeing are the: monobasic aluminium acetate, aluminium formoacetate, normal aluminium formate, and basic aluminium formate.[87] The water-proofing effect is achieved by dipping the textiles into a solution of a water-soluble soap then into an aluminium salt to form a water resistant aluminium soap $[Al(OOCR)_n(OH)_{3-n}]$ on the surface of the fibres.

Other uses for the acetate include:

[81] G. G. Merkl, US Patent No. 2325018, 1973.
[82] G.G. Merkl, US Patent No. 2957598, 1976.
[83] R. W. Jones and J. W. Clusky, *Cereal Chem.*, 1963, **40**, 589.
[84] Takeda Chemical Industries, Japanese Patent No. 15353, 1963.
[85] R. C. Mehrattra and A. K. Rai, *J. Indian Chem. Soc.*, 1962, **39**, 1.
[86] N. Shoji and M. Hideo, Japanese Patent No. 10476, 1961.
[87] A. L. McKenna, 'Aluminium Carboxylates', in 'Kirk–Othmer Encyclopaedia of Chemical Technology', 4th edn., Vol. 2, 1991, p. 273.

- monobasic aluminium acetate $[(HO)Al(OOCCH_3)_2]$ for cross-linking acid-hydrolysed collagen in the food and photographic industries;
- aluminium diacetate as a nucleating agent in conjunction with polypropylene in the manufacture of printing plates with good printing characteristics;
- aluminium diacetate $[(HO)Al(OOCCH_3)_2]$ as an esterification catalyst in the manufacture of dimethyl ester;
- dibasic aluminium monoacetate $[(HO)_2Al(OOCCH_3)]$ as a heat stabilizer and to prevent yellowing in the glyoxal–glycol durable press finish; and
- monobasic aluminium formoacetate $[(HO)Al(OOCH)(OOCCH_3)]$ to impart water repellency and in the tanning of collagen tape for surgical sutures.

The higher molecular weight aluminium carboxylates which are commercially important are the stearates, palmitate, and 2-ethylhexanoate.

The stearates of commercial interest are:

dihydroxyaluminium monostearate	$[(HO)_2Al(OOC(CH_2)_{16}CH_3)]$
monohydroxyaluminium distearate	$[(HO)Al(OOC(CH_2)_{16}CH_3)_2]$
aluminium tristearate	$[Al(OOC(CH_2)_{16}CH_3)_3]$

The monostearate is made by reacting solutions of chlorodihydroxyaluminium with sodium stearate. The distearate is prepared by reacting aqueous solutions of aluminium salts such as the sulfate or chloride with sodium stearate and adjusting the pH. Aluminium tri-stearate can be made from aluminium isopropoxide and stearic acid.

The aluminium stearates are insoluble in water, alcohol, and ethers and are used for a broad range of thickening applications because of their capacity to gel vegetable oils and hydrocarbons. Typical uses are:

- thickeners for synthetic grease;
- water-proofing agents for wood, bricks, masonry, *etc.*;
- fibre lubricants;
- mould release agents;
- thickener for cosmetic formulations;
- additives to enhance fuel efficiency;
- coating additives; and
- antifoam agents.

The monostearate has also found many uses in the pharmaceutical industry as a gellant.

Considerable use is made of poly(oxo-aluminium stearate) $([-O-Al-(OOC-(CH_2)_{16}CH_3)-]_n)$ for the treatment of bricks to prevent rising damp. This may be manufactured by reacting aluminium isoproproxide and stearic acid, adding water and then heating.

Aluminium 2-ethylhexanoate finds applications as a gellant for petrol and inks. Aluminium palmitate is used as a gloss coating for paper and together with aluminium oleate is used as a drier for paints and varnishes.

A number of drugs are converted to their aluminium salt form to reduce acidity and make them more palatable. Typical examples are:

- aluminium acetylsalicylate, an alternative to aspirin;
- 2-(3-phenoxyphenyl) propionic acid in the monohydroxy aluminium or dihydroxy aluminium salt for arthritis treatment; and
- ibuprofen as the aluminium salt as an analgesic and antipyretic.

The standard method of making poly(oxo-aluminium stearate) is by reacting together approximately equimolar amounts of aluminium isopropoxide, stearic acid, and water in the presence of isopropyl alcohol, as a solvent for the stearic acid; and white spirit as a diluent for reducing the viscosity of the mixture during the vigorous stirring which is necessary in order to avoid gel formation. A reaction temperature of 140 to 180 °C is used in order to remove as a distillate the isopropanol that is formed during the reaction.

12 ALUMINIUM SULFATE

Some 50 acidic, basic, and neutral hydrates of aluminium sulfate have been reported;[49] only a few of these, however, have been well characterized. Traditionally the formula $Al_2(SO_4)_3.18H_2O$ has been written for commercial dry aluminium sulfate but the material is generally amorphous and probably does not represent a distinct crystalline species. There is considerable confusion concerning the name and this material is referred to as 'paper maker's alum', 'coke alum', 'patent alum', or simply 'alum'. The double salts potassium aluminium sulfate and ammonium aluminium sulfate are frequently called 'common alum' or 'ordinary alum'.

Aluminium sulfate is sold as either 'commercial' purity (with an iron level of up to 0.5%) or 'iron free' aluminium sulfate. It is marketed both as a solid (Al_2O_3 content 17–17.5%, 16%, and 14%) and as a liquid (typically 6–8% Al_2O_3 content).

The world-wide market of 'commercial' grade aluminium sulfate is over 2 million tonnes per annum whilst the iron-free grade is about a quarter of this total.

12.1 Production of Aluminium Sulfate

Commercial grade aluminium is usually produced by the reaction of sulfuric acid (30–60%) with aluminium hydroxide, bauxite, or clay.[88] Before use, the clays are normally calcined to destroy the organics and to increase the availability of the alumina for extraction. The optimum conditions for roasting the clay and the optimum strength of the sulfuric acid depend upon the particular source of the clay, its mineralogy, and impurity content. The solid residue from the dissolution reaction is removed by sedimentation followed by a final polishing filtration, although the neutral aluminium sulfate solution may have to be diluted first. The digestion takes place at a temperature of 100–120 °C.

Iron removal is difficult and 'iron free' grade (Fe_2O_3 <0.005%) is normally produced by the reaction of sulfuric acid with aluminium hydroxide; no heat input is necessary to sustain the reaction. As concern grows about disposal of wastes, the percentage of alum manufactured from aluminium hydroxide is growing rapidly.

88 K. V. Darragh, 'Aluminium Sulfate', in 'Kirk–Othmer Encyclopaedia of Chemical Technology', 3rd edn., Vol. 2, Wiley, New York, 1978.

12.2 Applications of Aluminium Sulfate

Formerly the use of aluminium sulfate was evenly divided between its application in paper making and its use as a coagulant for treating water. Usage in paper making, however, has progressively declined because of the replacement by neutral or alkaline size systems in place of acid rosin size systems. This has been driven by the growing use of calcium carbonate as a filler in paper instead of clays. In paper making the principal use of the aluminium sulfate is to promote the combination of rosin and cellulose fibres to improve both the strength of the paper and its resistance to water and ink. In addition the aluminium sulfate is used to neutralize the acidity of the wood pulp, soften hard water, and to treat the resulting effluent.

Effluents from a wide variety of other industries can also be successfully treated with aluminium sulfate. Aluminium sulfate can be utilized to remove both coarse suspended particles and colloidal particles. It functions by reacting with the carbonate present in the water to produce a voluminous gelatinous precipitate of aluminium hydroxide which coagulates and carries down the suspended matter present. Additionally, the positive aluminium hydroxide species are attracted to the colloidal impurities which are normally negatively charged, thereby forming larger and heavier flocs. Combinations of aluminium sulfate and polyelectrolytes are particularly effective in treating oil-contaminated effluents; in these cases aeration is used to form flocs on the surface of the effluent which can be removed by skimming.

The most important use of aluminium sulfate is the treatment of potable water where its excellent coagulating characteristics are used to remove suspended solids and reduce colour. It is also effective in reducing the presence of micro-organisms in the water, and trihalomethanes. Other uses of aluminium sulfate include: water-proofing textiles, dyeing, tanning, photography, for the production of colloidal aluminium hydroxide for gastric disorders, as a fire retardant in cellulosic fibre loft insulation, slug killing products, and in fire extinguishers.[88]

13 ZEOLITES

Synthetic zeolite manufacture (see also Chapter 12) is a very large and rapidly growing market for aluminium hydroxide. The main growth area has been in washing powders where zeolites have been substituted for phosphates. Significant growth is also occurring in new catalyst systems and selective scavenging in pollution control processes, while other uses include gas-drying, separation processes, cracking and reforming catalysts for petroleum products, and hydrocarbon conversion.

There are over 150 types of zeolite differentiated by the Si: Al ratio in their anionic framework. Their general formula is:

$$M_{2/n}O.Al_2O_3.xSiO_2.yH_2O$$

where M is an alkali or alkaline earth metal whose valency is n. The most important zeolites industrially are A, X, and Y with Si: Al ratios of 1, 1.5, and 3 respectively.[92,93]

The manufacture of the zeolites industrially is carried out by reacting sodium silicate with sodium aluminate to form a sodium alumino silicate gel. This gel is then aged, filtered, washed, dried, and reactivated by heating to *ca.* 400 °C.

14 ALUMINIUM FLUORIDE

Aluminium fluoride is a crystalline solid which sublimes without melting at 1272 °C forming monomeric AlF_3 in the gas phase. It has a density of $2.88 \, g \, cm^{-3}$ and a ΔH^o_f of $-1504 \, kJ \, mol^{-1}$. Three well-defined hydrates are formed: $AlF_3.9H_2O$ which crystallizes from aqueous solutions of aluminium fluoride at temperatures less than 8 °C; $AlF_3.3H_2O$ which exists in metastable γ-form and is stable at room temperature; $AlF_3.H_2O$ which is obtained by dehydration of the above. Non-stoichiometric hydrates, $AlF_3.nH_2O$ can also be produced by different heating regimes.[94]

Two different routes are used to manufacture anhydrous aluminium fluoride on an industrial scale: one starts from feldspar and the other involves the utilization of fluorosilic acid obtained from scrubbing waste gases produced from phosphate rock fertilizer plants.[95]

In the former process, calcium fluoride containing minerals are reacted in a rotary kiln with fuming sulfuric acid, $H_2SO_4.SO_3$, to produce anhydrous HF gas and gypsum. The HF is then fed into the bottom of a three-stage fluid-bed reactor and aluminium hydroxide is introduced at the top stage where it is converted into activated alumina. In the middle stage, rising HF contacts activated alumina to form AlF_3.

The second and alternative route involves reacting aluminium hydroxide at about 100 °C with fluorosilic acid obtained from scrubber gases. The silica is filtered and $AlF_3.3H_2O$ crystallizes out of solution. This trihydrate is then calcined at 500–550 °C to give anhydrous AlF_3.

Aluminium fluoride is a very important constituent in the manufacture of aluminium metal as it is added together with cryolite, calcium fluoride, and sometimes other fluorides to the alumina electrolyte in order to lower its freezing point. Some 700 000 tonnes are used annually in this application; minor uses include the addition to ceramics, glasses, glazes, enamels, welding rods, and catalysts.

15 FLUOROALUMINATES

Many fluoroaluminates occur naturally, *e.g.*

[89] Taki Fertilizer Manufacturing Company Ltd., UK Patent No. 1222561, 1971.
[90] Handy Chemicals Ltd., US Patent No. 5069893, 1991.
[91] Handy Chemicals Ltd., US Patent No. 5149400, 1992.
[92] W. Büchner, R. Schliebs, G. Winter, and K. H. Buchel, 'Industrial Inorganic Chemistry', VCH, Weinheim, 1989.
[93] A. J. Dyer 'An introduction to Molecular Sieves', Wiley, New York, 1988.
[94] K. Wade and A. J. Banister 'The Chemistry of Aluminium, Gallium, Indium and Thallium', in 'Comprehensive Inorganic Chemistry', Chapter 12, Pergamon Press, Oxford, 1973.
[95] W. C. Sleppy, 'Aluminium Compounds', in 'Kirk–Othmer Encyclopaedia of Chemical Technology', 4th edn., Wiley, London, 1991, p. 252.

cryolite   – Na_3AlF_6
chiolite   – $Na_5Al_3F_{14}$
cryolithionite – $Na_3Li_3(AlF_6)_2$
hagemannite   – $NaCaAlF_6.H_2O$
jarlite   – $NaSr_3Al_3F_{16}$
weberite   – Na_2MgAlF_7
pachnolite   – $CaAl(F,OH)5.H_2O$
prospite   – $CaAl_2(F,OH)_8$

The most important fluoroaluminate commercially is cryolite (sodium hexafluoroaluminate) which is used together with other fluorides to dissolve alumina in the electrolytic process to manufacture aluminium.[94] A typical electrolyte might contain:

80–85% cryolite
5–7% calcium fluoride
5–7% aluminium fluoride
2–8% alumina

Minor additions of magnesium fluoride and lithium fluoride are also sometimes incorporated as well, in order to reduce the liquidus temperature even further.

Naturally-occurring cryolite has historically been the major source of cryolite, especially from Ivigtut in Greenland, but increasingly synthetic cryolite is now used. Cryolite is industrially manufactured from HF, alumina, and sodium carbonate or hydroxide.

16 ALUMINIUM BROMIDE

Aluminium bromide, $AlBr_3$ forms colourless deliquescent crystals which melt at 97.5 °C and boil at 256 °C. It has a density of 3.01 g cm^{-3}. It reacts violently with water, vigorously with alcohols, and dissolves in most organic solvents.[94]

Aluminium bromide may be prepared from metallic aluminium and bromine; however, manufacture is carried out on only a small scale.

Since anhydrous aluminium bromide is more soluble than anhydrous aluminium chloride it is a more versatile Friedel–Crafts catalyst. Its greater cost, however, limits its use to small-scale speciality use.

Two hydrates of aluminium bromide are known: $AlBr_3.6H_2O$ crystallizes out of solution at *ca.* 20 °C and $AlBr_3.15H_2O$ crystallizes out at −9 °C.

17 ALUMINIUM IODIDE

Aluminium iodide is a deliquescent solid with a melting point of 190 °C. It reacts violently with water and oxidizes to alumina and iodine when heated in air. It has a density of 3.98 g cm^{-3}.[94] Aluminium iodide can be prepared by the reaction between iodine and an excess of aluminium powder in a sealed tube at 300 °C for 24 hours and then purified by vacuum sublimation. Two hydrates, $AlI_3.6H_2O$ and $AlI_3.15H_2O$ have been reported.

18 ALUMINIUM ALKOXIDES

Many aluminium alkoxides are known but the most important ones commercially are the isopropoxide $[Al(OP_r^i)_3]$ and t-butoxide $[Al(OBu^t)_3]$. They are manufactured by reaction between the appropriate alcohol and metallic aluminium in a finely divided from such as powder or foil; the reaction may be activated by the use of mercuric chloride or iodine. Higher alcohols may be prepared by alcohol displacement starting from aluminium isopropoxide.[94,96]

The solubilities of aluminium alkoxides in organic solvents make them convenient intermediates for carboxylates and glycinates. Aluminium glycinate is used in the production of antiperspirants. Higher alcohol aluminium alkoxides have been shown to be very effective paint driers for alkyl based paints.[97-99] Aluminium alkoxides have also been used to manufacture high purity (99.99%) alumina powders for high performance ceramic applications. Spherical aluminas can be produced under carefully controlled conditions.

19 ORGANOALUMINIUM COMPOUNDS

Aluminium alkyls and aryls are colourless liquids or low melting solids which generally react violently with water and often burst into flames on contact with air. Two of the most frequently used aluminium alkyls are the triethylaluminium $(Et_3Al)_2$ and triisobutylaluminium, $(Bu_3Al)_2$. They are manufactured from aluminium metal, hydrogen, and ethylene or isobutylene respectively.[94,95]

The alkylaluminium halides, $RnAlX_{3-n}$ (where X is Cl, Br, or I and R is methyl, propyl, or isobutyl) have been well characterized and are prepared from aluminium metal and RX.

Organoaluminiumhydrides such as dimethylaluminium hydride (Me_2AlH), and diisobutylaluminium hydride (Bu_2^iAlH) are also well characterized.

Organoaluminium compounds are important commercially and used on the thousands of tonnes per annum scale as catalysts or starting materials for the production of plastics, elastomers, detergents, and organometallics containing tin, zinc, or phosphorus. Of particular importance is the polymerization of olefins by Ziegler–Natta catalysts. These are heterogeneous catalysts prepared by mixing organoaluminium compounds with transition metal halides (*e.g.* titanium chloride and triethylaluminium and alkoxides).

This type of catalyst absorbs ethylene readily at room temperature to yield linear polyethylene. The interesting feature is that the ethylene inserts itself into the aluminium–carbon bond as the polymer chains grow.

[96] D.C. Bradley, *Prog. Inorg. Chem.*, 1960, **2**, 303.
[97] J. Turner, US Patent No. 4631087, 1986.
[98] J. Turner, US Patent No. 4622072, 1986.
[99] J. Turner, UK Patent No. 1434191, 1973.

20 ALUMINIUM NITRIDE

Aluminium nitride, AlN, has been known for nearly 100 years and was the subject of a considerable amount of work in the first 20 years of this century; it seems to have been 'rediscovered' in the mid 1970s. Its key properties which have excited the interest of ceramists have been its exceptionally high thermal conductivity combined with its low coefficient of thermal expansion, high temperature stability, good thermal shock resistance, and high electrical resistivity.

Restrictions to its use have arisen because of:

- problems in densification of components;
- high cost; and
- difficulty in metallization of substrates made of AlN.

An important factor in obtaining an AlN ceramic with good thermal properties is the oxygen content; the lower the better but materials in the range 0.5–2% oxygen are readily available.

Theoretically its thermal conductivity could be as high as $320 \, W \, m^{-1} \, K^{-1}$ but the highest found in practice to-date is $230 \, W \, m^{-1} \, K^{-1}$ with 160–$180 \, W \, m^{-1} \, K^{-1}$ for most commercially available AlN powders. This compares with 15–$20 \, W \, m^{-1} \, K^{-1}$ for Al_2O_3, 230–$260 \, W \, m^{-1} \, K^{-1}$ for beryllia and 270 for SiC (all at room temperature). One of AlN's particular advantages, however, is that the thermal conductivity remains relatively constant with temperature whilst that of beryllia falls. For example, at $400 \, °C$ AlN has a thermal conductivity of 120–$140 \, W \, m^{-1} \, K^{-1}$ whilst the value for beryllia has fallen to $75 \, W \, m^{-1} \, K^{-1}$.

The density of AlN is $3.28 \, g \, cm^{-3}$, its dielectric strength 14–$15 \, kV \, mm^{-1}$, its dielectric constant 8.6–9.0, and its volume resistivity 10^{13}–10^{15} ohm cm^{-1}.

20.1 Production of Aluminium Nitride

The most commonly adopted route to manufacture aluminium nitride is direct nitridation of aluminium metal; catalysts are frequently employed to expedite the reaction.

A number of commercially used processes involve a carbothermal route which entails heating to $1100 \, °C$ an intimate mixture of alumina and carbon; unreacted carbon is then removed by controlled oxidation at 600–$700 \, °C$ in air. The oxygen pick-up can then be reduced by heating to at least $1400 \, °C$ in nitrogen or a vacuum. Carbon contamination by this method remains a problem but can be reduced by the use of high-purity sub-micron alumina and sub-micron lampblack. Fine γ-alumina is sometimes used as the aluminium source.

Other routes to manufacture aluminium nitride include:[9]

- plasma arc using consumable aluminium electrodes;
- pyrolysis of poly(iminoalane) precursors to AlN;

- pyrolysis of films formed by chemical vapour deposition (CVD) from gaseous mixtures of aluminium chloride and ammonia; and
- pyrolysis of dimethylaminoalane [$AlH_2N(CH_3)_2$].

20.2 Use of Aluminium Nitride

The intense interest in high performance ceramics in the mid 1980s attracted many companies into the development of aluminium nitride: by the end of the 1980s there were at least thirty offering samples for development, but this has now fallen back to perhaps half of that number. The annual consumption of aluminium nitride is currently in the range 70–100 tonnes annum^{-1} but it has been projected to grow to thousands of tonnes by the end of the century.

The major growth area is seen to be in substrate and electronic packages. Alumina is currently the most frequently used ceramic material in this application but there is a requirement for materials with improved thermal conductivities and better dielectric performance as electronic devices become both smaller and more sophisticated. It is the exceptionally good thermal conductivity described above and the electrical characteristics of aluminium nitride that enable it to fulfil the requirements of this application. The aluminium nitride must have high purity, low oxygen content, and good sintering behaviour. The finer the aluminium nitride, the greater the sintering ability, but then the more difficult it is to reduce the oxygen content. The target product for this application is a sub-micron powder with an oxygen level of <0.75% and other impurities of <300 p.p.m.

Alternatives to alumina or aluminium nitride for substrates are beryllia, silicon nitride, co-fired glass-ceramics (borosilicate or cordierite glasses), or combinations of materials, *e.g.* aluminium nitride films on alumina, aluminium nitride on a glass substrate, or aluminium nitride/silicon carbide composites.

Another potentially large market for aluminium nitride is ceramic armour: particular niches are body armour, law enforcement vehicles, helicopter, aircraft, and military vehicle protection. This is an area of considerable research[101] and there are many competing materials such as alumina, boron carbide, silicon carbide, titanium diboride, or various composites, *e.g.* alumina/silicon carbide whiskers, nickel/titanium carbide, titanium diboride/boron carbide, titanium diboride/silicon carbide, borosilicate glass reinforced with silicon carbide, or carbon fibres, and Lanxide^{TM} composites.

Another area where aluminium nitride's thermal conductivity offers attractions is in the filling of epoxy or silicone polymers which are used for encapsulating electronic components for protection from the environment. The competing materials for this application are alumina and boron nitride.

Miscellaneous uses for aluminium nitride include:

- titanium diboride/aluminium nitride composites for vacuum metallizing boats, crucibles, insulators;

[100] L. Sheppard, *Ceramic Bull.*, 1990, **69**, 1802.
[101] D. J. Viechnicki, M. J. Slavin, and M. I. Kliman, *Ceramic Bulletin*, 1991, **70**, 1035.

- silicon nitride/aluminium nitride composites for high temperature engine applications;
- silicon nitride/aluminium nitride composites for high temperature resistant refractories;
- composites with Lanxide™ material; and
- packing/bedding powder for inert gas sintering of components made by a powder metallurgy route.

A major problem in the utilization of aluminium nitride powders is hydrolysis of the surface of the particles which causes the evolution of ammonia and oxidation of the surface layers. Careful handling techniques or the use of surface-treated aluminium nitride grades are therefore necessary. Recently, development effort has produced grades which allow aqueous processing techniques to be employed.

CHAPTER 12

Silicas and Zeolites

P. KLEINSCHMIT

1 INTRODUCTION AND GENERAL SURVEY

With respect to the nature and size of this publication the subject 'Silicas and Zeolites' requires some limiting definition. The chapter will exclude natural and synthetic crystalline and non-crystalline quartz and quartz products as well as 'Kieselguhr' (siliceous materials gained from diatomic sediments).[1] Also water-soluble silicates and water-dispersed colloidal silicas[2] produced from them commercially by ion exchange methods are beyond the scope of this account. It will concentrate on:

- some economically important classes of synthetic amorphous silicas which frequently (which is slightly misleading) are summarized as fillers; and
- a few examples of the microporous class of crystalline alumosilicates which are well known as 'zeolites' or 'molecular sieves'.

1.1 Silicas

The synthetic silicas referred to below are in the majority pure, sometimes ultrapure, higher-priced specialities, and therefore not 'fillers' in the actual sense of the word, which implies normally cheap materials being added to higher-priced products such as paper, paints, coatings, and polymers, including plastics and rubbers, for the purpose of reducing the manufacturing costs of the end products and which, at best, do the job without any loss of quality. The silicas reviewed in the following account are, in many cases, auxiliary materials which give special and remarkable effects to the formulations they are added to in various, often small, quantities and which therefore pay their costs. In other applications they are used as bulk materials. Table 1 summarizes important uses of synthetic silicas.

The basic raw material for the production of synthetic silicas is sand or quartz. There are two main routes to these products: 'wet' processes starting from sodium

---

[1] A survey of these products is given by O.W. Flörke and B. Martin, in 'Ullmann's Encyclopaedia of Industrial Chemistry', Vol. 23A, VCH Publishers, Weinheim, 1993, p. 583; L. Benda and S. Paschen, *ibid.*, p. 607.

[2] H. E. Bergna and W. O. Roberts, in 'Ullmann's Encyclopaedia of Industrial Chemistry', Vol. 23A, VCH Publishers, Weinheim, 1993, p. 614.

Table 1 *Selected Uses of Synthetic Silicas*

Effect	Field of Application	Concentration (wt. %)
(1) *As additives*		
– reinforcement	rubber, silicone rubber, plastics	20–40
– thixotropy, thickening	lacquers, paints, coatings, printing inks, cosmetics	1–5
– matting	lacquers, paints, coatings	1–3
– free running, free flow	sticky solids, liquids to be transferred into powders	1–5
– anti-blocking	plastic foils	1–2
(2) *As bulk materials*		
– thermal insulation	(vacuum) panels for storage heaters, refrigerators	70–100
– adsorption	desiccating of gases, liquids, chromatrography	100
– carrier function	catalysts, enzymes	100

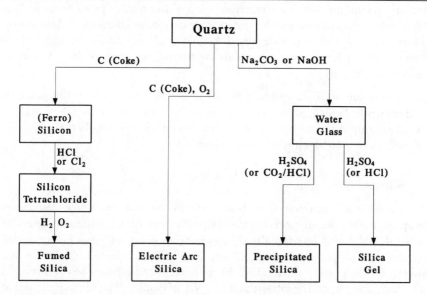

Figure 1 *Basic Processes for the Production of Synthetic Silicas*

silicate; and 'thermal' processes, using high-temperature reactions in which, in the first step, volatile silicon compounds such as the tetrachloride or the monoxide are prepared as intermediates, the latter being subsequently hydrolysed, or rather oxidized, also at elevated temperatures, to give the highly dispersed end products (Figure 1).

Despite the apparent simplicity of the chemistry involved in these processes, the properties of the silica products can vary within a large range (Table 2), and the number of different types sold on the market amounts to well over one hundred. Unlike the zeolites, these products are amorphous colloids, which therefore do not

Table 2 *Properties of Synthetic Silicas*

Type	Diameter[1] of Primary Particles/ (nm)	Specific Surface Area[2] ($m^2 g^{-1}$)	Tapped Density[3] ($g\,L^{-1}$)	Silanol Group Density ($SiOH\,nm^{-2}$)	Porosity
Fumed	7–40	50–500	50–120	2–4	Non porous ($<300\,m^2 g^{-1}$)
Gel	2–10	300–1000	100–800	5–6	Very high
Precipitated	7–100	50–800	100–350	5–6	Low

[1]Evaluation of TEM photographs; [2]Acc. to BET; [3]Acc. to DIN ISO 787/XI.

cause silicosis on inhalation. They are characterized and specified mainly by physical–chemical criteria such as specific surface area (*e.g.* BET-method), average particle size, particle size distribution, average pore diameter, pore diameter distribution, tapped density, moisture content, ignition loss, silanol group density, pH-value in aqueous suspension, surface charge, sieve residue, dibutylphthalate or oil adsorption number, chemical impurities, whiteness, and refractive index. However, many of the essential quality criteria as far as application is concerned cannot be judged by these data. A great number of standardized tests – results of which are also part of the product specifications – have been developed to give information on the performance of these 'fillers' in the systems they are added to.

Before a short survey is given on the industrial importance of the main classes of silicas, on the manufacturing processes, and on some trends of development, it is useful to mention a class of non-siliceous reinforcement fillers which historically, application-wise and with respect to some newer developments, have to be seen in close context to the silicas: the carbon blacks.[3] They have been by far the most important group of reinforcement fillers for rubber since 1910 when it was found that carbon black filled tyres had a much higher abrasion resistance and therefore a longer life than the previously used tyres filled with mineral powders. In 1989 the world-wide production capacity for carbon blacks exceeded 6.5 million tonnes annum^{-1}, about 95% of which was used in the rubber industry. Compared to that the synthetic silica business with its approximately 1.0 million tonnes annum^{-1} total production capacity is still a relatively small one (Table 3).

The fact, however, that the carbon black industry is dependent on petrochemical raw materials has been – at least for countries without domestic oil sources – a continuing problem, not only since the 'energy crises' of the 1970s and 1980s. This challenge and the desire to further improve the performance of tyres and other mechanical rubber goods has been a stimulant for researchers to look for adequate or even superior substitution products for carbon blacks as reinforcement fillers.

Some early attempts to find suitable 'white carbon' fillers for the rubber industry resulted in the development of highly dispersed calcium carbonates and silicates and special grades of alumina gels. The products are useful for special purposes, but their reinforcing power is poor compared to carbon blacks. At

³ M. Voll and P. Kleinschmit, in 'Ullmann's Encyclopaedia of Industrial Chemistry', Vol. 5A, VCH Publishers, Weinheim, 1986, p. 140.

Table 3 *Synthetic Silicas: Production Capacities and Suppliers 1991*

Product Group	Capacity	Main Suppliers
Pyrogenic Silicas	$100\,000\,\mathrm{t\,a^{-1}}$	Degussa, Cabot, Wacker, Tokuyama Soda
Silica Gels	$90\,000\,\mathrm{t\,a^{-1}}$	Grace, Akzo, Asahi, Crosfield
Precipitated Silicas	$800\,000\,\mathrm{t\,a^{-1}}$	Degussa, Rhône-Poulenc, Akzo, PPG, Huber, Tokuyama Soda, Nippon Silica

present, the topic is still of great significance, and in fact it seems as if a breakthrough has recently been achieved (see Section 3.2.3).

1.2 Zeolites

Unlike the *amorphous* synthetic silicas discussed in this paper, zeolites are *crystalline* solids with a regular porous lattice structure normally built up from SiO_4- and AlO_4-tetrahedra in which the negative charges are compensated by (exchangeable) mono- or multivalent cations. Approximately 40 different natural zeolites have been discovered since the Swedish minerologist Cronstedt found the first such mineral in 1756 and named it 'zeolite', derived from the Greek, meaning 'boiling stone' because he had observed that, when rapidly heated, these minerals seemed to boil (due to their capability to adsorb substantial amounts of water).

Nevertheless the zeolites are the 'newcomers' among the synthetic silicas and silicates. Up to now about 100 different structural types have been synthesized and almost monthly new ones are reported and characterized in the scientific literature.[4] They differ in their crystal structures, the size of the pores ('cages'), the diameter of the pore openings, and the SiO_2/Al_2O_3-ratio ('modulus') which can vary in the range $2-\infty$ and which also determines the number of exchangeable cations. It should be mentioned that in the last decade many novel materials with zeolite structure have been developed in which Si and/or Al are partially or completely substituted by other elements such as P, As, Ti, Ge, Ga, and/or Fe.[5] They cannot be treated here.

The main fields of applications for zeolites are:

- ion-exchange (*e.g.* as detergent builders);
- adsorption (*e.g.* as desiccants, molecular sieves); and
- catalysis (*e.g.* as cracking catalysts in oil refineries)

2 SILICAS PRODUCED BY THERMAL PROCESSES

2.1 Fumed (Pyrogenic) Silicas

The insufficient reinforcement properties obtained with different precipitated products (see Section 1.1) stimulated Dr Harry Kloepfer, working for the German

[4] W. M. Meier and D. H. Olson, 'Atlas of Zeolite Structure Types, 3rd Edn.', in *Zeolites*, 1992, **12**, 450.
[5] For example, E. M. Flanigan *et al.*, *Pure and Appl. Chem.*, 1986, **58**, 1351.

company DEGUSSA, in the late 1930s to try a different approach to finding 'white carbon'. Earlier, he had developed a manufacturing process for channel blacks using, instead of natural gas, coal tar oil as feedstock which was vaporized in a stream of hydrogen and partially burnt in diffusion flames. Using a similar kind of equipment and procedure he exchanged the feedstock tar oil for a volatile silicon compound, silicon tetrachloride, which was burnt in an oxygen–hydrogen torch to give – by a process of 'flame hydrolysis' – highly dispersed silica and hydrogen chloride. A new class of products with remarkable properties had been found. They were called 'Aerosils': 'airborn' silicas; in Anglo-Saxon countries they are known as pyrogenic or 'fumed' silicas.[6,7]

$$SiCl_4 + 2H_2 + O_2 \rightarrow SiO_2 + 4HCl \qquad (1)$$

Considering the different steps involved in the production of fumed silicas it is obvious that these materials are not exactly cheap: sand or quartz is first reduced in an electric arc process to silicon metal, then the latter is chlorinated with chlorine or hydrogen chloride to silicon tetrachloride which, finally, is burnt in the presence of expensive hydrogen to give SiO_2 again [equation (1)].

And indeed, as a substitute for carbon blacks as reinforcement fillers in rubbers, fumed silicas and other fumed oxides were and are too expensive. At present, they cost about £4.00 kg^{-1}, whereas common tread blacks are purchased at about £0.50 kg^{-1}. While fumed silicas are excellent reinforcement fillers for silicone rubber, they are not at all suitable for normal rubber, mainly because the unvulcanized raw mixes show a poor processability. For example when being mixed, calendered, or extruded, the viscosity of the raw mix is too high; and there is, besides other disadvantages, a detrimental reduction in pre-vulcanizing time.

However, these ultra-pure and ultra-fine fumed silicas found, despite their high price, a continuously growing technical application in different industries. Some information on the production capacity for the different classes of synthetic silicas including fumed silicas is given in Table 3.

2.1.1 Properties and Applications. Fumed silicas are fluffy white powders of amorphous structure. The average diameter of the spherical primary particles can be varied by using different reaction conditions during the flame hydrolysis; in commercial grades within the range of approximately 7–40 nm, corresponding to BET-surface areas between 380 and 50 m^2 g^{-1} (Table 2). Normally up to a few hundred primary particles are fused to three-dimensional aggregates, thus forming a 'secondary structure' with void volume in between the primary particles (see Figure 2 and Figure 5). Only coarser types (Figure 3), which are obtained at higher reaction temperatures in oxygen-enriched flames show little of this structure and mainly consist of discrete spherical particles. Fumed silicas with surface areas up to 300 m^2 g^{-1} are non-porous; grades with higher surface areas can contain a certain amount of micropores.

[6] H. Kloepfer, Patent No. DE 762 723 (Degussa), 1942, E. Wagner and H. Brünner, *Angew. Chem.*, 1960, **72**, 744.
[7] M. Ettlinger, in 'Ullmann's Encyclopaedia of Industrial Chemistry', Vol. 23A, VCH Publishers, Weinheim, 1993, p. 635.

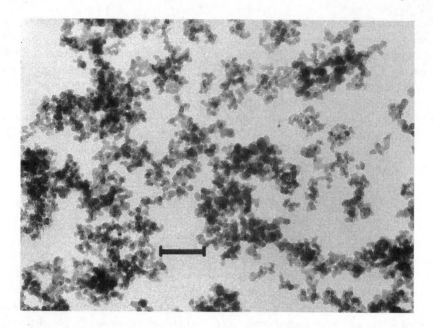

Figure 2   *Primary Particles of Fumed Silica (bar = 0.1 µm)*

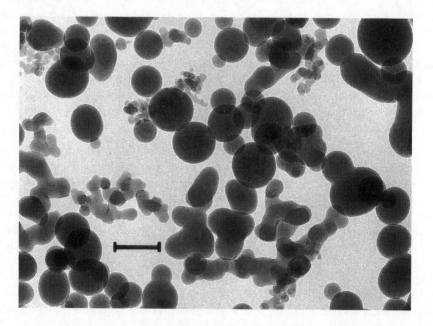

Figure 3   *Coarse Particles of Fumed Silica (bar = 0.1 µm)*

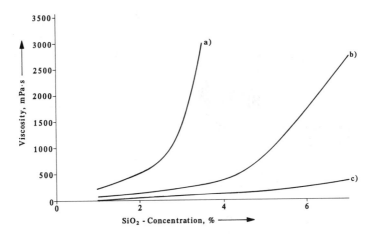

Figure 4 *Thickening Effect Caused by Various Amounts of Fused Silica* $(200m^2g^{-1})$ *in Different Solvents:* (a) *xylene;* (b) *butyl acetate;* (c) *butanol*

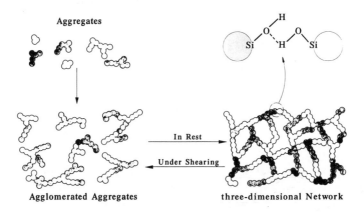

Figure 5 *Thickening and Thixotropic Effect of Liquids Caused by Fumed Silicas (Schematic)*

Very important for the outstanding properties of fumed silicas as thickening, thixotropic, and reinforcing agents in liquid, plastic, and elastic systems are the evenly distributed silanol groups on the silica surface and their ability to form hydrogen bonds. Generally, the silanol group density of fumed silicas is in the range of 2–$4\,SiOH\,nm^{-2}$. When fumed silica is dispersed in liquids the silanol groups of adjacent particles can interact via hydrogen bonding. In this way a vast three-dimensional network is formed which is extended throughout the liquid and causes, especially in non-polar liquids, a substantial increase of the viscosity (Figure 4). This effect is reversible: by means of shearing forces the network can be broken down again and the viscosity thus reduced. This means the composition is thixotropic. These relations are demonstrated schematically in Figure 5. Technically important systems where fumed silica is used in this way as a thickening and thixotropic agent are, *e.g.* coatings, printing inks, gelcoats, sealants, cosmetics, and toothpastes. For this purpose, normally small amounts (a

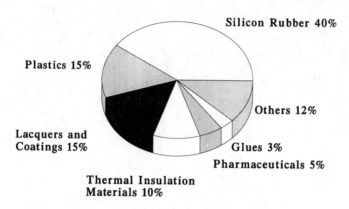

Figure 6   *The Division of the Main Fields of Application of Fumed Silicas*

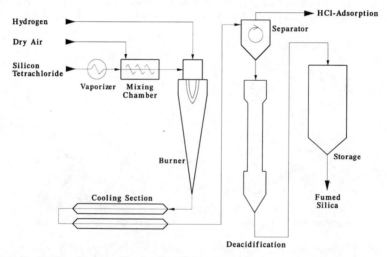

Figure 7   *A Flow Diagram Showing the Modern Fumed Silica Production Process*

few percent) are sufficient to give an optimal effect. This also applies to the application as free-flow and anticaking agents. Reinforcement of silicones with fumed silica, which creates among other positive effects a substantial increase of tensile strength, requires filling grades up to 30 wt.%. In high-performance insulation composites, where silica very often is the main or sole component, advantage is taken of the microporous structure formed by the void volume between the silica aggregates which hinders gas diffusion (see Figure 12).

A division of the main fields of application of fumed silicas is given in Figure 6.

2.1.2   Manufacturing Process. A flow sheet of a modern fumed silica production process is given in Figure 7. Whereas Kloepfer initially used a reactor with many small diffusion flames, modern plants are very similar to furnace black plants: the flame hydrolysis takes place in one big flame at temperatures within the range

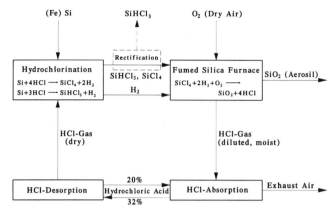

Figure 8 *Process Scheme of a Combined SiCl₄/Fumed Silica Plant*

750–2000 °C, depending on the grade. The ratio of the reactants silicon tetrachloride, hydrogen, and oxygen (in form of dry air, if necessary oxygen-enriched), which can be varied within a substantial range, determines the properties of the silica, *e.g.* particle size, specific surface area, and 'structure'. The reaction mixture then passes through a coagulation and cooling device before the silica particles are separated by means of cyclones or filters. Accompanying HCl is removed, *e.g.* in a fluid bed reactor, with steam or hot air. Bagging is normally combined with some form of compaction in order to raise the bulk density, which originally is well below $20\,\mathrm{g\,l^{-1}}$.

Economically as well as ecologically it is advantageous to run fumed silica plants in combination with a silicon tetrachloride plant, which produces a good deal of the necessary hydrogen and makes use of the by-product HCl from the silica plant. A simplified flow sheet for such a closed loop operation is shown in Figure 8. Instead of, or in addition to, $SiCl_4$ other silanes, *e.g.* methyltrichlorosilane or other by-products of the silicone manufacture can be used as raw materials for fumed silicas.

In the past a lot of effort has been made to find alternative – and possibly cheaper – methods for the production of fumed silicas. A promising approach was made by Flemmert[8] who developed the 'Fluosil-Process', which uses SiF_4 as a raw material:

$$SiF_4 + 2H_2 + O_2 \leftrightarrow SiO_2 + 4HF \tag{2}$$

Due to the fact, however, that the equilibrium of the hydrolysis reaction lies – in contrast to the case of $SiCl_4$ – only at very high temperatures completely on the right-hand side, the 'Fluosil Process' is more difficult to run, apart from material problems caused by the handling of HF and SiF_4. On the other hand the by-product HF can, theoretically, be recycled in order to produce in reversion of the above reaction equation (2), SiF_4 from sand; which then, apart from the

[8] G. L. Flemmert, 'Hydrogen Fluoride and Pyrogenic Silica from Fluosilicic Acid', The Fertilizer Society, Proceedings No. 163, London, 1977.

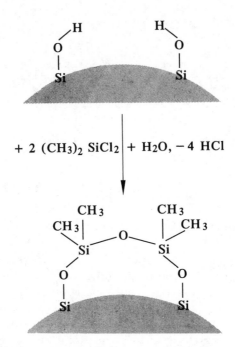

Figure 9 *Hydrophobation of Silica with Alkylchlorosilanes*

necessary compensation for losses, would be the only (and very cheap) raw material.

Nynäs Petroleum in Sweden ran for some years a small commercial plant for the production of 'Fluosil'. However, the technical (*e.g.* corrosion) problems plus, probably, environmental difficulties could apparently not be mastered and the plant was closed down. The latest attempt to commercialize this technology was undertaken by the W.R. Grace Company in Belgium four years ago. They constructed a huge production plant which was due to be tied into a phosphate fertilizer facility using the latter's waste by-product SiF_4 as a raw material and to produce simultaneously fumed silica and hydrofluoric acid. The fact that this plant also never went on-stream successfully seems to indicate that the technical and ecological hurdles for the realization of Flemmert's concept are indeed too high.

2.1.3 Modification with Organosilanes. Fumed silicas are hydrophilic. For some applications, however, a hydrophobic behaviour is required. Hydrophobic silicas are obtained by surface modification with alkylchloro- or alkylalkoxy-silanes.[9] As shown schematically in Figure 9 the silanol groups at the silica surface react (at elevated temperatures) under elimination of HCl (or H_2O) with alkylsilanes. The dimethylsilyl-modified silica (Figure 9) – trade name 'Aerosil' R972® – is a commercially-important product which is no longer wettable by water. It is used in the silicone rubber industry and, *e.g.* for improving the flowability of sticky powders. Other alkyl-modified silicas are obtained with hexamethyldisilazane,

[9] H. Brünner and D. Schutte, *Chem.-Ing. Technol.*, 1965, **89**, 437.

Table 4 *Some Oxides Obtained by Flame Hydrolysis*

Oxide	Melting Point (°C)	Starting Compound	Boiling Point (°C)
Al_2O_3	2015	$AlCl_3$	183 (subl.)
Cr_2O_3	2266	CrO_2Cl_2	117
Fe_2O_3	1565	$FeCl_3$	315
		$Fe(CO)_5$	103
GeO_2	1115	$GeCl_4$	84
Nb_2O_5	1485	$NbCl_5$	254
		$NbOCl_3$	400 (subl.)
NiO	1984	$Ni(CO)_4$	43
SiO_2	1610	$SiCl_4$	58
TiO_2	1830	$TiCl_4$	136
WO_3	1473	WCl_6	347
		$WOCl_4$	228
ZrO_2	2715	$ZrCl_4$	331 (subl.)

octylalkoxy-, octadecylalkoxysilane, or polydimethyl-siloxanes. Organofunctional groups can also be anchored on the silica surface if difunctional silanes, *e.g.* methacryloxypropyltrialkoxysilane:

$$CH_2 = \overset{\underset{\displaystyle CH_3}{|}}{C} - \overset{\overset{\displaystyle O}{\|}}{C} - O - CH_2 - CH_2 - CH_2 - Si(OR)_3 \qquad (3)$$

are used. A fumed silica modified in this way finds application as active reinforcement filler chemically bonded to dental materials based on poly(methylmethacrylate) because the double bond of the methacrylic group takes part in the polymerization process of the matrix.

2.1.4 Other Fumed Oxides. It should be mentioned that the method of flame hydrolysis for synthesizing highly-dispersed materials can be extended to other volatile metal compounds provided that the equilibrium of the hydrolytic reaction is on the side of the oxide, and the melting point of the latter is higher than the flame temperature. In Table 4 a selection of highly-dispersed fumed oxides is listed that have been synthesized by flame hydrolysis, mainly on the laboratory scale. Produced on a technical scale and commercially available are, in addition to the silicas, the oxides of aluminium, titanium, and iron which are used, *e.g.* as antistatic agents for plastic powders and in the textile industry (Al_2O_3), as heat stabilizers in hot curing silicone rubber, as UV-absorbers in sun protection creams (TiO_2), and as transparent red pigments for metallic coatings (Fe_2O_3).

2.2 Electric Arc and Plasma Silicas

In the past, several commercial plants for the production of highly dispersed silicas by electric arc processes have been run. The principle of this method is as follows.

At temperatures $> 2000\,^{\circ}$C, quartz is reduced with coke to – at that temperature – volatile SiO which, in a second step, is oxidized with air or steam to silica again:[10]

$$SiO_2 + C \rightarrow SiO + CO \tag{4}$$

$$SiO + CO + O_2 \rightarrow SiO_2 + CO_2 \tag{5}$$

The energy-intensive procedure and the general performance of these products cannot compete with the fumed silicas. Their technical importance is low. Efforts, *e.g.* by a research group of Lonza Switzerland,[11] to gain products equivalent to fumed silicas by the reduction of quartz in an alcohol-stabilized plasma torch never materialized in a commercial plant.

3   SILICAS BY WET PROCESSES

Belonging to this section, in accordance with Section 1 and Figure 1, are the amorphous product groups of the precipitated silicas and the silica gels. To complete the picture, it has to be mentioned that 'silica sols' (colloidal aqueous suspensions) are produced on a large commercial scale. They are obtained mainly by de-ionizing dilute sodium silicate solutions by means of cation exchangers, followed by concentration (up to 20–50 wt. % SiO_2) and stabilization with small amounts of alkali. They find application as binders, for example in the manufacture of cast mouldings and insulation composites, in granulation, and as polishing agents for silicon wafers.

In 1991, the estimated world-wide production capacity (Table 3) of solid silicas gained by wet processes was approximately 900 000 tonnes annum^{-1} (800 000 t a^{-1} precipitated silicas; 90 000 t a^{-1} silica gels). This is about ten times as much as the fumed silica capacity. It has to be mentioned, however, that, with the exception of some higher-priced products for special applications, the majority of these silicas are standard materials with a price level in the range of £0.80–1.00 kg^{-1}.

Historically, the silica gels, although today less important, are older than the precipitated silicas. Commercial gel production was started in about 1920 by Davison,[12] now part of the W.R. Grace Company which is still the biggest supplier world-wide. Precipitated silicas, today by far the biggest group of synthetic silicas, went commercial only in the early 1940s,[13] then gradually replacing as the 'third generation' of synthetic white reinforcement fillers their predecessors calcium carbonates and alumina gels; and calcium silicates and aluminium silicates. Nevertheless the latter products are still made and used as less active fillers in rubber, as extenders for titania pigments, in the paper industry, and as flame retardants in plastics.

[10] H. Biegler *et al.*, Patent No. DE 1180 723 (Degussa), 1963.
[11] E. R. Schnell *et al.*, *Powder Technol.*, 1978, **20**, 15.
[12] W. A. Welsh, in 'Ullmann's Encyclopaedia of Industrial Chemistry', Vol. 23A, VCH Publishers, Weinheim, 1993, p. 629.
[13] D. Kerner, P. Kleinschmit, and J. Meyer, in 'Ullmann's Encyclopaedia of Industrial Chemistry', Vol. 23A, VCH Publishers, Weinheim, 1993, p. 642.

3.1 Silica Gels

3.1.1 Properties and Applications. Silica gels, like their 'relatives' fumed and precipitated silicas, are X-ray amorphous and pure products (SiO_2-content, dry basis >99 wt.%) which are, depending on their respective applications, supplied as finely-ground powders, broken granulates, beads, or shaped monoliths. Their primary particles are spheroids called micelles with diameters that can be varied between about 2–10 nm and which are crosslinked together to form a random amorphous structure of high porosity. Specific surface area (~ 300–1000 m^2 g^{-1}), pore volume (~ 0.3–2.0 cm^3 g^{-1}), and the average pore diameter (~ 2–25 cm) can be adjusted over wide ranges according to the desired product profile (see also Table 2).

Due to their high porosity the classical application for silica gels is water adsorption from moist air or other gases. After removing the water included in the pores by activation, especially narrow pore gels have a high affinity to traces of water vapour as effective desiccants. Recently, however, they have met increasing competition from zeolites in this field of application (see Section 4.1). They are also used as adsorbents in the liquid phase, *e.g.* in the purification and stabilization of beer or in chromatography. Other uses include matting agents in lacquers and coatings, catalyst carriers (*e.g.* in form of beads), fillers for dental composites (competing with fumed and precipitated silicas), and thermal insulation, for instance in the form of aerogel monoliths.

3.1.2 Manufacturing Process. In synthesizing a silica gel, a (still liquid) hydrosol is formed as an intermediate by mixing alkali (normally sodium) silicate with (preferably) sulfuric acid at an acid pH-value and at temperatures within the range of 30–80 °C [equation (6)]. The mixture is then allowed to set in vessels or on conveyer belts. After an ageing process of several hours the resulting transparent rigid mass is broken and washed salt free with water, for example in counter-current reactors.

$$Na_2O \cdot 3.3SiO_2 + H_2SO_4 \rightarrow 3.3SiO_2 + Na_2SO_4 + H_2O \qquad (6)$$

If gels with higher pore volume and bigger pore diameters are required the washing is carried out under basic conditions. The drying step that follows also influences substantially the properties of the resulting end-product. Slow drying leads to shrinkage of the gel network and small pore gels, flash-drying to products with high pore volumes and bigger pore diameters. Similar effects are obtained by removing the water before drying with organic solvents and subsequent normal or supercritical drying. The latter products are known as 'aerogels'. After drying as required, a formulation step (*e.g.* breaking, classifying, milling, beading) and packaging complete the production process (for which a simplified flow sheet is given in Figure 10).

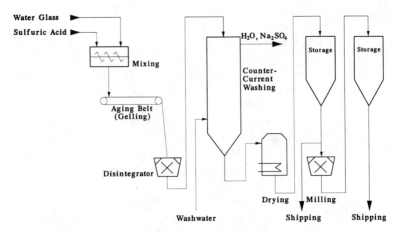

Figure 10 *A Simplified Flow Diagram Showing Production of a Silica Gel*

3.2 Precipitated Silicas

3.2.1 Properties and Applications. The synthesis of the third and nowadays most important class of amorphous synthetic silicas, the 'precipitated silicas', is based on the same chemical reaction and the same raw materials as for silica gel production [equation (6)]. As, however, by applying different reaction conditions the setting of the reaction mixture is avoided, the properties of the resulting products differ. Unlike fumed silicas they show a certain amount of porosity, but much less than the gels, whereas silanol group density is comparable (5–6 SiOH nm^{-2}) as well as the range in which the specific surface area can be adjusted (see Table 2). The thickening effect in liquids, as is the case of silica gels, cannot compete with that of fumed silicas.

Similar to the latter, their primary particles are fused to three-dimensional aggregates whose size, form, and amount can be varied by altering the production process within much wider limits than is possible with fumed silicas. The 'structure' can be measured quantitatively by special adsorption methods (*e.g.* dibutylphthalate-adsorption: 'DBP-number'). Specific surface area and 'structure' are, among others, key parameters for the characterization of the 'reinforcement strength' of fillers in rubber which are well established in the rubber and carbon black industry.

Reinforcement of elastomers accounts for >60%, today by far the most important application for precipitated silicas (see Figure 11), even if the total amount used is still well below 10% of the carbon black consumption for the same purpose. The fact that most of these 'rubber silicas' are used for the fabrication of shoe soles demonstrates that the ambitious target of substituting carbon black as high performance reinforcement filler in the rubber, especially in the tyre industry, has not yet been reached. Nevertheless this market segment recently showed a rapid increase, due to a development described in Section 3.2.3.

Another promising new application for silicas – fumed and precipitated – is high and low temperature insulation (see Section 2.1.1 and Figure 12). Whereas panels

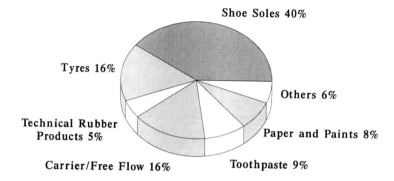

Figure 11 *Worldwide Consumption and Uses of Precipitated Silicas in 1990*

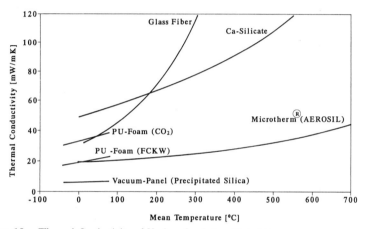

Figure 12 *Thermal Conductivity of Various Insulation Materials*

('Microtherm®') obtained by pressing a mixture of fumed silica and some additives are used as high performance microporous insulation materials at elevated temperatures, 'vacuum insulation panels' based on precipitated silica which is sealed in impermeable evacuated foils have been developed as most effective substitutes for CFC-containing polyurethane foams in refrigerators and other appliances. Figure 12 shows the superior performance of both silica-based materials.

Figure 11 gives a survey of the most important applications for precipitated silicas and their estimated shares in the total consumption.

3.2.2 Manufacturing Process. In contrast to silica gels which are produced under acidic conditions, precipitation is carried out in neutral or alkaline medium. The properties of the precipitated silica can be influenced by the design of the plant equipment and by varying the process parameters. The production process consists of the following steps: precipitation, filtration, drying, milling, and, in some cases, compacting and granulation (Figure 13).

So far only batch precipitation has attained technical importance. Generally,

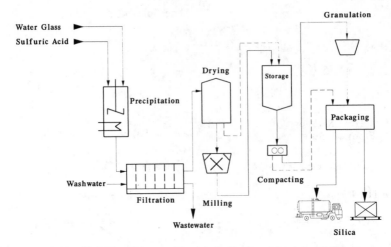

Figure 13 *A Simplified Flow Diagram Showing Production of Precipitated Silica*

mineral acid (preferably sulfuric acid) and sodium silicate solution are fed simultaneously into a water-containing vessel, while the reaction mixture is stirred. At the beginning, a hydrosol with silica seeds is formed. When precipitation starts the viscosity increases, indicating that a three-dimensional network is formed which later disintegrates, then the reaction proceeds and the silica particles grow under the influence of continuous stirring and temperature increase.

The separation of the silica from the reaction mixture and the washing out of salts contained in the precipitate is carried out in filter equipment such as rotary filters, belt filters, or filter presses. The solids content of the washed filter cake is 15–25% depending on the type of silica and filter. Drying is carried out in turbine, plate, or rotary driers. Special product properties can be achieved by spray drying the filter cake after redispersion in water or acid. To adjust the desired fineness of the particles, mills and separators are used. Special grades can be obtained by air-jet or steam-jet milling.

A small amount of the commercially available precipitated silicas is of the hydrophobic type. They are obtained by chemical reaction of silanol groups with certain silanes (as described in Section 2.1.3). Others are coated, *e.g.* with waxes, in order to improve dispersability in certain organic systems.

3.2.3 Recent Developments. In Section 3.2.1 it was mentioned that the performance of silicas in rubber as reinforcement fillers is inferior – compared to that of carbon blacks. This has chemical reasons: the surface carbon black is hydrophobic and has functional groups (for example, olefinic or conjugated double bonds) which can react or at least physically interact with double bonds of the rubber matrix, thus giving good reinforcement. This is not the case with the hydrophilic silica surface with its silanol groups, which, moreover, disturb the vulcanization process and give a limited reinforcement to the vulcanized rubber mainly because of the small particle size.

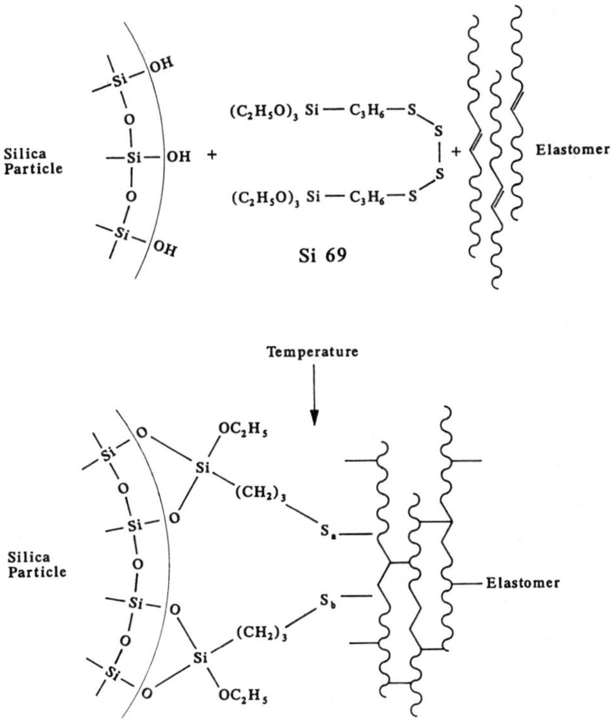

Figure 14 *Si 69® as a 'Reinforcement Agent' for a Silica-filled Rubber, before and after Vulcanization*

A promising approach to improve the performance of silicas in rubbers substantially – both the processability of the unvulcanized raw mixes and the mechanical properties of the cured material – was made some time ago by the development of certain bifunctional organosilanes, which are used as 'reinforcement agents' in silica-containing rubber mixtures. A very effective compound in this respect is the bis-(3-[triethoxysilyl]-propyl)-tetrasulfane (Si 69®, Degussa):

$$(C_2H_5O)_3Si—(CH_2)_3\text{-}S_4\text{-}(CH_2)_3\text{-}Si(OC_2H_5)_3 \tag{7}$$

which can on the one hand react with the silanol groups on the silica surface, forming stable siloxane bonds, and on the other hand, with the polymer network via the tetrasulfane groups, thus enabling a chemical fixation of the filler particle within the polymer (Figure 14).[14,15]

Although it had been known for about ten years that with the aid of this coupling agent silica-filled tyres could be made, only in the production of special tyres (*e.g.* for 'earth movers', and of highly-stressed technical rubber goods) was any advantage taken of this development, due to the higher price of the system silica/organosilane compared to carbon black.

[14] S. Wolff, *Kautsch. Gummi, Kunstst.*, 1979, **32**, 760.
[15] U. Deschler, P. Kleinschmit, and P. Panster, *Angew. Chem.*, 1986, **98**, 237.

In 1993, however, the French tyre company Michelin[16] introduced a new automobile tyre with outstanding properties: it combines low rolling resistance (which means less gasoline consumption), high abrasion resistance (long life-time), and good adhesion on wet and dry roads (safety!). Therefore they called it 'the green tyre', although it looks black. The tread of this tyre consists mainly of a newly-developed (solution-)polymer, a new type of precipitated silica, and the reinforcement agent 'Si 69'. Carbon black is not an essential ingredient of this tread mixture. Of course the production of this tyre will be more expensive but as it seems today it will repay its costs due to its superior properties, especially the energy savings it enables. This development could also be the beginning of a new era in silica technology.

4 ZEOLITES

4.1 Properties and Applications

Although zeolites differ in many respects from the synthetic amorphous silicas there are similarities too, for instance in certain applications (in the adsorption sector they are increasingly successful in competition with silica gels) and in production technology: in zeolite A synthesis, at least partially, equipment is used that is known from precipitated silica technology.

As described in the introductory remarks (Section 1.2), zeolites are crystalline microporous materials containing cavities and channels of molecular dimensions. Molecules that fit in these uniform cavities whose size is characteristic for the special type of zeolite can be selectively adsorbed and in this way separated from others that are too big to be adsorbed. This effect, widely used in chemical technology (*e.g.* for separation, 'form-selective' catalysis), is the reason for the zeolites' second name: molecular sieves.

Figure 15 shows schematically the structures of four different types of industrially important zeolites: (a) zeolite A (produced in huge tonnages as detergent builder and, in smaller quantities, for 'static' and 'dynamic' adsorption, *e.g.* for drying of gases and liquids); it has a three-dimensional pore and channel system, as has (b) Faujasite (a large pore zeolite used as catalyst in 'FCC'-oil-cracking units); (c) zeolite ZSM-5 (developed by Mobil as catalyst, *e.g.* for the 'MTG' – '*M*ethanol-*T*o-*G*asoline-Process') with a two-dimensional channel system; and (d) mordenite (catalyst for various chemical reactions) with a single-dimensional channel system.

The primary building units for zeolites are SiO_4-tetrahedra, which are connected to each other by shared oxygen atoms. Substitution of a SiO_4-unit by an AlO_4-tetrahedron results in a negative charge of the framework which is compensated by a non-framework cation located in the cages or channels and which is exchangeable by other cations (Figure 16). This ion-exchange capability is the basis for the application of zeolite 4 A (Na^+-form) as a detergent builder because the Ca^{2+}- and Mg^{2+}-ions which are responsible for water hardness easily substitute the Na^+-ions. If the ions are exchanged by protons, highly acidic sites

[16] R. Pauline, Patent No. FR 91 02433, 1992; A. Trono, *Tyre Technol.*, 1992, Pap. 5, p. 1.

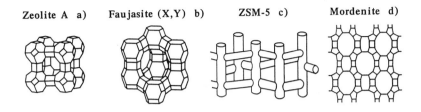

Figure 15 *The Structures of Four Industrially Important Zeolites*

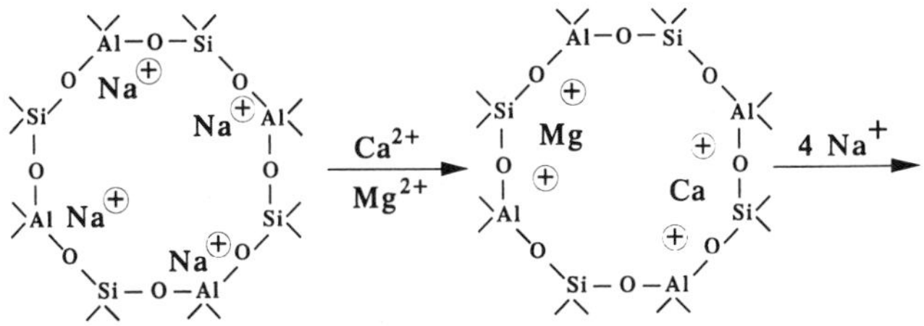

Figure 16 *Structure Detail and Ion Exchange Mechanism of Zeolite 4A*

are formed which determine or at least influence the catalytic properties.

After synthesis the channels and cages of the zeolites are normally filled with water. By calcination (heating to 400–500 °C) the water can be removed. In this 'activated' form zeolites, especially A-type zeolites, are highly efficient adsorbents.[17] The pore size opening can be modified within certain limits by the choice of the non-framework cation (K^+ → zeolite 3A; Na^+ → zeolite 4A; Ca^{2+} → 5A; the respective number stands for the pore size opening measured in Ångstroms). The adsorption capacity for water can be as high as 30 wt. %. Zeolites are technically applied also for a lot of other separation and purification processes such as air separation (*e.g.* 'PSA-' = *p*ressure *s*wing *a*dsorption), or n- and iso-paraffin-separation (Parex, and Molex process). Special silica-rich ('high modulus') Faujasite-type zeolites ('US-Y' = *u*ltra *s*table *Y*-zeolites) which can be prepared either by special synthesis or by suitable after-treatment ('de-alumination') are hydrophobic and therefore able to selectively remove organic substances, for instance solvents, in low concentrations from moist air or wastewater.

Zeolites are normally white powders with high (inner) specific BET-surface areas (up to $1000 \, m^2 \, g^{-1}$). In powder or slurry form they are used for detergent formulation. For application as adsorbents and catalysts they normally have to be shaped with the aid of suitable binders to extrudates, granules, or micro-beads (the latter being obtained by spray-drying).

When the zeolites left the laboratory scale in the late 1950s, they were at first used as ion exchangers and adsorbents.[17] Later, their outstanding catalytic properties were discovered which resulted in a fast growing-market for special

[17] D. M. Ruthven, *Chem. Eng. Prog.*, 1988, **84** (2), 51.

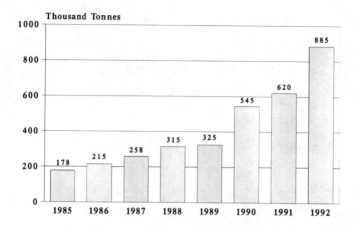

Figure 17 *Production Capacity for Detergent Grade Zeolite A from 1985–1992 in Europe*

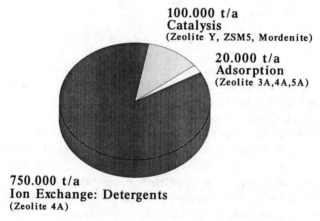

Figure 18 *An Overview of the Most Important Technical Applications of Zeolites (1991)*

grades, such as the already mentioned (rare earth exchanged) X- and Y-types as ingredients of cracking catalysts.[18]

As far as tonnages are concerned, however, a new dimension was added to the synthetic zeolites business when the sodium-A-type turned out to be a suitable non-eutrophic and comparatively cheap substitution product for sodium tripolyphosphate (STPP) as a detergent builder in household washing formulations. In Europe, production capacity for zeolite A grew in the period of 1985–1992 by not less than five-fold to almost $900\,000\,t\,a^{-1}$ (Figure 17).

Figure 18 gives an overview of the most important technical applications of zeolites and their shares in the total consumption.

[18] N. Y. Chen and T. F. Degnan, *Chem. Eng. Prog.*, 1988, **84** (2), 42.

4.2 Manufacturing Processes

The industrial manufacturing[19,20] of zeolite powders is carried out in aqueous suspensions and commonly starts by preparing an alumosilicate hydrogel (less often from a modified clay mineral, *e.g.* metakaolin) which is subsequently crystallized at higher temperature. The hydrogel can be prepared from various sources of silica and alumina, preferably sodium silicate and sodium aluminate solutions. Depending on the nature of these starting materials and the method of mixing them as well as on the desired properties of the end product (*e.g.* structure, particle size, and size-distribution), addition of bases or mineral acids may be necessary. For the synthesis of special, more 'sophisticated' zeolites so-called 'structure-directing agents' or 'templates' (amine compounds such as tetra-propylammonium salts or hydroxide) must be added, and the synthesis may have to be carried out at higher temperatures under elevated pressure ('hydrothermal conditions').

After powder synthesis, in many cases additional after-treatment steps are necessary to modify the properties of the zeolite by, for example, calcination (to destroy the template), ion-exchange (to alter pore size and acidity), chemical modification (to increase the modulus by de-alumination), and, last but not least, shaping to pellets, beads, or extrudates for the application as adsorbents or catalysts.

4.2.1 Zeolite A. Due to its world-wide use as a detergent builder, zeolite 4A [sodium form, Equation (8)], is volume-wise by far the most important single product of the large and rapidly growing molecular sieve family:

$$Na_{12}(AlO_2)_{12}(SiO_2)_{12}.27H_2O \qquad (8)$$

Although it is insoluble in water, zeolite A nevertheless has – at least when the average particle size is in the order of 4 μm or less – a good dispersability and a low sedimentation tendency; and, furthermore, a high calcium-binding capacity (> 160 mg Ca g^{-1}, anhydrous basis) and a low abrasiveness. The combination of these properties was essential for the commercial success of this 'heterogenous inorganic builder concept' which in the beginning was regarded as rather unusual. Of course, high purity of the material is also required.

A typical method of synthesis used in industrial 4A-production starts from commercial sodium silicate solutions (35–40 wt. %; Na$_2$O/SiO$_2$ ratio 1:3.3 or 1:1.9). For the preparation of sodium aluminate, alumina hydrate is dissolved by caustic soda.

A suitable composition which meets the requirements of the practice in terms of product quality, yield, and synthesis time is defined by the molar ratio given in Equation (9); in this case an excess of alumina is employed.

$$2.11Na_2O: Al_2O_3: 1.74SiO_2: 77.6H_2O \qquad (9)$$

[19] D. E. W. Vaughan, *Chem. Eng. Prog.*, 1988, **84** (2), 25.
[20] E. Roland, in 'Zeolites as Catalysts, Sorbents, and Detergent Builders', ed. H.G. Karge and J. Weitkamp, 'Studies in Surface Science and Catalysis', Vol. 46, Elsevier Publishers, Amsterdam, 1989, p. 645.

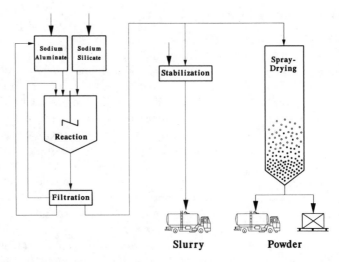

Figure 19 *A Simplified Flow Diagram for the Preparation of Detergent NaA Zeolite*

As mentioned above, the average particle size is a very critical point. It is mainly influenced by the parameters chosen in gel preparation, *e.g.*:

- the Na_2O/H_2O ratio in the reaction mixture;
- the mixing sequence and the addition rate of the different components;
- temperature;
- the stirring/shearing energy applied during the gel preparation;
- seeding.

In Figure 19 a flow diagram for the preparation of detergent NaA zeolite is shown. Usually, the gel precipitation temperature is between 50 and 80 °C, the temperature during the crystallization being slightly higher (80–90 °C). The time required for the gel preparation may vary from 0.5–1.5 hours, then another 1.0–2.0 hours are necessary for the zeolite crystallization. When crystallization is complete, the zeolite suspension is filtered with filter presses, rotary filters, or band filters. The filtrate which usually contains certain amounts of sodium hydroxide can be recycled to the aluminate-preparation unit.

The entire process is normally run semi-continuously. Whereas it is possible to prepare the gel continuously, crystallization is carried out batch-wise in reaction vessels having a size of sometimes more than $100 \, m^3$.

Many detergent manufacturers prefer a zeolite slurry instead of a zeolite powder. Water and a suspension stabilizer, which prevents rapid sedimentation, are added in this case to the filtered zeolite. Otherwise, the zeolite is spray-dried.

4.2.2 Preparation of Other Zeolites. The synthesis of Faujasite-type zeolites X and Y in their Na^+-form resembles in principle that of zeolite A, with differences in gel preparation including molar ratios of the starting compounds, reaction and crystallization time, and conditions, *etc.* The SiO_2/Al_2O_3 modulus should be at least 5, preferably higher, when the zeolite is to be used as a component of FCC

(= fluid catalytic cracking) catalysts. In this case subsequent ion exchange steps with ammonium and rare earth salts followed by calcination have to be undertaken.

Higher SiO_2/Al_2O_3-ratios are obtained by 'de-alumination', which can be done, among other methods, by steam-treatment of NH_4-Y zeolite and, if extremely high moduli (>50) corresponding to strong hydrophobicity are required, by treatment with gaseous $SiCl_4$.[20] The latter method leads to products with excellent crystallinity and outstanding temperature stability.

The famous high-silica 'pentasil' zeolite 'ZSM 5' that was first developed in the Mobil laboratories in the late sixties proved to be an excellent catalyst for a number of important petrochemical processes which include methanol conversion to gasoline (MTG-process), octane enhancing in fluid catalytic cracking, xylene isomerization, and benzene alkylation to ethylbenzene.

Each of these processes requires a catalyst tailor-made for the respective application, with many parameters having to be optimized, such as modulus, number, and strength of acid sites, crystal size and morphology, specific surface area, nature, and amount of binder amongst others.

The preparation, therefore, is much more sophisticated than for zeolite 4 A synthesis and requires much longer reaction times. Best results in powder synthesis, which is performed in autoclaves 'hydrothermally' at temperatures between 1 and 180 °C and the corresponding autogenous pressures, are obtained with 'templates' as 'structure-directing' agents (*e.g.* tetrapropyl-ammonium bromide).

A bright future in catalytic oxidation reactions with hydrogen peroxide is predicted for a modified ZSM-5-type called 'Titanium Zeolite TS 1' in which part of the Si-tetrahedra in the lattice structure are substituted by TiO_4-units.[21] The first applications carried out on a technical scale are the direct hydroxylation of phenol with high selectivity to hydroquinone, and the direct conversion of cyclohexanone with NH_3 and H_2O_2 into cyclohexanonoxime, which is an intermediate for the preparation of caprolactam and polyamide plastics respectively.

Only a few examples of commercially relevant applications of zeolites could be given here. It is a class of materials whose potential has by no means yet been exhausted.

[21] B. Notari, *Catal. Today*, 1993, **18**, 163.

CHAPTER 13

Titanium Dioxide Products

T. A. EGERTON AND A. TETLOW

1 INTRODUCTION

Titanium (atomic number 22) is widely distributed and is the ninth most abundant element in the Earth's crust. Its most important oxide is the dioxide which crystallizes in three structural forms: brookite, anatase, and rutile. All three exist as minerals, though brookite is rare. Because of their high refractive index, lack of absorption of visible light, ability to be produced in the correct size range, chemical stability, and non-toxicity, both anatase and rutile are produced commercially and have become the world's predominant white pigments. (Brookite can be synthesized hydrothermally but has no technological importance as it has no advantages to compensate for the more difficult synthesis.) This chapter reviews the uses, manufacture, and properties of titanium dioxide pigments and describes some new non-pigmentary applications.

Selected properties of anatase and rutile are summarized in Table 1. The more open crystal structure of anatase is reflected in its lower density and lower refractive index. Both anatase and rutile are electrical insulators at room temperature. However, since an empty conduction band, derived from the empty Ti^{4+} 3d orbitals, lies only $ca.$ 3 eV above the filled valence band derived from the O^{2-} 2p orbitals, electrons may be excited to the conduction band by UV radiation of wavelengths shorter than 410 nm (rutile) and 385 nm (anatase). In normal daylight there is sufficient UV to excite electrons to the conduction band and induce photoconduction and photocatalytic activity. However, since the human eye is sensitive only to wavelengths longer than 400 nm, both anatase and rutile are white. Rutile is the thermodynamically favoured form at all temperatures.

A family of oxides of general formula Ti_xO_{2x-1} can be produced from rutile but these, and also the sesquioxide, Ti_2O_3, are dark blue/black. Consequently, they cannot be used as white pigments and will not be considered further here.

For general information on the chemistry of titanium see Barksdale[1] and Whitehead.[2]

[1] J. Barksdale, 'Titanium. Its occurrence, chemistry and technology', The Ronald Press Company, New York, 1966.
[2] J. Whitehead, in 'Kirk-Othmer: Encyclopaedia of Chemical Technology', Volume 23, 3rd Edition, John Wiley and Sons Inc., New York, 1983, p. 131.

Table 1   *Some Physical Properties of Anatase and Rutile*[3]

	Anatase	*Rutile*
Crystal structure	Tetragonal	Tetragonal
a, nm	0.3785	0.4593
c, nm	0.9514	0.2959
Density, $kg\,m^{-3}$	3890	4260
Refractive Index	2.54* (550 nm)	2.75* (550 nm)
Dielectric Constant	48*	114*
Band Gap, eV	3.25	3.05
Melting Pt, °C	converts to rutile	1830–1850

Because of their tetragonal symmetry, both anatase and rutile are anisotropic. Figures marked with *
are mean values, *e.g.* the refractive indices, μ, are mean values calculated from $(\mu_1 + 2\mu_2)/3$ where μ_1 is
the value in the 'c' direction and μ_2 is the value in the 'a' and 'b' directions. See also Reference 1.

2   COMMERCIAL DEVELOPMENT

Commercial production of titanium dioxide pigments began in 1918–1919, almost
simultaneously in North America and Western Europe. The pigment then was
anatase (*ca.* $100\,t\,a^{-1}$) and subsequently composites of anatase and barium/calcium
sulfate. Production rose to *ca.* $10\,000\,t\,a^{-1}$ by 1929 and *ca.* $100\,000\,t\,a^{-1}$ by 1937.
Rutile production commenced in the period 1946–1949 and, because of superior
optical properties and durability, steadily progressed to account for more than
80% of production.

Annual production reached 1 million tonnes in 1963, 2 million tonnes in 1976,
and 3 million tonnes in 1989. Within this timescale the most significant trend has
been the transfer from the older pigment production route (Sulfate Process) to the
newer Chloride Process, and today more than 50% of total pigment production is
from this route.

Table 2 shows the current major producers of titanium dioxide pigments and
the countries where they, or their subsidiaries, operate. Of the 3 million tonnes
produced, approximately 57% goes into paints, 17% into plastics, 14% into
paper, 3% into inks, and the remainder into miscellaneous applications.

The largest TiO_2 users, by area, are North America (1.1 million tonnes),
Western Europe (0.9 tonnes), and Asia–Pacific, 0.6 million tonnes) although the
last is potentially the fastest growth area.

In the last 2–3 years the world TiO_2 pigment market has remained quite flat at
ca. 3 million tonnes annum^{-1} due to world recession. However, assuming
moderate economic growth and no major perturbations (*e.g.* energy costs) some
experts predict a demand close to 4 million tonnes by the year 2000.

Because of increasing efficiencies of the industry TiO_2 prices (in 1992 £'s) have
dropped by 25–30% over the last 40 years. However, the actual price is very
sensitive to world economic conditions. In the mid 1980s prices increased from
£1200 tonne^{-1} to a peak of £1700 tonne^{-1} from which it has now (1992) fallen
markedly.

[3] F. A. Grant, *Rev. Modern Physics*, 1959, **34**, 646.

Table 2 *Principal Titanium Dioxide Pigment Producers*

Company	Countries
DuPont	USA, Mexico
Tioxide, (ICI)	UK, France, Spain, Italy, Canada, Australia, South Africa, Malaysia
SCM Chemicals Ltd., (Hanson Industries)	USA, UK, Australia
Kronos Inc., (NL Chemicals Inc.)	USA, Canada, Belgium, Norway, Germany
Kemira, (Finnish Gov.)	Finland, USA
ISK Group, (Ishihara SK)	Japan, Singapore, Taiwan
Bayer	Belgium, Germany, Brazil
Kerr McGee	USA
Thann et Mulhouse, (Rhône Poulenc)	France

In addition to pigmentary sales, *ca.* 100 000 tonnes of TiO_2 are now used annually in non-pigmentary applications. These are described in Section 7.

3 THE PRODUCT ATTRIBUTES OF TITANIUM DIOXIDE

The processes used for the manufacture of titanium dioxide white pigments must yield a product with specific and demanding specifications. What are these requirements and how do they influence the production routes?

Colour is of greatest importance. Mineral sources of titanium dioxide almost invariably appear black; the whole production process is based on the need to remove impurities in order to obtain a product which appears white. Even parts per million of transition metal ions can cause observable and unacceptable discoloration of the resulting white pigment. Therefore, a prime requirement in the selection of raw materials and manufacturing processes must be elimination of colour-inducing impurities, *e.g.* copper and vanadium.

Titanium dioxide is sold for its opacity. Because of this opacity it is used even in non-white systems, *e.g.* coloured paints. Without it the paint would be coloured but transparent – more like a stain. The opacity results from the scattering of light by the titanium dioxide pigment particles. Light scattering theory shows that Q, the scattering efficiency of an isolated, spherical, optically isotropic particle, is dominated by two parameters. The first is the ratio, m, of the refractive index of the sphere to that of the surrounding medium; $m = n_p/n_m$ where n_p and n_m are the refractive indices of the particle and the medium respectively. The second parameter is the ratio of the particle size to the wavelength of the incident radiation. For convenience this ratio is often expressed in terms of $x = 1.5\pi d/\lambda$, where λ is the wavelength of the light in the surrounding medium and d is the particle diameter. In Figure 1 values of Q are plotted against x for different values of m (fine structure in the plots is omitted for clarity). The higher the refractive index ratio, m, the higher the scattering, Q, – which is why titanium dioxide is the predominant white pigment and why rutile is generally preferred to anatase. The figure also demonstrates that for each value of m there is a corresponding value of

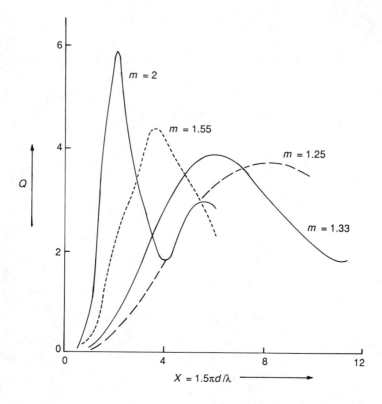

Figure 1 *Light Scattering Efficiency versus x = 1.5πd/λ for particles of different refractive index. Note The total scattering by a given particle is Qπd²/4 and the total scattering per unit volume of particle is 1.5 Q/d*

particle size, x, at which the scattering, Q, is maximized. The particle size of the pigment must therefore be controlled. A third point is that the first maximum in the curves occurs for lower values of x at higher values of m – a pigment with a high refractive index must be small. In a real system, *e.g.* a high-gloss oil paint, in which the pigment concentration may be 15 v/v%, the 'infinite dilution' approximation is not valid and only 25% of the particles will be formed of individual TiO_2 crystals. Theoretical calculations must take these facts into account. However, despite the approximations implicit in its application to real pigmented systems, the theory has strongly guided practical pigment development and the general conclusion that size control is extremely important remains true. For rutile in a gloss film the refractive index ratio is typically 1.8 and the corresponding maximum scattering efficiency for visible light is calculated to occur in the size range 0.18–0.30 μm, depending on the system. A variation of only 10% in the particle size can cause significant changes in the opacity of practical systems. It is therefore extremely important that during pigment production the size of the titanium dioxide is controlled to very tight tolerances.

Once the titanium dioxide has been made it must be dispersed in the product – paint, plastic, paper, *etc.* – to which it is giving opacity. There are two aspects to this. First the pigment must be speedily wetted and dispersed in the medium of use – to allow, *e.g.* a printing ink producer to obtain the highest throughput on his machinery. Second, it must stay well dispersed, *e.g.* during application and drying of a paint, in order that the maximum opacity is obtained – so that the paint covers well. Figure 2(a) is an electron micrograph of titanium dioxide dispersed in a paint film and Figure 2(b) shows a corresponding picture of pigment in a nylon fibre. These are just two applications in a spectrum which includes relatively polar printing inks, non-polar poly(ethylene) or coalescing emulsion paint. In order to achieve the necessary degree of dispersion the surface chemistry of the pigment must be controlled to give colloid stability appropriate to the final use. This control is effected by coating the titanium dioxide surface with white hydrous oxides. Alumina, silica, titania, and zirconia are commonly used, different coatings being used for different applications. Figures 3(a) and 3(b) are electron micrographs of uncoated and coated titanium dioxide pigments.

Further control of the dispersion process may be obtained by organic surface treatment. A large number of organic compounds has been used, mainly polyols and alkanolamines. For pigments designed for incorporation into plastics, treatment with dimethylsiloxane is common.

Coating also reduces the photocatalytic activity of the titanium dioxide – it improves the durability of the pigment – and hence minimizes the heterogeneous photocatalytic oxidation of the organic media in which the pigment is dispersed. Both anatase and rutile are semiconductors and both act as photocatalysts, though anatase is more active. Under the influence of UV radiation an electron can be excited from a filled valence band to an empty conduction band. The electron vacancy that remains in the conduction band may be regarded as being positively charged relative to the original, unexcited, environment and is described as a positive hole. The excited-electrons and positive-holes can migrate to the surface of the crystal and there participate in a series of reactions that lead to the formation of hydroxyl and perhydroxyl radicals. It is these radicals that are believed to lead to oxidative breakdown of the surrounding organic media. This breakdown can be minimized in two ways. First the recombination of the electrons and excited holes can be encouraged – the equivalent in organic photochemistry is enhancing non-radiative transfer of an excited electron to the ground state – by doping the titanium dioxide crystal. Second the interaction between the hydroxyl/perhydroxyl radicals and the surrounding organic can be reduced by coating the pigment surface. Hence, coating of pigment is not only used to control dispersion, it is also used to minimize photocatalytic activity – to increase the durability of the pigment.

4 RAW MATERIAL SOURCES

Although titanium comprises *ca.* 0.6% of the Earth's crust there are only a few minerals in which it occurs in significant amounts – the TiO_2 polymorphs rutile, anatase, and brookite, and ilmenite ($FeTiO_3$) and its weathered alteration products, including leucoxene, sphene ($CaTiSiO_5$), and perovskite ($CaTiO_3$). Of

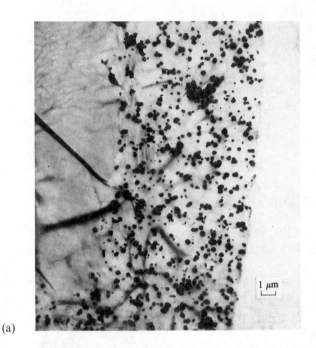

(a)

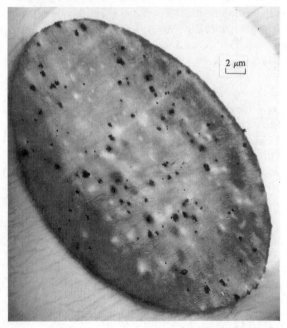

(b)

Figure 2   (a) *An electron micrograph of titanium dioxide dispersed in a paint film. The photograph shows a thin cross-section of a paint film which has been embedded in a resin (at the left of the photo) and then sectioned. The pigment particles appear black; (b) An electron micrograph of titanium dioxide dispersed in a nylon fibre. The photograph shows a thin cross-section of the fibre*

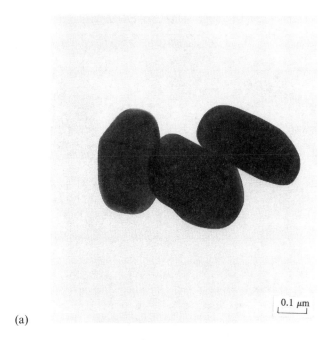

(a)

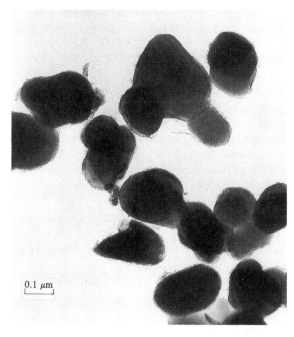

(b)

Figure 3 (a) *Uncoated titanium dioxide pigment;* (b) *Coated titanium dioxide pigment*

Table 3   *Titanium Minerals – Reserves and Production (1989)*

Country	Reserves ($\times 10^6$ mt. TiO_2)	Production ($\times 10^3$ mt. ilmenite)
S. Africa	36.0	
Norway	32.0	930
India	31.0	150
China	30.0	150
Canada	27.0	
Australia	24.0	1692
USA	7.8	300
USSR	5.9	460
Sri Lanka	3.6	107
Brazil	1.6	144
Finland	1.4	
Malaysia	1.0	520

these minerals ilmenites and mineral rutile are the main ones exploited commercially, although mineral anatase may emerge in the future.

Ilmenites from rock deposits typically contain 44–46% TiO_2 and in excess of 40% iron oxides (mainly FeO) and are readily attacked by sulfuric acid. (Dissolution of these ores in sulfuric acid is the first step in the Sulfate Process for titanium dioxide production.) Beach sand ilmenite deposits, however, due to oxidation/hydrolysis and expulsion of Fe_2O_3, contain more TiO_2 ($\geq 55\%$) and less iron oxides but a higher ratio of Fe_2O_3:FeO. They are widely used in the Sulfate Process although extreme weathering ($>60\%$ TiO_2) gives reduced reactivity with sulfuric acid and hence less usefulness. There is a balance, therefore, between maximizing TiO_2 input from the ore and maximizing the TiO_2 extraction efficiency with sulfuric acid. In addition, because of their deleterious effect on colour, ilmenites containing high levels of transition metal impurities are not used in the Sulfate Process. Also, for radiological reasons, strict limits are imposed on thorium and uranium levels.

Despite these restrictions, ilmenites form the most important direct, or indirect, input to the TiO_2 industry and Table 3 (for 1989) shows the principal reserves and producing countries.[4]

Whilst ilmenites are also used in the Chloride Process, high grade ores with increased TiO_2 content are more commonly used, *e.g.* mineral rutile (which is not readily attacked by sulfuric acid) in order to minimize wasteful iron chloride by-products. Again there are operational restrictions on the feedstocks and, for example, high levels of manganese and alkaline earths are not acceptable.

Both processes, Sulfate and more recently Chloride, also utilize up-graded ilmenite (slags). Slags are formed by the high temperature electrothermal reduction of ilmenites. A high proportion of the iron oxide content is reduced to metallic iron and is separated as a saleable by-product and the resultant slag, which is a complex solid solution of iron titanate, magnesium iron titanate, and

---

[4] Roskill, 'The Economics of Titanium 1991', Roskill Information Services Ltd., London, 7th Edition, 1991.

reduced TiO_2, may contain 70–90 + % TiO_2. The more enriched slags have been targetted for Chloride processing primarily due to the limited reserves of mineral rutile and the opportunities for adding value to the mineral products.

$$2FeTiO_3 \xrightarrow{C} FeTi_2O_5 + Fe(metal) + CO \qquad (1)$$

(Mg impurity may also be incorporated)

$$3FeTi_2O_5 \xrightarrow{C} 2Ti_3O_5 + 3Fe(metal) + 5CO \qquad (2)$$

Numerous other ilmenite beneficiation processes have been pursued primarily to provide alternative Chloride Process feedstocks. Such processes include:

(i)   leaching of ilmenite (often reduced) with acid

$$\xrightarrow{2HCl} FeCl_2 + TiO_2 + H_2O \qquad (3)$$

$$Fe_2TiO_5 \xrightarrow[900-1000\,°C]{C} FeTiO_3$$

$$\xrightarrow{H_2SO_4} FeSO_4 + TiO_2 + H_2O \qquad (4)$$

and
(ii) reduction of ilmenite followed by aeration and leaching

$$Fe_2TiO_5 \xrightarrow[1100-1200\,°C]{2C} 2Fe(metal) + TiO_2 + CO_2 + CO \qquad (5)$$

$$2Fe(metal) + 3H_2O + 3/2O_2 \rightarrow 2Fe(OH)_3 \qquad (6)$$

In summary, the overall raw material supply to the TiO_2 pigment industry may be represented by Table 4 and again the most recent, comprehensive data are for 1989.[4]

5   MANUFACTURING PROCESSES

Titanium dioxide pigments are manufactured by two processes: the Sulfate Process (extraction with sulfuric acid); and the Chloride Process (extraction with chlorine). Apart from the common objective of economic viability, both processes, as outlined in Section 3, must provide high-quality products with high purity (colour), stringent mean size and narrow size distribution (opacity), and media protection (durability). These parameters for the base pigments are influenced to some extent by the extraction process used. However, the majority of TiO_2 pigments are coated with other oxides, *e.g.* Al_2O_3 and SiO_2, which can enhance application properties such as dispersion, dispersion stability, gloss, and durability.

Table 4 *Mineral Supplies to the TiO₂ Industry*

Feedstock	% of TiO$_2$ input
Ilmenite	
Australia	22.2
Norway	9.8
Malaysia	6.7
Others	15.2
Slag	
Canada	15.6
S. Africa	13.9
Norway	1.9
Mineral Rutile	
Australia	5.2
Sierra Leone	2.9
Others	1.5
Synthetic Rutile	
Australia	5.1

Such end-treatment also minimizes differences between Sulfate and Chloride route pigments.

In the 1940s, when the Chloride Process was being developed, all production was by the Sulfate route. By 1948 there was less than 1% Chloride production which has subsequently increased to *ca.* 33% in 1982 and *ca.* 50% in 1990 and, of course, annual production has increased to 3 million tonnes. This trend to Chloride production is likely to continue, primarily due to environmental pressures.

5.1 The Sulfate Process

As described in the previous section, the Sulfate Process mainly uses ilmenites (selected) and slags as feedstocks, and typical analysis of four feedstocks commonly used is shown in Table 5.

A schematic diagram of the Sulfate Process is given in Figure 4. Reaction of the mineral with sulfuric acid converts metal oxides into soluble sulfates, primarily of titanium and iron:

$$FeTiO_3 + 2H_2SO_4 \rightarrow TiOSO_4 + FeSO_4 + 2H_2O \tag{7}$$
$$Fe_2TiO_5 + 4H_2SO_4 \rightarrow TiOSO_4 + Fe_2(SO_4)_3 + 4H_2O \tag{8}$$

After the resultant liquor has been allowed to settle to remove unreacted gangue minerals, a proportion of the iron in solution is removed by cooling (crystallization)/centrifugation leaving a relatively clean titanium solution for the hydrolysis stage. The product from this is a microcrystalline oxide of titanium (pulp) which is converted into the required crystalline state by calcination. In theory it is a relatively simple process but some stages merit further comment.

In the digestion procedure, which may be batch or continuous, ground ore is

Table 5 *Analysis of TiO$_2$ Process Feedstocks*

%	Ilmenites		Slags	
	Norway	Australia	Canada	South Africa
TiO$_2$	44.2	54.1	78.2	85.2
Ti$_2$O$_3$	—	—	16.2	27.6
FeO	34.3	21.5	9.7	10.5
Fe$_2$O$_3$	11.0	19.5	—	—
Fe(metal)	—	—	0.10	0.10
Cr$_2$O$_3$	0.07	0.03	0.17	0.18
V$_2$O$_5$	0.21	0.15	0.57	0.43
SiO$_2$	1.5	0.5	2.7	2.1
Al$_2$O$_3$	1.0	0.65	3.2	1.3
MgO	3.0	0.15	5.4	1.1
CaO	0.3	0.01	0.5	0.16
MnO	0.3	1.6	0.2	1.7
ZrO$_2$	0.01	0.1	0.06	0.23
Nb$_2$O$_5$	0.01	0.14	0.02	0.12
P$_2$O$_5$	0.03	0.02	0.01	0.01
p.p.m. ThO$_2$	<10	70	15	40
p.p.m. U$_3$O$_8$	<10	<10	10	10

reacted with strong sulfuric acid (85–95%) and whilst the reaction is stoichiometric it is encouraged by using excess acid –1.5 to 1.6 × (ilmenite), 1.7 × (slag). This excess acid also stabilizes resultant titanium solutions and, at a later stage, influences hydrolysis (precipitation). Each ore also requires an optimum acid strength (85–96%) for maximum titanium extraction of about 95%.

Although the digestion reaction is very exothermic, heat is needed for initiation. This 'set-off' may be achieved by an addition of water/dilute waste acid to the blend of ore/strong acid (98%). The reaction temperature can rapidly then attain 200 °C with copious evolution of off-gases (not with continuous digestion) – mainly steam. The charge solidifies and is baked to complete reaction. The solidified charge (the cake) is then dissolved. During this process the complex iron/titanium sulfates present in the cake dissolve and it is convenient to consider the system as a mixture of metal sulfates. Scrap iron is added at dissolution to reduce any ferric iron to ferrous, since ferrous is more easily removed later in the process, though this is not necessary for slag feedstocks.

$$Fe_2(SO_4)_3 + Fe \rightarrow 3FeSO_4 \qquad (9)$$

Flocculation, settling, and filtration are used to remove the majority of waste/unreacted mineral (sludge) and the hot liquor is then cooled to allow separation and removal of copperas (FeSO$_4$.7H$_2$O). Further filtration and then concentration by evaporation yields a clarified liquor for the important hydrolysis stage. At this point the liquor composition, in terms of TiO$_2$, Fe, acid, and water, is carefully monitored to allow strict control of the thermal hydrolysis during a programmed time/temperature cycle.

Within this cycle nucleation seeds may be added directly or generated *in situ*

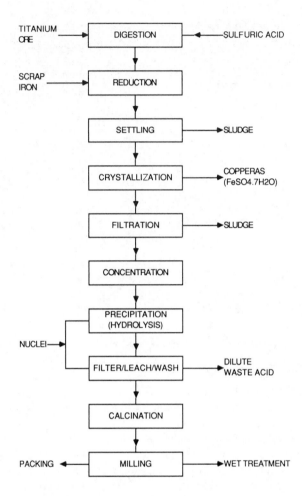

Figure 4 *The Sulfate Process*

and, in the latter case, more may be added post-precipitation. The precipitated titania is microcrystalline but the nuclei control the direction of subsequent crystallization (to anatase or rutile) upon calcination. The nuclei also influence precipitation yield and crystal size and, therefore, are key process parameters.

Leaching, washing, and filtration steps then follow to remove residual trace metal impurities, *e.g.* Fe < 30 p.p.m. It is at this stage that the majority of dilute waste acid is generated. The feed pulp for calcination may typically also retain 7–8% acid. Treatment chemicals are also added at this point to modify crystallization, crystal growth, colour, dispersibility, and durability.

Calcination (Figure 5) is normally carried out in large rotary oil or gas-fired kilns with feed-end temperatures *ca.* 400 °C and discharge temperatures *ca.* 900 °C. Residence times may be 18–24 hours and within this timescale several operations may be identified – water removal, decomposition of acid/sulfates, crystallization,

Figure 5 *Sulfate Process Calcination*

and crystal growth. Not too surprisingly this is a further key process area and the time/temperature cycle is carefully controlled.

After calcination the cooled product is milled and may be packed directly or end-treated (coated) for improvement to application properties. To a large extent the pigmentary characteristics are well established at calcination and, from ore input through calcination, some 80–85% TiO_2 has been recovered.

5.2 The Chloride Process

As with the Sulfate Process, the Chloride Process (Figure 6), in theory, seems relatively straightforward, *i.e.* extraction of titanium by chlorine from a titaniferous ore, *e.g.* mineral rutile, followed by oxidation of the titanium tetrachloride so produced to titanium dioxide (pigmentary size) and chlorine (for recycle).

$$\text{`}TiO_2\text{'} + 2Cl_2 \xrightarrow{C} TiCl_4 + CO_2 \tag{10}$$

$$TiCl_4 + O_2 \rightarrow TiO_2 + 2Cl_2 \tag{11}$$

Figure 6   *A Chloride Process Plant*

Again, however, several stages merit further comment particularly with reference to Figure 7, a schematic process diagram.

The potential feedstocks, in terms of decreasing TiO_2 content (and cost) are mineral rutiles, synthetic rutiles, high grade slags, leucoxenes (highly altered ilmenites), and ilmenites. Typical analysis of some of these, from Australia, is shown in Table 6.

In a typical chlorination the majority of metal oxides, including titanium, chlorinate and for every tonne of $TiCl_4$ produced there is a higher chlorine requirement with the lower grade feedstocks (although they are cheaper anyway) and insufficient chlorine is recovered from the $TiCl_4$ oxidation stage. The majority of this deficit should be recovered from iron chlorides for economic and environmental reasons.

With mineral rutile, whilst there is still multi-element chlorination, the dominant reaction is chlorination of titania:

$$2TiO_2 + 4Cl_2 + 3C \rightarrow 2TiCl_4 + 2CO + CO_2 \qquad (12)$$

This reaction takes place at about 950 °C typically in a fluid bed chlorinator and at this temperature the reaction is autothermal. The ore and coke characteristics are most important, both in terms of analysis (*e.g.* low alkaline earth content to minimize bed sintering) and size (reactivity/fluidizing properties). Other elements can be chlorinated (mainly iron) and exit with the titanium tetrachloride but

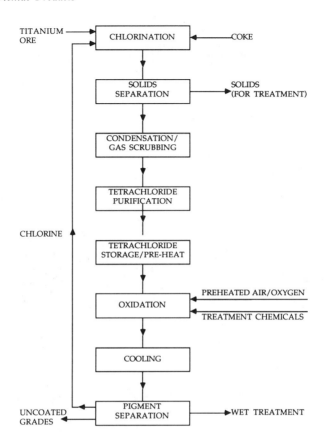

Figure 7 *The Chloride Process*

some, such as zirconium and silicon do not react readily at this temperature and these 'inerts' are purged periodically from the bed.

The chlorination exit gases/vapours comprise titanium tetrachloride, carbon monoxide, carbon dioxide, and impurity metal chlorides/oxychlorides. These impurities are largely removed by partial cooling and may then be further processed for by-product/chlorine recovery. Further condensation/gas scrubbing of the titanium tetrachloride stream removes more impurities and the crude titanium tetrachloride then passes to primary purification – distillation. Remaining impurities are reduced to very low levels at this stage. Vanadium oxychloride, which condenses at a similar temperature to titanium tetrachloride, may be reduced at this stage and converted into a non-volatile sludge. The final, distilled product ('tetra') is extremely pure – a key feature in the production of base pigment with good colour.

Good quality base pigment, in terms of colour, durability, mean size, and size distribution, is produced at the next critical stage – the oxidation reactor or 'burner'. Here pre-heated tetra is reacted with pre-heated oxygen or oxygen-enriched air at 1000–1200 °C in the presence of nucleation/growth additives.

Table 6 *Analysis of Feedstocks to the Chloride Process*

%	Mineral Rutile	Synthetic Rutile	Ilmenite
TiO_2	96.0	93.0	58.0
Ti_2O_3	—	11.0	—
FeO	—	3.2	11.5
Fe_2O_3	0.7	—	25.8
Fe(metal)	—	0.1	—
Cr_2O_3	0.16	0.14	0.07
V_2O_5	0.65	0.24	0.21
SiO_2	0.7	1.0	0.8
Al_2O_3	0.3	1.2	1.0
MgO	0.03	0.3	0.18
CaO	0.03	0.03	0.02
MnO	0.1	1.0	1.1
ZrO_2	0.7	0.1	0.1
Nb_2O_5	0.4	0.2	0.12
P_2O_5	0.02	0.03	0.03
p.p.m. ThO_2	60	200	100
p.p.m. U_3O_8	10	10	10

Additional heat (if necessary) to pre-heat may be provided by flame burners/plasma arc, *etc.*

Unlike the slow calcination stage in the Sulfate Process the nucleation, growth, rutilization, and quench operations are very rapid and the design of reactor is crucial to effective and reliable operation and is critical to achieving quality. Not too surprisingly burner designs, whether impinging jets, slot, or concentric flow systems, tend to be proprietary, guarded features. The patent literature illustrates many examples.

Hot gases leaving oxidation are cooled and recycled to chlorination after removal of the solid base pigment which may remain untreated for some applications or post-treated (coated) for others. Post-treatment by a wet, aqueous route has many similarities to that for the Sulfate Process although the Chloride Process also allows for vapour phase, non-wet, end treatment.

Being essentially a gaseous operation, *e.g.* at chlorination and oxidation, the Chloride Process also affords increased throughput, and improved economics, via higher pressures. Such benefits are not achieved easily or cheaply, however, bearing in mind the aggressive nature of reactants (Cl_2, $TiCl_4$) and the technological problems presented.

Finally, despite the potential advantages, the Chloride Process too is not without environmental problems (see later) and, to date, the Process has also not produced commercially acceptable anatase pigment.

5.3 Wet Treatment (Coating)

The majority of titanium dioxide pigments sold today are coated to enhance the pigmentary characteristics of the base pigment, *e.g.* dispersibility, dispersion

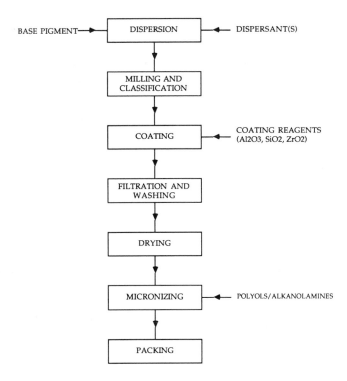

Figure 8 *Pigment end Treatment (Schematic)*

stability, opacity, gloss, and durability. Such end-treatments are tailored to the requirements of the final application and are very important, therefore, to the paint, plastics, and paper industries.

Coating techniques have been developed over many years and to some extent, using wet route coatings, are common to Sulfate and Chloride base pigments. Figure 8 is a schematic flow sheet.

The base pigments are generally dispersed in water with phosphate, silicate, and/or organic dispersants (*e.g.* mono-isopropanolamine) and the slurries may then be milled and/or classified to remove oversize material. The dispersed particles are then coated by selective precipitation of (usually) small quantities of colourless oxides, *e.g.* P_2O_5, SiO_2, Al_2O_3, TiO_2, ZrO_2, deposited via controlled time/temperature pH cycles. A detailed knowledge of surface/colloid chemistry is important at this stage to ensure that the required solution species are present and are subsequently deposited at the pigment surface. By this procedure the surface characteristics are markedly altered to give colloid stability appropriate to end use.

After the coating stage the coated product is filtered, washed, and dried prior to fine milling (micronizing), usually with steam, and packing. At the micronizing stage there may be further control of the dispersion process by organic surface treatment using, for example, alkanolamines, polyols, silicones, and siloxanes.

6 ENVIRONMENTAL ISSUES

Wastes from the TiO_2 Industry have received special, and unfavourable, attention from Environmental and Governmental Agencies especially in the USA and Western Europe. The quantity (as well as quality) of wastes has received very special attention despite monitoring/assurance from numerous independent, authoritative groups including Water Authorities.

With the Sulfate Process, for example, and a factory producing $100\,000\,t\,a^{-1}$ of pigment from ilmenite the wastes could be as high as:

	$t\,a^{-1}$
Copperas $(FeSO_4.7H_2O)$	60 000
Strong acid $(>0.5\%)$	130 000 (as 100% acid)
Weak acid $(<0.5\%)$	80 000 (as 100% acid)
SO_x	5 000

EC legislation has continued since initial recommendations in 1978 and the current requirements for discharges to water and air are shown in Table 7.

To meet these requirements manufacturers of TiO_2 by the Sulfate route have embarked on a variety of process options and, where possible, look to off-set the expense with in-process savings and production of further useful and saleable by-products. Some of the possibilities include:

(a) Concentration and recycle of waste acids.
(b) Roasting of metal sulfates – SO_x and metal oxide values.
(c) Neutralization/land dumping, where feasible – and subject to tight control.
(d) Production/sale of more copperas for water treatment and agricultural purposes.
(e) Iron oxide pigments.
(f) Speciality iron products.
(g) White gypsum production, *e.g.* for wall boards.
(h) 'Red' gypsum production, *i.e.* less pure.
(i) Alternative by-products, *e.g.* ammonium sulfate.
(j) Sale of CO_2 generated during gypsum production.

Although the Chloride process produces less waste, especially when using mineral rutile as a feedstock, it too does not escape environmental legislation. The limits imposed by EC environmental legislation are presented in Table 8.

Table 7 *Environmental Requirements (Sulfate Process)*

Waste	Permit Limit	Proposed Compliance Date
Strong acid	None permitted	31.1.93
Solid waste	None permitted	31.1.93
Weak acid plus	800 kg $SO_4\,t^{-1}\,TiO_2$*	31.12.93
Neutralized waste		(or delayed to 31.12.94)
SO_x	10 kg $t^{-1}\,TiO_2$	1.1.95
Dust – major	50 kg/n M^3	31.12.93
– minor	150 kg/n M^3	31.12.93

*Interim 1200 kg $SO_4\,t^{-1}\,TiO_2$

Table 8 *Environmental Requirements (Chloride Process)*

Waste	Permit Limit	Proposed Compliance Date
Strong acid	None permitted	31.1.93
Solid waste	None permitted	31.1.93
Weak acid plus	130–450 kg t^{-1} TiO$_2$	31.1.93
Neutralized waste	(depends on feedstock)	
Chlorine	5 mg/n M^3 (average)	31.1.93
	40 mg/n M^3 (inst.)	
Dust – major	50 mg/n M^3	31.1.93
– minor	150 mg/n M^3	

Table 9 *Non-pigment Applications of Titanium Dioxide*

Vitreous enamels	Glass ceramics
Coloured pigments	Electroceramics
Welding fluxes	Electrical conductors
Catalysts	Photocatalysts
UV blocks for cosmetics	UV absorbers for plastics
Ultrafine for novel optical effects	

Factories sited outside the EC may be expected to experience similar constraints. Chlorine and titanium tetrachloride production, storage and transport are also closely regulated. Chlorination of ilmenites leads to iron chloride wastes and the consequence need to replenish chlorine inventories. In addition the iron chlorides must be disposed of. In the USA deep wells are used but this may not always be allowed. All of this puts extra pressure on cleaner feedstock supplies. However, chlorination of ilmenite could continue via selective chlorination, recovery of chlorine values and sale of iron by-products.

7 NOVEL TiO$_2$ APPLICATIONS

In addition to the pigmentary uses of TiO$_2$, about 100 000 t a^{-1} are sold for non-pigment applications. Some of these are listed in Table 9. Many of the uses have been described in an earlier publication[5] and, therefore, rather than describe all of them, we will concentrate on those which have developed significantly in recent years. These include ultrafine titanium dioxide for UV blocks in cosmetics and plastics, and novel optical effects, and high purity titanium dioxide for electroceramics and catalysts.

7.1 Ultrafine Titanium Dioxide

If the crystal size of titanium dioxide is reduced from *ca.* 0.2 μm to 20–50 nm, described as 'ultrafine', the light scattering properties change dramatically. Since scattering depends on the ratio λ/d, scattering in the visible region is minimal. Instead the ultrafine TiO$_2$ has high attenuation in the UV region. This

[5] G. F. Everson, 'The Preparation and Uses of Inorganic Titanium Compounds', ed. R. Thompson, Special Publication No. 40, The Royal Society of Chemistry, London, 1981, pp. 226–247.

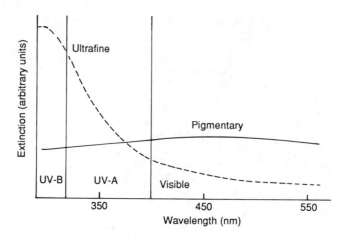

Figure 9 *A Schematic Representation of the Extinction of Light Passing Through Aqueous Suspensions of Ultrafine and Pigmentary Titanium Dioxide*

difference in optical properties is shown schematically in Figure 9.

The shift in scattering to shorter wavelengths allows ultrafine TiO_2 in cosmetic products, such as suncreams, to provide protection from UV radiation without the skin whitening encountered with pigmentary TiO_2. In the past, most suncreams contained organic UV absorbers aimed at affording protection from UV-B (290–320 nm) radiation, which causes sunburn and erythema (skin reddening). Efficacy of sun creams with respect to UV-B radiation is usually expressed in terms of a Sun Protection Factor (SPF). For an SPF of *ca.* 15, a product would contain 5% TiO_2. More recently, the knowledge that UV-A (320–400 nm) radiation causes wrinkling and ageing has led to the use of broad spectrum sun creams. Ultrafine TiO_2 is a broad spectrum UV attenuator and, therefore, protects from both UV-A and UV-B.

Ultrafine titanium dioxide may also be used as a UV protector to extend the life of plastics, and as a transparent UV barrier in applications such as food packaging. For food packaging applications, the resistance to migration of the inorganic absorber is particularly desirable and the product is approved for such applications by, *e.g.* the US Food and Drugs Administration (FDA).

For both the cosmetic and food applications high levels of chemical purity are essential. As with pigment, the dispersion properties of ultrafine titanium dioxide are very important since if the fine crystals are allowed to aggregate the optical properties shift towards pigmentary TiO_2. Surface treatments are used to modify the hydrophilic/hydrophobic nature of the surface and thus allow dispersion in both aqueous and non-aqueous systems and, *e.g.* consequent introduction into either the oil or the water phase of a cosmetic emulsion.

7.2 Ceramic Applications of Titanium Dioxide

The traditional uses of titanium dioxide in ceramics are in vitreous enamels and in thread guides for the fibres industry. These applications have traditionally used pigmentary type titanium dioxide. For use in enamels, impurities such as niobium and tungsten, which can affect colour, must be controlled.

In recent years electroceramic applications have imposed increasingly demanding specifications on titanium dioxide. These applications are based mainly on alkaline earth titanates, *e.g.* barium titanate usually made by co-calcination of titanium dioxide and barium carbonate. Barium titanate ceramics can have exceptionally high dielectric constants and are used as high performance capacitors capable of withstanding high voltage surges. Suitably doped titanate ceramics have large positive temperature coefficients of resistance and are widely used in positive temperature coefficient (PTC) thermistors which are incorporated into heaters, power controls, and to control currents, *e.g.* in motor starters. Since these electrical properties are extremely sensitive to small quantities of impurities it is essential that the impurities are controlled to a very low and very consistent level. Requirements for a titanium dioxide content in excess of 99.9% rule out the possibility of using material extracted from a pigment stream. Typical products are made by the hydrolysis of high purity titanium tetrachloride.

7.3 Catalytic Applications of Titanium Dioxide

Although titanium dioxide does not have the widespread catalytic uses of, *e.g.* alumina and silica, it does have important and growing applications. An early application was in selective oxidation catalysts, *e.g.* the oxidation of *o*-xylene to phthalic anhydride by vanadia/titania. In the last decade, growing environmental pressures have led to catalysts for the Selective Catalytic Reduction (SCR) of nitrogen oxides (NO_x) in gaseous emissions from power stations. In these SCR systems controlled amounts of ammonia are reacted with the NO_x to form nitrogen and water. Tungsten oxide on TiO_2 is a favoured catalyst. A third use, hovering on the edge of commercial exploitation, is application of the photocatalytic properties of TiO_2. Again, the focus is on the environment and the need to remove trace amounts of organics, *e.g.* pesticides, from water. Since anatase is generally catalytically and photocatalytically more active than rutile, most of these applications are based on anatase. The starting material may be derived from pigment production but careful control of sulfate and phosphate levels and dedicated calcination facilities are usually necessary. The growing demands for more efficient catalysts justify the consequent costs.

8 CHALLENGES FOR THE FUTURE

Future process development must continue to pursue lower cost processes whilst ensuring that high quality is maintained and that environmental impact is negligible. Wastes must be recyclable, convertible to useful by-products, or rendered completely harmless.

Activation of ores may benefit selective attack by conventional reagents or allow attack by simpler, cheaper reagents. Selective biological extraction also merits attention.

At the research level there are two major applications for titanium dioxide, both related to solar power. In the first of these titanium dioxide is used as an anode in photo-assisted electrochemical splitting of water to form hydrogen and oxygen.

$$4OH^- \rightarrow 2H_2O + O_2 + 4e \qquad (13)$$
$$2H^+ + 2e \rightarrow H_2 \qquad (14)$$

The hydrogen can then be used as a non-polluting fuel. This research began 20 years ago but, at the time of writing, prospects do not seem good. The challenge is to develop systems which allow the four-electron reaction leading to generation of oxygen to be sustained without the use of toxic sacrificial compounds such as methyl viologen. Secondly, in 1991 this whole area was given new impetus by the announcement of a photovoltaic device based on absorption of visible radiation by a ruthenium dye followed by electron transfer to a transparent titanium dioxide electrode. Quantum efficiencies of *ca.* 10% were reported for this device in which the electrode was made by hydrolysis of titanium alkoxides. These alkoxides are made by reacting titanium tetrachloride – from the chloride process – with the appropriate alcohol. If the present aim of reliably doubling quantum efficiencies is achieved, these devices could make a significant contribution to the world's need for economical, renewable energy sources.

Acknowledgement

The Authors thank the Directors of the Tioxide Group for the opportunity to publish this article.

CHAPTER 14

Elemental Phosphorus, Industrial Phosphoric Acid, and Their Derivatives

J. C. McCOUBREY

1 INTRODUCTION

This industry shares a common root with the larger fertilizer industry for, although it uses less than 10% of the annual world production of phosphate ores, both are serviced by broadly the same mining and ore transportation systems and increasingly by a common primary phosphoric acid technology. Indeed, for environmental reasons, it also lives somewhat in the shadow of world agronomy.

Like the early fertilizer industry, its origins in the Nineteenth century were in bones; these being rich in calcium phosphates provided a raw material for a crude phosphoric acid which could be reduced to a pure but hazardous white elemental phosphorus P_4, largely in demand for the making of matches.[1]

Requirements for safer products predicated on P_4 grew quite slowly from 1850–1890, being supplied by P_4 made by a retort-based technology until the advent of the electric smelting furnace introduced by Redman in 1888. This technique was acquired and exploited early in the UK by Albright and Wilson in the 1890s on a scale of 80 kW furnaces, and by its subsidiary the Oldbury Electrochemical Company with cheap power in the USA in 1896 on a scale of 6×50 kW.[1] Later technological advances owe much to the Tennessee Valley Authority projects of the 1940s[2] and to a small group of manufacturers in North America and Europe from the 1950s.

Today's international industry offers a capacity world-wide of more than 1 million tons of P_4, the value of phosphorus being increased 6–10 fold from that in the rock. It also makes a provision of more than four million tons of pure phosphoric acid, for a wide variety of applications, whose value is about three time per ton that of equivalent fertilizer acid.

[1] R. E. Threlfall, '100 Years of Phosphorus Making 1851–1951', Albright and Wilson, Oldbury, 1951.
[2] R. B. Burt and J. C. Barber, Tennessee Valley Authority, Chemical Engineering Report No. 3, Alabama, 1952.

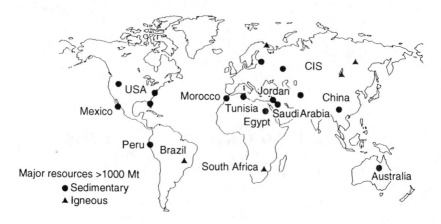

Figure 1   *World Distribution of Major Phosphate Rock Resources*
(Reproduced with permission from A. J. G. Notholt and D. E. Highley, *Trans. Inst. Min. Metall.*, 1986, **95**, p. 127.)

2   OCCURRENCE OF PHOSPHATE ROCKS

Since elemental phosphorus is very reactive, it only occurs in combined form in nature. Although phosphorus compounds of many kinds are essential for life in plants and animals, it is overwhelmingly the metal phosphates which today represent useful raw materials, mainly in the form of apatite minerals.[3,4]

Apatites are generically $A_{10}(XO_4)_6Z_2$. Hard igneous phosphate rocks have a greater preponderance of fluorapatite $Ca_{10}(PO_4)_6F_2$. More common sedimentary rocks are usually carbonate apatite often occurring with structures of the Francolite type or even as fluorapatite mixed with Dahllite $Ca_{10}(PO_4)_{6-x}(CO_3)_x(OH,F)_{2+x}$.

The map in Figure 1 shows the broad world distribution, the most commercially exploited regions USA, North Africa, and the CIS having huge sedimentary rock deposits which account for a substantial proportion of the workable reserves of rock which can readily be up-graded to the useable levels of more than 30% P_2O_5 ($>13\%$ P).[5]

Much phosphate is recovered by open-cast mining, but much larger world resources $200–300 \times 10^9$ tons are available given a greater capability for deep underground mining and extensive beneficiation. The US Bureau of Mines in 1983 indicated[6] that 14×10^9 tons was recoverable at less than \$35 ton^{-1} and a further 34×10^9 tons recoverable at less than \$100 ton^{-1}. This taken against an annual usage of about 160 million tons at \$20–40 ton^{-1} suggests that resources are distinctly finite.

Figures for 1989–90 show the USA exporting about 5 M out of its 48 M ton y^{-1} production, Africa 16 M of its 37 M, and the Middle East 8 M of its 12 M, with

[3] D. E. C. Corbridge, 'Phosphorus', Elsevier, Amsterdam, 1990.
[4] J. R. van Wazer, 'Phosphorus and its Compounds', Interscience, New York, 1958, Vol. 1, 1961, Vol. 2.
[5] A. J. G. Notholt and D. E. Highley, *Trans. Inst. Min. Metall.*, 1986, **95**, 125.
[6] H. Kotabe, *Phosphorus Potassium*, 1987, **149**, 16.

China using the whole of its 19 M ton annual output and the old USSR supplying Eastern Europe at 39 M tons y^{-1}.

Commercial phosphate deposits vary in P_2O_5, in organic content and in many secondary elements. Most rocks require beneficiation to remove other undesirable minerals and clays and are ground, washed, screened, deslimed, and magnetically separated. Flotation is used to separate apatite from silica, and calcination to reduce carbonates and organics.

Modern fertilizer acid technology calls for an optimum of iron, magnesium, and aluminium, and low levels of organics and carbonates for good processability.

For industrial phosphoric acid, control of secondary elements means that arsenic, cadmium, vanadium, or uranium may become part of rock specifications together with P_2O_5, sizing, excess lime, moisture, *etc.*

3 WHITE ELEMENTAL PHOSPHORUS

3.1 Production

A prime constituent of the rock used to produce P_4 is fluorapatite $Ca_9(PO_4)_6CaF_2$ since rock is further intensively calcined. The chemistry of the furnace operation may thus notionally be expressed into a primary reduction:[7]

$$2Ca_3(PO_4)_2 + 6SiO_2 + 10C \rightarrow 6CaSiO_3 + 2P_2 + 10CO - 3MJ \qquad (1)$$

The strongly endothermic reduction of tricalcium phosphate by carbon in the presence of silica requires temperatures around 1200 °C to proceed vigorously but about 1400–1500 °C to expedite with the simultaneous removal of molten slag; only about half of the energy input is used in the thermodynamics of equation (1).[8]

Much of the CaF_2 remains 'unreacted' in the furnace, the slag being broadly a mixture of a silicate mineral wollastonite and a fluorine containing mixed mineral cuspidene. About 25% of the fluorine is reacted, coming over the top via the secondary reaction:

$$2CaF_2 + 3SiO_2 \rightarrow SiF_4 + 2CaSiO_3 \qquad (2)$$

Figure 2 shows a prototype furnace of the American TVA developments of some years ago. State of the art technology of phosphorus production is today optimized in furnaces of electrical loadings of 60–80 MW.[9] These are typically constructed as a tall cylindrical steel shell of around 12 m diameter, lined with refractory, and having a carbon hearth, normally covered with molten slag in operation. Power is supplied by three large consumable carbon electrodes vertically mounted passing, via special gas-tight (telescopic) seals, through holes in the arched refractory roof. These electrodes can be renewed either by fitting new screw-in sections (typically

[7] D. R. Peck, in 'Mellors Comprehensive Treatise on Inorganic and Theoretical Chemistry', Longmans, London, 1971, Vol. VIII, Supp. III, p. 111–148.

[8] H. A. Curtis, *J. Electrochem. Soc.*, 1953, **100**, 81.

[9] D. C. deWitt, 'Phosphorus, Encyclopedia of Chemical Processing and Design', Vol. 36, ed. J. J. McKetta, Marcel Dekker, New York, 1990, p. 1.

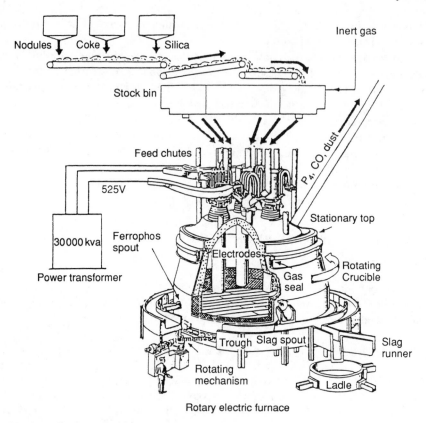

Rotary electric furnace

Figure 2   *Drawing of TVA's* 25 000 kW h *Circular Furnace*
(Reproduced with permission from a paper by P. Ellwood in *Chem. Eng.*, 1965,
February, p. 54; courtesy McGraw Hill, Inc., New York, NY 10020.)

1.4 × 2.75 m, weighing 10 tons in North American furnaces) or continuously by
baking graphitic paste, in the Soderberg electrode construction of German
furnaces and those used in the massive installations of Kazakhstan (different roof
designs also being needed).

The reaction profile of a phosphorus furnace is complex; the electrodes provide
power in some combination of resistive and arc heating, and are sustained on
hydraulic rams which permit (via control technology) the electrical profile to be
optimized. The furnace burden has to be sufficiently open to permit reasonable
flow and counter-current heat transfer of off-take gases. The sizing and
preparation of coke, silica, and especially rock (which has to be pelletized or
nodulized and calcined above 1000 °C) is vital. Breakdown of reactor solids can
result in sintering in the furnace with irregular performance and with channelling
and high gas velocities which entrain solids that stabilize colloidal behaviour in the
condensing system producing 'phosphorus mud'.

Gases leaving the furnace top at about 300 °C consist of P_4, CO, H_2, and SiF_4.
These are passed through dust-removing stages and the P_4 is largely condensed by

warm (70 °C) water sprays, followed by a colder finishing condenser (P_4 melts at 44.1 °C). Phosphorus mud has to be removed in a settler before the pure P_4 is pumped to store, and fluosilicic acid in the aqueous streams has to be neutralized. The CO/H_2 stream can be a source of fuel after cleaning.

At the bottom of the furnace the reducing conditions are such that iron oxide (1–2%) in the rock goes to iron, which combines with some phosphorus (Fe_3P) and a little silicon (C/SiO_2) to form a small heavy molten phase on top of which sits the major burden of molten calcium silicate slag (1400–1500 °C). These molten phases are tapped (the metal has some metallurgical value) and the 8–10 tons of slag per ton P_4 have to be cooled and broken for uses, or landfilled.

3.2 Environmental Aspects of P_4 Production

(i) The preparation of rock feed (*e.g.* pelletizing) can result in the calcination off-gases containing significant amounts of hydrogen fluoride which has to be absorbed, neutralized, and either disposed to waste or economically recovered.

(ii) The substantial amounts of water used in condensing, settling, and storing phosphorus (known collectively as 'phossy water') can contain dissolved, and suspended P_4, acids, and silica, *etc.* This water has to be carefully neutralized with lime, forming precipitates, and finally oxidized to remove any residual phosphorus before it can be discharged.

(iii) Since most sources of phosphate rock contain minute quantities of radioactive oxides which remain unreduced in the furnace, the slag will be a repository of these as complex 'silicates'. Levels are no greater than those of some natural stones and may not pose a problem depending upon geography and application.

3.3 Phosphorus Capacity

Table 1 shows the current position which has changed considerably in the last ten years. In the USA, capacity for P_4 production has declined and consolidated largely into Idaho where electric power costs are lowest. In the old USSR, particularly in Kazakhstan under the banner of Kazfosfor, large furnaces with an effective capacity close to half a million tones per year are operated with much of the output dedicated more to agriculture than to industrial products, and with exports to Western Europe and elsewhere. In Canada, production formerly for the UK and captive requirements has ceased, and in Western Europe one large and two small producers continue. In China, both large and small furnaces appear to contribute to the substantial output which is partly exported to other Far East countries.

3.4 Handling Phosphorus

White phosphorus is handled as a liquid for processing, though it may be transported solid under water in drums, in iso-tanks, or in rail cars which are steam

Table 1 *Phosphorus Production Estimated, 1992*

Company	Capacity/(tons)
FMC, Pocatello, USA	120 000
Monsanto, Soda Springs, USA	100 000
Rhône–Poulenc, Silverbow, USA	30 000
Hoechst, Vlissengen, Holland	90 000
Atochem, Epierre, France	15 000
Enichem, Cretoni, Italy	15 000
Kazfosfor, Dzambul, Kazakhstan	150 000
Novo Dzambul, Kazakhstan	140 000
Chimkent, Kazakhstan	190 000
Togliatti, Kuibyshev	45 000
China, (Yunan and elsewhere)	150 000
	1 045 000

melted for use. Exposure to air gives rise to rapid oxidation with dense fumes, and pumping and plumbing have to take account of that.

Transfers of liquid phosphorus within factories are normally effected using a centrifugal pump the body of which is immersed in water. Transfer lines are heated and purged.

Transfers to storage and to road or rail cars are effected by displacement with water, phosphorus, and warm water circuits being linked. Water used in transferring phosphorus from tank cars to customers may be returned to the phosphorus producer or handled under close conservation. Careful control of blanket waters to pH < 8 to avoid formation of gaseous phosphine is done with buffers.

Safeguards for employees include availability of emergency baths or showers, heavy safety clothing, breathing apparatus, dental checks to avoid necrosis, and antidotes for severe phosphorus burns.

4 PHOSPHORIC ACID

4.1 Production of Thermal Phosphoric Acid

About 80% of the phosphorus produced is converted into phosphoric acid. The reaction of phosphorus with oxygen in air is a highly exothermic process $\Delta H \sim 3\,\text{MJ}\,\text{mol}^{-1}\,P_4O_{10}$. Addition of phosphoric oxide P_4O_{10} to water is substantially less exothermic, and this reaction in an excess of re-cycle phosphoric acid is normally coupled with the oxidation in compact modern processes to give 'thermal' acid; the whole being carried out in a pair of connected vessels of stainless steel or lined with carbon or refractory (Figures 3).[10]

To contain this hot, corrosive system it is not always practicable to get significant heat recovery though some older acid processes did so with more

[10] D. R. Gard, 'Phosphoric Acid and Phosphate, Encyclopaedia of Chemical Processing and Design', Vol. 35, ed. J. J. McKetta, Marcel Dekker, New York, 1990, p. 350.

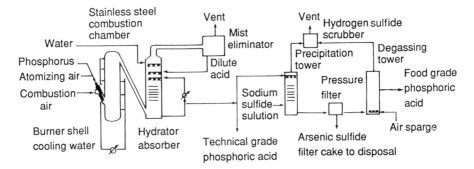

Figure 3 *Flow sheet for the manufacture of thermal phosphoric acid from elemental phosphorus. This schematic shows a stainless steel combustion chamber and a packed hydrator/absorber. Sodium sulfide is added for the removal of low concentrations of arsenic in the production of food grade acid*
(Reproduced with permission from Article on 'Phosphoric Acid and Phosphates' by D.R. Gard in 'Encyclopaedia of Chemical Processing and Design', Vol. 35, ed. J.J. McKetta, Marcel Dekker, Inc., New York, 1990, p. 467.)

cumbersome designs. Hoechst have a modern design where heat is recovered. A major requirement of the 'direct' hydration of phosphorus combustion products is the ability to collect acid mists which are recycled.

Thermal phosphoric acid is usually produced directly at higher P_2O_5 concentrations than agricultural wet acid. Traditionally pure acids of up to 62% P_2O_5 final concentration can be made readily, even after processing to remove arsenic as sulfide by addition of sodium sulfide. The level of heavy metals in such acids is very low since the feed phosphorus is very pure (99.9%). Extremely pure phosphoric acid to provide electronic grades can be achieved by fractional crystallization of the 62% P_2O_5 acid.

Thermal phosphoric acid can also be produced at higher P_2O_5 levels up to 84% (polyphosphoric acids). Due to handling problems such acids are not normally de-arsenicated.

4.2 The 'Wet Process' as a Primary Source of Industrial Phosphoric Acid

Because wet process acid is mainly used for fertilizer production in many countries, the scale and diversity of the technology is considerable. Single-stream plants of 600–1000 tons day^{-1} of P_2O_5 are common and an installation of 8 × 600 ton day^{-1} in Morocco is probably the biggest in the world. In a number of cases units are coupled with sulfuric acid plants and the substantial heat requirements are provided from these. Plants produce either brown acids from uncalcined phosphate rock or green acid from calcined rock.

In Europe the logistics of materials and residues handling is bringing about a situation in which primary phosphoric acid is being bought from the very large plants erected by the phosphate rock producing countries of North Africa and the

Middle East. In the USA technical and logistic rationalization occurs continuous-ly.[11]

Although the core chemistry can be simply written as:

$$Ca_3(PO_4)_2 + 3H_2SO_4 \rightarrow 3CaSO_4 + 2H_3PO_4 \tag{3}$$

the process is substantially more complex due both to impurities and to solid phase effects which determine much of the engineering of the plants used.

All commercial phosphate ores contain significant amounts of silica and about 3.8% of fluorine which combine to produce gaseous fluorine products which have to be dealt with. In addition carbonate and organic contents combine to give rise to stable froths and foams.

The basic principle of the process is that rock is digested with recycled phosphoric acid to give a solution of monocalcium phosphate. The calcium is then precipitated by addition of sulfuric acid. The resultant hydrated calcium sulfate is separated from the phosphoric acid (see Figure 4) and washed, the washings being added to the recycle. Finally, the acid from the filter is concentrated by evaporation. There are many guides to equipment for reaction, crystallization, filtration, *etc.* and for process engineering available, and turn-key plants are on offer to designs by Prayon, Rhône-Poulenc, Norsk Hydro, or Nissan internationally.[12]

Since calcium sulfate can occur in two hydrated forms, variations can be played. The process can be tuned to give $CaSO_4.2H_2O$ at lower temperatures (dihydrate process), or $CaSO_4.\frac{1}{2}H_2O$ at higher temperatures (hemihydrate process); or it may be run as a double-crystallization with the second stage adding further sulfuric acid and giving dihydrate (HDH process), which improves the overall recovery of P_2O_5 (see Figure 5). Changes from dihydrate to hemihydrate may alternatively be used when it is desired to clean up the calcium sulfate for use as a filler, *etc.*

The principal process objectives are to achieve large uniform calcium sulfate crystals thereby speeding up filtration and overall throughput, and to obtain the highest possible yield by solubilizing as much P_2O_5 as possible. Because the acid phosphate ion HPO_4^{2-} is almost the same size as SO_4^{2-} both ions compete for Ca^{2+}; this means that HPO_4^{2-} can easily disappear with the calcium sulfate.

Choice of technology for a particular plant is a complex function of the value of the rock/product, and the cost of energy and capital; and it may depend upon location, secondary impurities in the rock, or on the product range to be offered; or on procedures to handle emissions of fluorides, the separation of cadmium (also controlled by choice of rock), or the disposal of large amounts of calcium sulfate. Conventional dihydrate plants produce acids of up to 30% P_2O_5 content at 94% yield on rock. A $250\,000\,t\,y^{-1}$ HDH process operated until recently by Albright and Wilson in England made 45% P_2O_5 acid at 97% yield based on ground Moroccan phosphate and discharged calcium sulfate to the sea. If calcium sulfate

[11] P. Becker, 'Phosphates and Phosphoric Acid, Raw Materials Technology and Economics', Marcel Dekker, New York, 1989.

[12] *Phosphorus Potassium*, 1990, **170**, 28; *Phosphorus Potassium*, 1991, **174**, 23; and *Phosphorus Potassium*, 1991, **176**, 26.

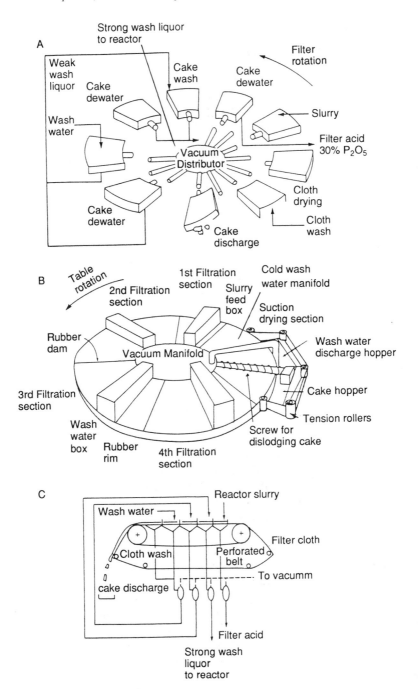

Figure 4 *Wet Acid Filter Designs:* (A) *Tilting Pan;* (B) *Rotating Table; and* (C) *Belt* (Reproduced with permission from *Phosphorus Potassium*, 1980, **106**, 41, courtesy of British Sulphur Corporation.)

The HDH process

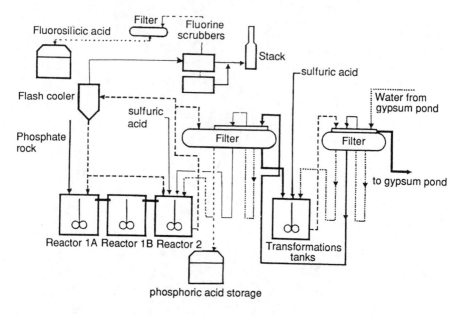

Figure 5 *The Hemihydrate–Dihydrate Phosphoric Acid Process*
(Reproduced from *Phosphorus Potassium*, 1990, **170**, 30, courtesy of British Sulphur Corporation.)

is to be recovered for cements or plaster board a process such as that offered by Prayon at Engis, Belgium may be used.[13]

In general wet process acids are concentrated to 54% P_2O_5 by vacuum evaporation with indirect heating and forced circulation. In most cases HF or SiF_4 is released, though hemihydrate processes with high temperatures of acidulation will have already released much fluorine. Fluorine values can be recovered using dilute fluorosilicic acid as a scrubbing liquor.

Production of 54% acid in the USA, and more recently in the countries of N. Africa and the Middle East, for shipments to many other locations is a growing practice. In 1990 26.7 M tons of P_2O_5 was produced as acid world-wide. Much of this was further reacted 'captively' with phosphate rock to give super phosphates – calcium phosphates of up to 50% P_2O_5 content, or ammoniated for example to diammonium phosphate.

About 15% of the phosphoric acid made is traded and a significant amount of this is beginning to form the basis of the purified phosphoric acid industry in developed countries.

[13] *Phosphorus Potassium*, 1991, **172**, 28 and *Phosphorus Potassium*, 1991, **174**, 42.

4.3   Purification of 54% P_2O_5 Wet Acid to Provide Technical- or Food-grade Phosphoric Acids

In the manufacture of detergent-grade alkali phosphates it has been known for many years to use wet process acid. In neutralizing the acid to make an alkali phosphate liquor, the insoluble basic phosphates and hydroxides of many metals will precipitate and a substantially purified liquor can be formed. Liquors purified broadly in this way, with some care, formed the basis of successful manufacture of sodium polyphosphates for some years. Such a process has limitations and cannot provide the flexibility needed to make a full range of technical phosphate or to approach the quality needed for food-grade products.

Wet process acids after evaporative defluorination, often in conjunction with addition of fine silica, still contain quite high sulfate levels and a host of heavy metal ions. Because chemical specifications were set over the years based on pure thermal acid, the targets involve very low residuals. Desulfation of green acids can be achieved at source by a further reaction with phosphate rock to provide a more suitable feed-stock for industrial acids.

Separation of individual or collective metallic impurities from phosphoric acid can be effected, without neutralization, by ion-exchange resins and by a variety of solvents both water miscible and immiscible. The use of precoat filters and ion-exchangers to remove magnesium can also lower some other metal impurities such as iron and aluminium. Solvent extraction processes using specific complexers have been designed to remove cadmium from phosphoric acid.[14]

In general the most developed commercial processes are the multi-element removals using water-immiscible solvents which are capable of dissolving phosphoric acid such as ketones, alcohols, ethers, or tributyl phosphate where the process becomes one of liquid–liquid extraction originally designed to be performed in mixer/settlers.

Such processes on various scales have been operated for some time in Japan, Israel, Belgium, France, Germany, and the UK. The technology is gradually asserting itself in the USA and in 1990 Texasgulf and Albright and Wilson began operating at Aurora in North Carolina a 100 000 ton P_2O_5 facility to manufacture a range of pure phosphoric acids.[15] The crudest acid containing metallic impurities (underflow) can be returned to the wet acid process for reprocessing or used in fertilizer applications; the pure acids range from industrial detergent-grades to fine food or beverage qualities. Figure 6 shows a diagram of the steps involved in such a process as that of the Aurora plant, where it can be seen that the solvent extraction process is carried out in continuous counter-current column extractors; these have rotating disc agitators and stator rings.

Probably the most technically versatile plant in Europe is that operated by Albright and Wilson at Whitehaven, England. The first unit was built in 1976 as a 40 000 tons y^{-1} P_2O_5 operation using methyl isobutyl ketone in mixer/settlers. To this was added a second unit of 90 000 tons P_2O_5 per year in 1979 still largely to

[14] *Phosphorus Potassium*, 1987, **147**, 31.
[15] *Phosphorus Potassium*, 1990, **169**, 12 and *Phosphorus Potassium*, 1990, **170**, 20.

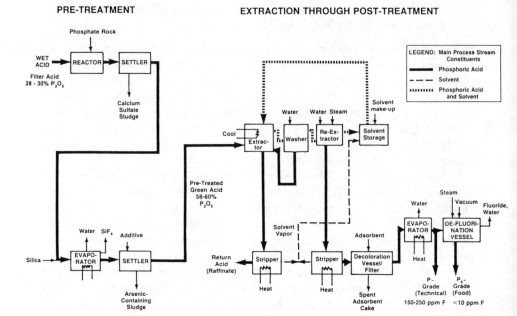

PRE-TREATMENT   EXTRACTION THROUGH POST-TREATMENT

Figure 6   *The Prayon Wet Acid Purification Process*
(Reproduced from paper by A. Davister and M. Peterbrook, *Chem. Eng. Prog.*,

make detergent grade acid. In 1984 a large column extractor was added to increase the overall P_2O_5 recovery (avoiding the underflow containing P_2O_5 traditionally used in fertilizer production). The acidic waste containing the metallic impurities then contained little of the original P_2O_5.

In 1987 a second column extractor was added to carry out the further purification required to permit production of a very pure food-grade acid from the final circuit of the overall process.

The decision between mixer settlers and columns for extraction is largely dictated by whether crude acids from different sources are to be used, when the flexibility of mixer/settlers is required; or if, as in Aurora, rock and wet acid come from one source, only then the simplicity of columns is preferred.

The Whitehaven pure acid complex has a capacity of more than $250\,000\ \text{t}^{-1}$ of technical- and food-grade acids at between 54 and 62% P_2O_5. In mid-1992 manufacture of green phosphoric acid ceased on this site, acid being imported, and a further environmental improvement was introduced with a process to neutralize to an insoluble residue the acidic wash containing metals for safe landfill disposal.

The relative purities of typical filter and typical pharmaceutical acids are given in Table 2.

Prime producers of green phosphoric acid have environmental problems ranging from the disposal of any residual radioactivities (uranium extraction technology developed in the 1970s is non-viable with the present low value of uranium) to the disposal of large excesses of phospho-gypsum produced at levels

Table 2   *Quality of Phosphoric Acids*

	Typical Filter Acid/%	*Typical Pharmaceutical Acid/%*
P_2O_5	27.4	62
H_2SO_4	2.9	0.005
HF	2.3	0.0003
Mg	0.05	0.0010
Ca	0.14	0.0060
As	0.001	<0.0001
Cd	0.004	0.0001
Zn	0.026	0.0001
Cu	0.003	0.0002
Al	0.21	0.0010
Fe	0.64	0.0010
Heavy Metal	~0.03	0.0010
Si	0.38	negligible
TOC	0.012	0.001
Solids	0.41	negligible

well in excess of demand for cement or plaster board, and containing about half of the cadmium which arises in the rock. Moves to lower cadmium rocks and to treatment processes are all parts of environmental strategies.

4.4   Uses of Industrial-grade Phosphoric Acid

The biggest proportion of this grade of acid is captively converted into metal phosphates by the prime producers. However, substantial quantities of acids are sold to be formulated in a variety of ways from soft drinks acidulants to foundry resin catalysts and de-rusting solutions; these acids are customarily 86% w/w H_3PO_4 (62% P_2O_5), 75% w/w H_3PO_4 (54% P_2O_5), and of high purity (Table 2). In addition, however, polyphosphoric acids of technical quality are sold having concentrations from 76% P_2O_5 to 84% P_2O_5 ('116% H_3PO_4') the former containing 50% of pyrophosphoric acid and the latter containing high levels of condensed acids averaging tetraphosphoric acid. The poly-acids are used as dehydrating agents, in acid phosphate ester making, as polymerization catalysts, in polyol manufacture, and in chemical and electrolytic polishing.

5   ALKALI PHOSPHATES

Orthophosphoric acid is a tribasic acid: $K_1 = \sim 7.10^{-3}$; $K_2 = \sim 6.10^{-8}$; and $K_3 = \sim 5.10^{-13}$ and the variety of salts produced on reaction with sodium, potassium, calcium, and ammonium alkalis including a range of hydrates offers a versatile chemistry of production and applications, spanning a wide range of buffering capacity.[10,16]

---

[16] 'Phosphorus *etc.*', in Kirk Othmer, 'Encyclopedia of Chemical Technology', 3rd Edition, 1980.

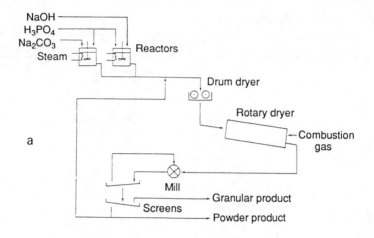

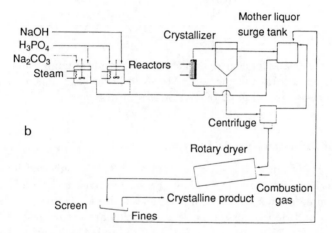

Figure 7   *Flowsheet for the Production of Technical- or Food-grade Orthophosphate Salts by* (a)
Solution Drying or (b) *Crystallization*
(Reproduced with permission from Article on 'Phosphoric Acid and Phosphates'
by D. R. Gard, in 'Encyclopaedia of Chemical Processing and Design', Vol. 35,
Ed. J.J. McKetta, Marcel Dekker, Inc., New York, 1990, p. 476.)

5.1   Orthophosphate Salts

Production of monosodium phosphate NaH_2PO_4, and disodium phosphate
Na_2HPO_4, is effected by reacting either anhydrous sodium carbonate or sodium
hydroxide solution with phosphoric acid to produce a liquor of the desired Na:P
ratio and a solution of the desired aqueous concentration. This is filtered and either
crystallized and centrifuged/filtered to give hydrated salts, or evaporated to
dryness in a spray or drum drier to produce anhydrous salts, see Figure 7.
Size-reduction and classification represent the final production stages before
packing or transfer for subsequent internal use. Mono- and dipotassium and

calcium salts are made in broadly similar ways depending upon phase diagrams, *etc.* with less difficulty in achieving anhydrous salts from solution. Anhydrous ammonium salts are readily obtained by crystallization.

The mono-phosphates of univalent cations have solutions with pH about 4.4 and are used as buffers, dispersants in food processing, and in pharmaceuticals; or specifically in the case of the potassium and ammonium salts, as plant or yeast nutrients.

The diphosphates typified by disodium phosphate are also used for buffering (pH about 9.2) and for their dispersant qualities; and this compound itself finds application in cheese processing, drilling fluids, and for preventing plumbosolvency in soft waters.

Diammonium phosphate is an important nutrient in brewing and has a special role as a fire extinguisher powder and fire retardant.

Monocalcium phosphate $Ca(H_2PO_4)_2$ is a long-established acidulant for self-raising flour, and a bread improver. Dicalcium phosphate dihydrate is a vital primary abrasive constituent in a number of toothpastes; the anhydrous salt is harder and can be used to provide secondary abrasion. Other important acid phosphates are those of magnesium and aluminium used as binders in cements and refractories and the mixed sodium aluminium phosphate an acidulant for chemically leavened flours, popular in the USA.

The trisodium and potassium phosphates (M_3PO_4), in their anhydrous form, are more difficult to produce, and have to be made by high-temperature calcination, the latter in a stainless steel rotary kiln, to complete reaction. Also of considerable commercial interest is the trisodium phosphate dodecahydrate made by crystallization. These very alkaline phosphates (pH solution about 12) are produced in large tonnage in granular or powder forms and their applications, mostly in solid or liquid detergents, are dictated by compatibility with other detergent constituents.

Other smaller volume salts and their applications are discussed by Toy[17] and by Corbridge.[3]

5.2 Formation of Polyanion Compounds – Polyphosphates Formed by Thermal Condensation of Orthophosphates

Just as in concentrated phosphoric acid where there is an equilibrium between orthophosphate and polyphosphates, as free water becomes scarcer, so orthophosphate salts can condense by eliminating water when heated sufficiently.

Two molecules of monosodium orthophosphate condense at about 220 °C to give an important compound $Na_2H_2P_2O_7$ – sodium acid pyrophosphate, a material ideal in mixtures with flour and bicarbonate baking powders to provide controlled CO_2 release and leavening for scones and cakes.

This compound is normally made from spray-dried orthophosphate which is converted by heating on a moving-band oven or agitated low-temperature

[17] A. D. F. Toy, 'The Chemistry of Phosphorus', Pergamon Texts in Inorganic Chemistry, Vol. 3, Pergamon, Oxford, 1975, p. 399.

calciner. The degree of reactivity in the product is critical and achieved by close control of temperature and residence time.

When NaH_2PO_4 is extensively heated into a molten state ($\sim 800\,°C$), hydrogen is almost completely eliminated and condensation occurs to give an extended polymer roughly equivalent to a metaphosphate $(NaPO_3)_x$. Products close to this composition can be represented by a variety of chains and rings. By careful control of Na:P ratios and temperatures, glassy products are manufactured in direct gas-fired high temperature refractory lined furnaces. These products are designed for use in water-treatment and in some food applications.

One such product is the traditional 'hexametaphosphate' made by quick chilling of the molten material and having a chain length about 12. It is used to prevent hard calcium scale build-up in the water side of boilers and heat exchangers. Quite small amounts of these polyphosphates can be effective by the mechanism known as 'threshold treatment' wherein an interruption of the coherent calcium scale structure is effected making it relatively easy to remove.

Other glassy polyphosphates can be used to give slow rates of dissolution to continuously treat hard waters, and some are effective dispersants finding many applications in polymer and food processing.

5.3   Alkaline Pyrophosphates

Disodium or dipotassium orthophosphates on dry heating in a kiln at 400–500 $°C$ give tetrapyrophosphates $M_4P_2O_7$. They can be made continuously or in multiproduct batch, taking spray-dried orthophosphates through bunkers to kilns. These materials give appropriate pH (>10) and sequestering power to remove calcium, strontium, or barium from hard process waters. They are used extensively in cleaning applications and in the case of the highly-soluble $K_4P_2O_7$ in liquid detergents as a builder, or in dispersant and complexing applications in metal and surface treatment.

5.4   Phosphates in Tooth Care

An interesting derivative of sodium metaphosphate is made by reaction with sodium fluoride to give the important dental fluoridating agent sodium monofluorophosphate Na_2PO_3F much used in modern toothpastes. This product together with calcium orthophosphates has played an important part in improving dental care particularly for children. The compatibility of dicalcium phosphate dihydrate abrasive with the fluorophosphate in toothpaste formulations aimed to effect *in situ* fluoridation is important and special quality products have been designed to maximize usefulness. Calcium pyrophosphate has also been used as an abrasive in sodium fluoride-containing toothpastes where its low solubility prevents the precipitation of calcium fluoride. Recently it has been found that pyrophosphates provide tartar control in toothpastes.[18] Systems based on 3–5% of tetrapotassium pyrophosphate have been claimed to be particularly effective in

[18] Colgate, US Patent No. 3927201/2.

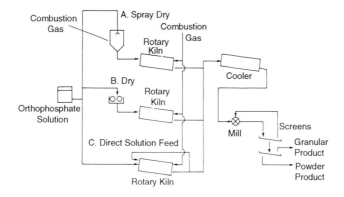

Figure 8 *Flowsheet for the Manufacture of Sodium Tripolyphosphate and Other Crystalline Sodium or Potassium Condensed Phosphates*
(Reproduced from Article on 'Phosphoric Acid and Phosphates' by D. R. Gard in 'Encyclopaedia of Chemical Processing and Design', Vol. 35, Ed. J.J. McKetta, Marcel Dekker, Inc., New York, 1990, p. 485.)

this respect. The effect appears to come from the inhibition of the crystallization of the naturally-arising amorphous phosphates which would otherwise convert bacterial plaque into tartar.

5.5 Triphosphates

The classic case of a large-scale condensed phosphate with very wide applications is illustrated by the continuous production of sodium triphosphate (tripolyphosphate) by the chemical condensation of a mixture of NaH_2PO_4 and Na_2HPO_4 in a sodium-to-phosphorus ratio of 5:3 to produce $Na_5P_3O_{10}$. This has to be produced with a controlled ratio of its two crystalline phases (Phase II which is produced at lower temperatures 350–400 °C and Phase I produced around 500 °C; conversions are also affected by water vapour concentrations).

An orthophosphate solution is made and primarily spray-dried to give an ortho solid which is fed to a large rotary calciner for reaction/conversion at between 400–500 °C leading to a product whose quality may be controlled by variation of temperature or humidity on line or rapidly off line. Tight controls are required by detergent manufacturers since the rate of hydration to the hexahydrate, and consequent temperature rise which varies between Phase I and Phase II, is an important part of detergent processing and this is simulated in the control testing. Another major factor in the product specification is its bulk density since different detergents require light or heavy material. This may be controlled by addition of some ortho liquor to the kiln where extra densification is required. Figure 8 illustrates the stages in the operation. The finished product is despatched in bags, tote bins, or larger containers.

Sodium tripolyphosphate is the best and still the major 'builder' used in the detergent industry. It sequesters or traps calcium and magnesium compounds whether from hard water or soiled clothes and prevents them from interfering with the cleaning action of the surfactants. Tripolyphosphate also makes the washing solution mildly alkaline which helps removal of soil, keeps the dirt in suspension in the washing solution, once removed from the fabric, and also has an advantageous effect on the micellar stability of the surfactants.

The amount of STPP in washing powders is probably optimum around 30% but great environmental pressure has been brought over the last 20 years to reduce the use of this compound since it hydrolyses to orthophosphate; it is claimed this contributes to the provision of nutrients for undesirable algae in slowly-moving waters. Although tertiary sewage treatment can effectively remove phosphate by precipitation, this has until recently not often been used and instead various substitutes for STPP have been sought and adopted in part or in whole depending upon location and view points. It is worth noting that in a number of cases agricultural wastes contribute considerably greater phosphate levels than detergents.

5.6   Ammonium Polyphosphates

These are manufactured from reaction of monoammonium orthophosphate, phosphoric acid, and urea followed by slow crystallization; or by direct neutralization of acid with ammonia under pressure to provide some polyphosphate in solution. Their uses range from valuable plant food, combining N and P nutrients with resistance to precipitation by hard water, to specialist flame retardants. Selected grades of ammonium polyphosphates are used as flame retardants for wood and as retardants and dispersants for paints and pigments.

5.7   Polyphosphates Are Important for Food

Using food-grade phosphoric acid, large quantities of polyphosphates, particularly sodium salts, are made into blends for processed food applications. Particularly in meat processing, *e.g.* in hams and bacons, they reduce oxidative spoilage, stabilize proteins, improve fat distribution, and give final products of improved colour and flavour. In frozen fish and poultry, injection of polyphosphate solutions prior to freezing reduces thaw and cook-out fluid loss with improved flavour and shelf-life. In burgers and sausages blends of phosphates aid salt solubilization of the proteins. Pyrophosphates are used extensively as emulsifiers in meat-based pet foods.

Dairy product uses includes emulsification in cheese processing, fat dispersion in condensed milks, and improved stability in desserts and certain ice-creams, all employing pyrophosphate or polyphosphate blends. Manufacturers produce literature describing many specific food applications.

5.8   Other Major Chemical Intermediates from White Phosphorus

Of the world's $1 \, M \, t \, y^{-1}$ production of P_4 probably no more than 15–20% is used to make derivatives other than via phosphoric acid. These other derivatives however are the vital raw materials of an important and far reaching organophosphorus industry.[19,20]

6 PHOSPHORIC ANHYDRIDE/ORGANIC ACID PHOSPHATES

Relatively little organophosphorus chemistry starts from orthophosphoric acid, but there is a significant industry to make acid phosphate esters from the anhydride P_4O_{10} (often referred to as P_2O_5) or in some cases from polyphosphoric acids.

The anhydride is made by burning phosphorus in dry air and condensing it in different ways to give variously reactive oxides with structures mainly in the hexagonal (H) form.

Reaction of P_2O_5 with C_4–C_{20} fatty alcohols or ethoxylated alcohols is the root of a phosphorus-based surfactant/solvent trade with many applications in textiles, resins, and metal extraction and in secondary chemicals manufacture. The chemistry of the primary reaction gives approximately equal quantities of mono- and di-acid esters:

$$3ROH + P_2O_5 \rightarrow RO.\overset{\displaystyle O}{\underset{\displaystyle OH}{\overset{\|}{P}}}{-}OH + (RO)_2\overset{\displaystyle O}{\underset{\displaystyle OH}{\overset{\|}{P}}} \qquad (4)$$

This ratio can be altered to higher mono- by incorporation of some polyphosphoric acid.

Since P_2O_5 is easily transported and handled, this chemistry is often conducted by non-captive producers. Apart from its general use as a desiccant the other significant industrial uses of P_2O_5 are, *e.g.*: one process to make phosphorus oxychloride where it reacts with PCl_5; and its reaction with by-product diethyl ether to give the plasticizer/solvent triethyl phosphate:

$$P_2O_5 + 3(C_2H_5)_2O \rightarrow 2(C_2H_5O)_3P{=}O \qquad (5)$$

This latter ester can be further reacted with P_2O_5 to give a condensed organic pyrophosphate, which was early used as an insecticide.

7 OTHER FORMS OF ELEMENTAL PHOSPHORUS

Various allotropic forms of phosphorus are known, but amorphous red phosphorus is the only one, apart from the common or α-white P_4, which has commercial significance. An early safety break-through in the match industry of the 19th century, it is still used in the strikers of the safety match box.

It is a polymer formed exothermally when white phosphorus is heated to reflux

[19] G. M. Kosolapoff, 'Organophosphorus Compounds', Wiley, New York, 1980.
[20] S. Trippett *et al.*, 'Organophosphorus Chemistry', Specialist Periodical Reports, The Royal Society of Chemistry, London, 1970, Vol. 1 onwards.

at about 280 °C. Typical reaction times of around 48 hours in static furnaces give a hard product which needs to be broken, wet-milled, and boiled in alkali before dressing, coating, or preparing for formulation with resins to give flame-retardant composites. It is also made in rotating ball mills.

Either white or red phosphorus have been used in pyrotechnic devices and to make metal phosphides. Phosphides such as those of copper (used for phosphor-bronzes), iron (used for low-spark brakes), or iridium (for hardening alloys) are relatively unreactive.

Reactive phosphides made by reaction of the metal with red phosphorus include aluminium phosphide AlP, used in formulations for its slow release of phosphine (PH_3) with moist air to effect fumigation of grain. Similarly made is zinc phosphide Zn_3P_2 used as an acute rat poison.

8 PHOSPHINE

The only reasonably stable simple hydride of phosphorus is phosphine, PH_3. It is easily oxidized, has wide combustion limits and even small traces of residual phosphorus can render it spontaneously inflammable. Phosphine boils at $-89\,°C$ under one atmosphere; it can be stored as a liquid or as a compressed gas, its critical temperature is 51 °C and its critical pressure 64 bars. It is very toxic. The technology of safe handling of phosphine plays a significant part in its industrial chemistry.

Phosphine is thermodynamically unstable with respect to its elements and has traditionally been available as a by-product gas mixed with hydrogen from processes in which phosphorus was reacted with alkali to make hypophosphite salts. It can be made somewhat purer by hydrolysing some metal phosphides or by pyrolysing phosphorus acid.

For the semi-conductor industry phosphine is extensively purified but used in only small amounts. On a large scale the preferred procedure is the two-stage process involving the acid hydrolysis of elemental phosphorus. This process[21] involves the conversion of the P_4 to a part polymer of high surface area, followed by controlled reaction of this polymer with orthophosphoric acid at a temperature of 280 °C(Figure 9). This corrosive system is usually contained in a carbon reactor with resistively heated carbon heat exchangers. By-product phosphoric acid is absorbed for general use. A little hydrogen is also produced in the reaction but the stoichiometry is largely:

$$2P_4 + 12H_2O \rightarrow 5PH_3 + 3H_3PO_4 \tag{6}$$

Handling of phosphine is done at near atmospheric pressure with a large gas holder being used to provide intermediate storage (see Figure 9). The plant involves a considerable degree of instrument control and automation.

Phosphine is used for:

[21] Albright and Wilson and Hooker Chemical Co., British Patent No. 990918, 1965.

Figure 9 *Photograph of Phosphine Plant (P_4/H_3PO_4) at Oldbury Works, Albright and Wilson Ltd., England (with permission)*

(i) production of phosphonium salts; and
(ii) reaction with olefins to produce di- or tri-alkyl phosphines and their derivatives, oxides, sulfides, and phosphinic acids.

Application (i) is typified by the reaction of PH_3 with formaldehyde and a strong acid to form a quaternary hydroxymethyl phosphonium salt:

$$PH_3 + 4CH_2O + HCl \rightarrow \begin{bmatrix} CH_2OH & CH_2OH \\ & P^+ \\ CH_2OH & CH_2OH \end{bmatrix} Cl^- \qquad (7)$$

Tetrakis hydroxymethyl
phosphonium chloride (THPC)

This compound and the analogous sulfate are produced in large quantities by carefully absorbing phosphine gas in formaldehyde/acid. THPC is the basis of manufacture of a flame-retardant system made by its reaction first with urea to give a pre-polymer, which is subsequently cross-linked on cloth with ammonia in the Proban® treatment of cellulosics and cellulosic mixtures.[22]

The compound THPS (sulfate) is made as a broad-spectrum biocide for water treatment in a number of different applications including control of sulfate reducing bacteria in oil recovery systems.

Application (ii) above is typified by the manufacture of trioctyl phosphine and its oxide used as an extractive solvent for a number of metal purifications:

$$PH_3 + 3CH_2{=}CH - C_6H_{13} \rightarrow P(C_8H_{17})_3 \overset{OX}{\rightarrow} O : P(C_8H_{17})_3 \qquad (8)$$

These products are made captively by PH_3 producers and no serious attempt has been made to move PH_3 in bulk, though it is available in cylinders.

Tertiary organophosphines are incorporated in important homogeneous catalysts but as we shall see later the preferred ligand is often triphenyl phosphine which cannot be made from phosphine.

9   HYPOPHOSPHITES AND HYPOPHOSPHOROUS ACID

Production of sodium hypophosphite monohydrate $NaH_2PO_2.H_2O$ is an industry of some thousands of tons, making product for several applications of which the largest is probably the electroless plating of nickel on metal or plastics. Hypophosphite made by disproportionation of P_4 has to be separated from phosphite and the end-product is of high purity:

$$3H_2O + P_4 + 4OH^- \rightarrow PH_3 + H_2 + HPO_3^{2-}\downarrow(Ca^{++}) + 2H_2PO_2^- \qquad (9)$$
$$\text{calcium phosphite}$$

Sodium hypophosphite is used via an acid exchange resin to make hypophosphorous acid, a significant fine chemical used as a catalyst, antioxidant, and in the production of special salts such as manganese hypophosphite catalysts or iron hypophosphites for pharmaceutical or food supplement uses.

10   PHOSPHORUS TRIHALIDES (PCl₃)

The exothermic reaction of elemental halogens with white phosphorus leads to the formation predominantly of reactive phosphorus P^{III} trihalides ranging from the gaseous trifluoride PF_3 (BP $-101°C$) to the solid PI_3 (MP 71 °C).

Although some hundreds of tons of PBr_3 (BP 106 °C) are made for specialist brominations it is the chloride PCl_3 which is far the most important member of the series. The US statistics in 1991 show a capacity of about 150 000 m tons and the

---

[22] 'Flame Retardants (Phosphorus Compounds)', in Kirk Othmer, 'Encyclopedia of Chemical Technology', 3rd Edition, 1980.

world production is perhaps 300–400 000 tons annually. Prices in the USA are stable around \$900 ton^{-1} and growth of demand there has been about 2% annum^{-1}.

The material is made in Britain by the controlled addition of chlorine to white phosphorus in a heel of PCl_3 in 2000–5000 gallons metal reactors (see Figure 10 and also Figure 11). The product so made is distilled continuously (BP 75.3 °C), using the heats of reaction. The purity of commercial PCl_3 is well above 99%, the main impurity, $POCl_3$, arising from entrained air and moisture in the phosphorus.

One of the major problems in handling phosphorus trichloride is its great reactivity with water: it fumes in moist air giving off HCl, reacts violently with small amounts of liquid water and with increasing quantities produces in turn diphosphorous acid and phosphorous acid. It can be handled cold in mild steel containers but, if there is any significant likelihood of hydrolysis, is better in lead-lined, glass, or nickel equipment. Well-sealed resin lined mild steel drums are also used commercially.

A major use for phosphorus trichloride is in the production of phosphorus oxychloride (see later). Perhaps an even greater outlet world-wide is in the manufacture of phosphorous acid and its derivatives. A third family of reactions is that with phenols or alcohols.

10.1 Organic Phosphites and Pesticides

$$2PCl_3 + 6CH_3OH + (\text{'base'}) \rightarrow \overset{\overset{\displaystyle O}{\displaystyle \|}}{PH}(OCH_3)_2 + P(OCH_3)_3 + CH_3Cl + \underset{\substack{\text{(base} \\ \text{hydrochloride)}}}{} \quad (10)$$

The notional reaction equation (10) indicates the labile chemistry of PCl_3 with alcohols, which can be run to form either or both of the important insecticide intermediates dimethyl phosphite (dimethyl hydrogen phosphonate) or trimethyl phosphite. If the stoichiometry is controlled and HCl is removed fast and effectively, trimethyl phosphite can be made alone.

PCl_3 can react with alkylene oxides to form useful intermediate chlorinated phosphites particularly those which subsequently give Arbusov rearrangements[19] to reactive phosphonates such as $(ClCH_2CH_2)P(O)(OCH_2CH_2Cl)_2$.

Di- and trialkyl-phosphites are the base of a long-established pesticides (mainly insecticides) industry in the hands of organic chemical developers like Bayer, Ciba-Geigy, Shell, *etc.* but partly also as downstream products made by basic phosphorus producers such as Monsanto, Hoechst, Rhône–Poulenc, and FMC.

Summaries of phosphorus-based pesticides are found elsewhere.[23] Typical compounds of this family are outlined in Equations (11) and (12) to illustrate the chemistry, that of trimethyl phosphite representing a traditional Perkow reaction, the active product being a *cis*-isomer.

[23] 'Pesticide Manual', British Crop Protection Council, London, 6th Edition, 1980 and later editions.

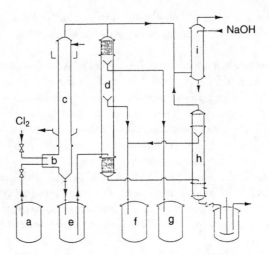

Figure 10 *German (Hoechst) manufacturing process for phosphorus trichloride.* (a) *Phosphorus storage:* (b) *Phosphorus–Chlorine reactor;* (c) *Condenser;* (d) *Fractionating column;* (e) *Phosphorus tri-chloride crude;* (f,g) *Phosphorus tri-chloride pure;* (h) *Residue re-cycle;* (i) *Scrubber.*
(Reproduced with permission from H. Harmisch and H. H. Saenger (Hoechst), *Chem. Technol. Band 1*, p. 412 (Anorganishe Technologie), Ed. Winnacker–Kuchler. (Courtesy of C. Hanser Gmbh and Co.)

$$(MeO)_2\overset{\displaystyle O}{\overset{\|}{P}}H + OHC - CCl_3 \rightarrow (MeO)_2\overset{\displaystyle O}{\overset{\|}{P}} \tag{11}$$

Chloral

$$CH - CCl_3$$
$$|$$
$$OH \quad \text{Trichlorfon}$$

$$(MeO)_3P + \underset{Me}{\overset{Cl}{\underset{\|}{C}}}\underset{\underset{O}{\|}}{\overset{H}{\underset{CNHMe}{C}}} \rightarrow \underset{MeO}{\overset{MeO}{\underset{\diagdown}{P}}}\overset{O}{\underset{\diagup}{\overset{\|}{\diagdown}}}\underset{O - C = C - CONH(Me)}{\overset{H}{|}} + MeCl \tag{12}$$
$$Me$$

2-chloro-*N*-methyl-3-oxobutryamide   Monocrotophos

Phenol derivatives do not experience the problem of acid conversion into disubstituted phosphites, and triaryl phosphites can be made by direct reaction of alkyl phenols with PCl_3. Trinonyl phenyl phosphite and other trialkyl aryl phosphites are made in large quantities as rubber or plastic stabilizers by intermediate manufacturers outside the mainstream phosphorus industry (Borg-Warner, Ciba-Geigy).

PCl$_3$ used to make phosphonates will be subsumed under reactions of H$_3$PO$_3$ later. Other important industrial chemistry of PCl$_3$ is to make acid chloride intermediates from fatty acids.

Triphenyl phosphine made by Wurtz reaction:

$$PCl_3 + 3PhCl + 6Na \rightarrow P(Ph)_3 + 6NaCl \tag{13}$$

is an important, though not-high volume, intermediate used as a ligand in homogeneously-catalysed hydroformylation or hydrogenation reactions. Probably the best known catalysts are those based on work by Wilkinson using RhH(CO)(PPh$_3$)$_3$.

Triphenyl phosphine is also important in fine organic chemicals production for its use in the Wittig ylide process[20] to make key olefin building blocks in the synthesis of vitamins, *etc.*

Other important industrial reactions, which include addition of chlorine to form PCl$_5$, are reaction with methane to form methyl dichlorphosphine CH$_3$PCl$_2$, an intermediate for the flame retardant Trevira®, and reaction with sulfur to form thiophosphorylchloride PSCl$_3$, an insecticide intermediate.

10.2 Phosphorous Acid

This material with structural formula P(O)H(OH)$_2$ is for most purposes a di-basic acid made commercially as a 70% solution or as a pure anhydrous solid (MP 70 °C). It is produced by hydrolysing PCl$_3$ with weak hydrochloric acid in a packed column and the HCl and water are progressively stripped. The acid is concentrated in both atmospheric and vacuum evaporators of special design and a final molten material is converted to solid on a rotating drum flaker.

10.2.1 Phosphonates A very important family of products derived from phosphorous acid is the group of aminoalkylene phosphonic acids made by a Mannich reaction which couples alkyl amines or diamines to the acid via formaldehyde, equation (14). *In situ* hydrolysis of PCl$_3$ may be used alternatively.

$$H_3PO_3 + \begin{matrix} H \\ \diagdown \\ C = O \\ \diagup \\ H \end{matrix} + \begin{matrix} R_1 \\ \diagdown \\ NH \\ \diagup \\ R_2 \end{matrix} - HCl \rightarrow \begin{matrix} R_1 \\ \diagdown \\ N \\ \diagup \\ R_2 \end{matrix} - \begin{matrix} H \\ | \\ C \\ | \\ H \end{matrix} - \begin{matrix} OH \\ \diagup \\ P = O \\ \diagdown \\ OH \end{matrix} + HCl + H_2O \tag{14}$$

These phosphonic acids, particularly in the form of their sodium potassium or ammonium salts, provide industrial sequestrants and scale and corrosion inhibitors whose stability under extremes of pH and temperature is substantially better than the traditional polyphosphates. They are however more expensive and are sold mainly under trademarks or in formulations.

Their applications range from use in detergents to sequester metals such as zinc or copper which can affect the stability of peroxy bleaches, to water treatment

Figure 11   *Photograph of Phosphorous Acid Plant at Oldbury Works, Albright and Wilson Ltd., England (with permission)*

under the extreme conditions of oil recovery where calcium-, strontium-, or barium-containing waters or brines would otherwise foul equipment which has very high volume water throughputs.

Large-scale production of these phosphonic acids in 1000–3000 gallons glass lined reactors by reaction and stripping of HCl/H_2O is done in a batch/continuous manner in highly instrumented plant. Products are controlled for strength, sequestering activity, and colour and a wide series of alkali salts is available, based mainly on products where ammonia, ethylene diamine, or diethylene triamine are the condensing amines.

A second type of phosphonic derivative readily synthesized by reacting phosphorous acid with acetic anhydride, equation (15), is a diphosphonic acid.

$$\begin{array}{c} CH_3CO \\ \diagdown \\ O + 2H_3PO_3 \\ \diagup \\ CH_3CO \end{array} \quad \begin{array}{c} HO \quad O \quad Me \quad O \\ \diagdown \quad \| \quad | \quad \| \\ P - C \ - P - OH \\ \diagup \quad | \quad | \\ HO \quad OH \quad OH \end{array} \quad + CH_3COOH \quad (15)$$

This sequestrant product, alternatively called acetodiphosphonic acid ADPA or 1-hydroxyethane-1,1-diphosphonic acid, has a particularly high association constant with Fe^{3+}. It or its salts have found a host of uses ranging from cleaning of iron oxides from railway carriages or stabilization of liquid fertilizers and denture cleaners to scale/corrosion inhibitor in desalination units.

A major use of phosphorous acid developed in the last fifteen years is for the manufacture of glyphosate, the essential ingredient of the widely used Monsanto herbicides. Many patents describe different procedures to make this material but the archetypal chemistry is shown in equation (16).

$$H_3PO_3 + CH_2O + \begin{array}{c} CH_2COOH \\ \diagup \\ NH \\ \diagdown \\ CH_2COOH \end{array} \rightarrow \begin{array}{c} CH_2COOH \\ \diagup \\ N - CH_2P(O)(OH)_2 \\ \diagdown \\ CH_2COOH \end{array}$$

iminodiacetic acid

chemical $\downarrow$ oxidation   (16)

$$\begin{array}{c} CH_2PO(OH)_2 \\ \diagup \\ NH \\ \diagdown \\ CH_2COOH \end{array}$$

Glyphosate (*N*-phosphono methyl glycine)

The isopropylamine salt of glyphosate is the basis of the well known Roundup® a very widely used post emergence herbicide.

11   PHOSPHORUS OXYHALIDES

Although both phosphoryl fluoride POF_3 (BP $-139\,°C$) and phosphorus oxybromide $POBr_3$ (MP $100\,°C$) are commercially available, neither has a substantial market. By contrast phosphorus oxychloride $POCl_3$ has been a major commercial chemical every since the need for organic phosphate plasticizers, solvents, and hydraulic fluids was appreciated in the 1940s. In the post war developments of PVC when 'tar acid' phenols were cheap, triaryl phosphates made from these and oxychloride were a major source of plasticizer.

Phosphorus oxychloride is manufactured in two major ways, the most widely used being the highly exothermic oxidation of phosphorus trichloride with oxygen or air [Equation (17)] sometimes in the presence of catalysts or anti-inhibitors in a column reactor, with or without continuous removal of volatile excess trichloride.

(A variant to this may be direct oxidation of elemental posphorus with a mixture of chlorine and oxygen.)

$$2PCl_3 + O_2 \rightarrow 2P(O)Cl_3 \tag{17}$$

A second process still used on a large batch scale is the reaction of slurry of phosphorus pentoxide with PCl_3 to which chlorine is added [Equation (18)]. The difference in boiling point between PCl_3 (75 °C) and $POCl_3$ (107.5 °C) makes for a relatively easy separation by distillation which helps in part to remove excess heat from the reactor which is controlled by the rate of addition of chlorine.

$$3PCl_5 + P_2O_5 \rightarrow 5POCl_3 \tag{18}$$

The purity of the finished product $POCl_3$ is high 99.8–99.9%.

Phosphorus oxychloride is a dense liquid (specific gravity 1.685 at 15 °C) which fumes in contact with moist air. The liquid can react with small amounts of water with explosive violence to form 'phosphoric acids' and HCl, sometimes after an 'induction' period. The violence of this reaction, like that of PCl_3, depends on the ratio of halide to water.

Applications of $POCl_3$ relate traditionally to the manufacture of trialkyl or triaryl phosphates since these can only be made from phosphoric acid with extreme difficulty. Tri-n-butyl phosphate has long been recognized as an important selective solvent and was used to separate uranium and plutonium. It is now used extensively to purify phosphoric acid and incorporated in hydraulic fluids.

11.1 The Role of Phosphorus Oxychloride and Other Phosphorus Chemicals in Flame Retarding

When PVC is loaded with conventional diester plasticizers (normally phthalates) its flame retardant properties become impaired. For many applications of plasticized PVC, traditionally mine belting, it is necessary to use a plasticizer which sustains flame-retarding character. The classical products to this end are triaryl phosphates, originally from cresol and xylenol, nowadays from a mixture of phenol and synthetic, isopropyl phenol or *p*-t-butyl phenol [see equation (19)]. These high boiling esters are made from $POCl_3$ in glass or nickel reactors and purified by vacuum distillation on a large scale.

$$POCl_3 + Pr^iC_6H_4OH + 2PhOH \rightarrow OP \begin{matrix} \diagup OPh \\ \\ - OC_6H_4Pr^i \\ \\ \diagdown OPh \end{matrix} + 3HCl \tag{19}$$

Modern products are made to be of very low toxicity and various levels of alkyl phenol are introduced depending upon formulation need. The same esters with some further antioxidant, *etc*. additives are used to provide fire retardant hydraulic fluids for large seals or machinery used in furnaces, steel works, *etc.* but are under considerable competition from 'high water' fluids.

A growing area of use for $POCl_3$ is in the production of chloroalkyl phosphates made from alkylene oxides. This is typified by the production of (tris-1,3-dichloro-2-propyl)phosphate [equation (20)] for flexible polyurethane foams. Other similar products are made for rigid polyurethanes and other plastics and rubbers.

$$POCl_3 + 3ClCH_2.\underset{\underset{O}{\diagdown\diagup}}{CH} - CH_2 \rightarrow OP\left(OCH\overset{\diagup CH_2Cl}{\diagdown CH_2Cl_3}\right)_3 \qquad (20)$$

epichlorohydrin   (tris-1,3-dichloro-2-propyl)phosphate

This range of flame-retardant chemicals together with those variously referred to above such as THPC, Proban®, for textiles; ammonium polyphosphates (often in conjunction with pentaerythrytol) in intumescent paint; and a variety of more complex diphosphates, phosphonates (Pyrovatex®), and phosphates form a substantial industry dedicated to produce fire-retardant materials for furnishing, work clothing, hospital use, and in many applications in aircraft, trains, *etc.*

The common mechanism of phosphorus intermediates as fire retardants involves appropriate early formation of phosphoric acids as the fire tries to spread giving rise to chars which act as barriers to further oxidation. All this is dealt with in various detailed references.[22] This technology marks a significant growth point for industrial fine organophosphorus chemistry.

12   PHOSPHORUS SULFIDES

Sulfur attaches to and interpolates in the molecular tetrahedron of P_4 to form a range of compounds from P_4S_3 to P_4S_{10}. Phosphorus sesquisulfide P_4S_3 is still an important compound for the match industry but its off-take volume is not high. The major commercial sulfide is P_4S_{10} (P_2S_5) known as phosphorus pentasulfide. This is a dense (specific gravity 2.09) crystalline, yellow solid with a melting point of 290 °C and a smell of H_2S caused by its slow hydrolysis in the atmosphere.

The manufacture of P_2S_5 brings together molten phosphorus and molten sulfur and involves exothermic reaction at temperatures ultimately above 300 °C in well-stirred reactors with good heat transfer. The application of the products depends upon its rate of reaction, usually with alcohols, and the literature contains many ways to alter the precise composition and crystal structure by small changes in the P:S ratio, *etc.* Defect structures involving P_4S_9 may arise. Methods of crystallization of the melt can also affect the specifications but generally the material is flaked on a drum under an inert gas blanket. A factor of some importance in manufacture of this product is the use of elemental phosphorus rigorously free from even tiny organic contaminants.

The capacity for this product in the USA is about 80 000 tons y^{-1} at a price of about \$1300 ton^{-1}. There has been a slight fall in demand over the last ten years.

12.1 Oil Additives

The biggest application of P_2S_5 is in the production of 'oil additives' commonly zinc, barium, or similar salts of short chain *O,O*-dialkyl dithiophosphoric acids (*e.g.* diisopropylthio acid). These can be made by addition of metal oxide to acids made by reactions with alcohols, Equation (21).

$$P_4S_{10} + 8ROH \rightarrow 4(RO)_2\overset{\displaystyle S}{\underset{\displaystyle SH}{\overset{\|}{P}}} + 2H_2S \tag{21}$$

Metal dialkyl dithio acid salts are long known to provide wear resistance, detergency, antioxidancy, and corrosion resistance under extreme pressure conditions in oils used for machinery or vehicles. Most large oil formulators and oil companies buy phosphorus pentasulfide in aluminium tote bins for delivery to their own plants for reaction and subsequent blending with oils. Sodium salts of dialkyl thio acids are quite extensively used as ore-flotation agents.

12.2 More Insecticides

The second biggest application of this product however is for two large families of insecticides[23] based on the dithio acids or on *O,O,*-dialkyl dithiophosphorochloridates made by reacting the acids with chlorine, equations (22)–(24).

$$(RO)_2\overset{\displaystyle S}{\underset{\displaystyle SH}{\overset{\|}{P}}} + 3Cl_2 \rightarrow 2(RO)_2\overset{\displaystyle S}{\underset{\displaystyle Cl}{\overset{\|}{P}}} + S_2Cl_2 + 2HCl \tag{22}$$

$$(MeO)_2\overset{\displaystyle S}{\underset{\displaystyle SH}{\overset{\|}{P}}} + \begin{matrix} CH-COOEt \\ \| \\ CH-COOEt \end{matrix} \rightarrow (MeO)_2\overset{\displaystyle S}{\underset{\displaystyle S-CH-COOEt}{\overset{\|}{P}}} \tag{23}$$

diethyl maleate

CH$_2$ – COOEt
Malathion
A low toxicity product of long
standing and wide application

An example of the second:

$$(EtO)_2P \quad + \quad \text{3,5,6-trichloro-2-pyridol (sodium salt)} \quad \rightarrow \quad \text{Chlorpyrifos} \tag{24}$$

3,5,6-trichloro-2-pyridol (sodium salt)

Chlorpyrifos
A more recent product used for
foliar crop pests

Phosphorus producers may offer dialkyl dithio acids and chloridates but progressively this is becoming the territory of specialists who buy P_2S_5 from phosphorus producers to make their intermediates and some end products, selling internationally.

Earlier higher-toxicity organophosphates products, such as Parathion, have been overtaken by more subtle organophosphorus compounds. Although more sophisticated insecticides like synthetic pyrethroids have taken some market share for certain crops, they are expensive and limited and organophosphorus insecticides continue to play a vital role in agronomy in the hands of large pesticide producers like Bayer, Ciba-Geigy, Rhône-Poulenc, and the phosphorus producers.

13 SUMMARY

Uses of white phosphorus in the Western world have decreased over the last ten years. This is partly because environmental pressures on the use of polyphosphates in detergents have been severe, and partly because cleaned up wet process phosphoric acid has replaced some of the acid made from P_4. Production technology for P_4 has however been fostered in Khazakhstan and in China. Outlets for industrial phosphoric acid have probably now stabilized and could begin to grow again. Wider international business in derivatives and improvements in economies of scale via rationalization of wet acid facilities will consolidate the industry.

New products in the next ten years could include phosphoric acid for fuel cells. phosphorus-based ceramics, agrichemical, water treatment, fire-retardant, or food chemical intermediates, but the industry looks most earnestly to cost reduction and to added value from quality and formulation in the next decade.

Subject Index